Technology Transfer and Small & Medium Enterprises in Developing Countries

Technology Transfer and Small & Medium Enterprises in Developing Countries

Editors
K. Ramanathan
Keith Jacobs
M. Bandyopadhyay

CENTRE FOR SCIENCE & TECHNOLOGY OF THE NON-ALIGNED AND OTHER DEVELOPING COUNTRIES
(NAM S&T CENTRE)

2011
DAYA PUBLISHING HOUSE®
Delhi - 110 035

ISBN 9789351242024

Centre for Science and Technology of the Non-Aligned and Other Developing Countries (NAM S&T Centre)
Core-6A, 2nd Floor, India Habitat Centre, Lodhi Road,
New Delhi-110 003 (India)
Phone: +91-11-24644974, 24645134, Fax: +91-11-24644973
E-mail: namstct@gmail.com
Website: www.namstct.org

Published by : **Daya Publishing House®**
A Division of
Astral International Pvt. Ltd.
– ISO 9001:2008 Certified Company –
4760-61/23, Ansari Road, Darya Ganj
New Delhi-110 002
Ph. 011-43549197, 23278134
E-mail: info@astralint.com
Website: www.astralint.com

Laser Typesetting : **Classic Computer Services**
Delhi - 110 035

Printed at : **Chawla Offset Printers**
Delhi - 110 052

PRINTED IN INDIA

Foreword

Present day main concern is that the poor and the powerless are getting more and more marginalised and are slowly disappearing from the main stream of society giving way to the rich and the powerful who are increasingly dominating their lives, their forms of livelihood, their lands and other assets they have, and even their traditional value systems. The introduction of new technologies should not in any way aggravate the already existing disadvantageous situation of these exploited people. After decades of working with the poor, I can see this tendency ever increasing at a faster rate. To alleviate the above concerns, I started a voluntary movement with my fellow teachers and students to help our village communities try and realise their full potential and attain a better quality of life. I still continue to be a student and community educator in the rural areas of this country. This is the fiftieth year of that endeavour and experience which is popularly known as the Sarvodaya Shramadana Movement of Sri Lanka.

Governments allocate large sums of money to bring new technologies to people but unfortunately due to issues related to poor governance it is largely not used for the purpose it is allotted for. The money allocated for technology transfer should not result in such counter-productive consequences which in fact largely affect the poor people. Most scholars and government administrators are aware of the numerous but appropriate technologies that the Sarvodaya Movement promoted in this country long before the governments even thought about the relevance of these so called 'second rate' technologies. On numerous occasions I have been ridiculed and accused as some crazy person trying to turn the clock back. In the early seventies I had the unique experience of being with Dr. E. F. Schumacher at his home in England and he casually gave me a manuscript to read that night. I was up almost the whole night as I got absorbed in it which was eventually published in 1973 under the title "Small is Beautiful". My concept of technology, as relevant to our people in economically exploited countries, was nurtured in my mind from that experience. Particularly the Chapter on Buddhist Economics is most relevant to our societies, even today, after 35 years of its publication. This is what he says – "While the materialist is mainly interested in goods, the Buddhist is mainly interested in liberation. But, Buddhism is the 'Middle Way' and is therefore in no way antagonistic to physical well-being. It is not wealth that stands in the way of liberation but attachment to wealth; not the enjoyment of pleasurable things but the craving for them. The key note of Buddhist economics, therefore, is simplicity and non-violence. From an economist's point of view, the marvel of the Buddhist way of life is the money utter rationality of its pattern – amazingly small means leading to extraordinary satisfactory results. The Sarvodaya Development Model is now promoted in over 15,000 villages in Sri Lanka. Its emphasis is not to create an affluent society but a no-poverty society where the satisfaction of basic needs, and the secondary needs of all, is given priority as opposed to self indulgence where people's desires are emphasized.

Thus, it is not our objective to create a consumerist market economy. Neither are we keen to introduce every new technological device that is introduced into put to the market. We don't believe that modernity is related to time but that it is more akin to its utility value, affordability and sustainability and so on. We have to be self-reliant as far as our food, energy, water and such other basic needs are concerned. They should not be allowed to be governed by price fluctuations as is happening today.

Not only appropriate technologies for production purposes, but also the methods and techniques on correct economic management and good governance should be innovated. The technologies Sarvodaya introduced to our rural communities were an extension and improvement of what they had earlier, and never replacing whatever was appropriate and functional. These included ecological farming, organic agriculture, use of effective micro-organisms (EM) instead of chemicals, construction of community gravity water systems, use of silos for preservation and storage of rice paddy and other grains, wind mills for irrigation and generating electricity, use of solar power, promotion of hydropower for rural electricity supply, fuel saving earthen stoves, micro credit programs to promote small and medium size enterprises and so on. Sarvodaya has tried them in our villages and succeeded.

We always had long term sustainability in our mind. I remember that somewhere in the seventies, at my invitation, Dr. Schumacher visited these Sarvodaya projects and gave us valuable suggestions about the further promotion of Intermediate Technology. To spread the word among interested development practitioners I organized a few public lectures and I still remember at one of these meetings, a local self-proclaimed intellectual getting up and ridiculing Dr. Schumacher in the most derogatory terms and condemning him even as a lackey of neo-imperialism. This intellectual was for promoting the most modern, highly expensive technology into our countries.

In fact the transfer of unsustainable technology has taken root in most countries of the South. I would rather call it in poor countries or, still more correctly, in poor and powerless sectors of all communities and countries. The real division is not based on geographical boundaries. It is a matter of who exploits whom economically, socially and politically. After fifty years of hard work we in Sarvodaya have come to believe that village or community self-governance is the objective we should pursue and this is the only path to freedom for millions of disadvantaged people the world over. Technology is only a means to an intended end. It is not an end in itself. Similarly small, medium-sized and large-scale enterprises are not ends in themselves. What is the ultimate goal of all these technologies and enterprises? Isn't it human happiness? If all these result in the acquisition of enormous wealth and power for a few, and lead to utter misery for the majority of people, of what use are all these? To get the real benefits of advances in science and technology we have to have a correct notion of the principles and ways in which new technologies should be introduced to the masses that are marginalized, exploited, dependent, indebted, poor and powerless. Technology per se is not going to improve their situation. It has to be an integral part of a total approach where there is a clear vision of the society they are going to create – for example a society where the well-being of all (Sarvodaya) is ensured; a society where non-violence is the hall mark of all human relationships; a society where participation of all members of the community with regard to human needs satisfaction as well as human rights preservation are guaranteed; a society where high priority is given to preservation of the environment; a society where non-renewable resources are conserved and no technology is introduced if it pollutes and destroys life-support systems such as air, water and soil; and a society where essential basic needs such as water, food, clothing, housing, health care, energy needs, total education, spiritual, moral, and cultural needs are satisfied. Under the guise of transfer of technology we should not encourage our people to get into debt particularly to financial institutions. Never allow people to mortgage their properties to acquire new technology. Micro-finance institutions should be established and village folk should be trained to run their enterprises efficiently. Self-governance methods and techniques should be developed and mastered. Minimal legal frameworks, enhanced by strong traditional constructive value systems, should be introduced for communities to be aware of their legal rights and responsibilities. While the community objective of economic justice for all should not be lost sight of, individual and family initiatives should also be encouraged.

Everything I said so far was based on fifty years of village development experience and the relevant technologies I referred to were all tried very successfully. Some of them we continue to practice in our village self-development activities and village self-governance programs. Others were taken over by government institutions, banks, rich NGOs, multinational corporations and other kinds of vested interests. Selfishness and greed are bad qualities. But when they get scientifically organized and technologically empowered they become social crimes. While the technologies we introduced liberated people from their economic bondage, vested interests imposed them on people to perpetrate their bondage. However we have to go on the path our conscience illumines. I would like to conclude my intervention once again quoting from the Epilogue of "Small is Beautiful."

"In the excitement over the unfolding of his scientific and technical powers, modern man has built a system of production that ravishes nature and a type of society that mutilates man. If there were more and more wealth, everything else, it is thought, would fall into place. Money is considered to be all powerful; if it could not actually buy nonmaterial values, such as justice, harmony, beauty or even health, it could circumvent the need for them or compensate for their loss. The development of production and the acquisition of wealth have thus become the highest goals of the modern world in relation to which all other goals, no matter how much lip-service may still paid to them, have come to take second place. The highest goals require no justification; all secondary goals have finally to justify themselves in terms of the service their attainment renders to the attainment of the highest."

A.T. Ariyaratne
President of the Sarvodaya Movement,
Moratuwa, Sri Lanka
E-mail: atariyaratne@gmail.com

**Based on the lecture given at the International Conference on South-South Cooperation for Technology Transfer and Development of Small and Medium Enterprises (SMEs), 18-22 August, 2008, Negombo, Sri Lanka*

Preface

The term technology transfer (TT) can be defined as the process of movement of technology from one entity to another. The transfer may be said to be successful if the receiving entity, the transferee, can effectively utilise the technology transferred and eventually assimilate it. The movement may involve physical assets, know-how, and technical knowledge. Technology transfer in some situations may be confined to relocating and exchanging of personnel or the movement of a specific set of capabilities or the movements of technology from the laboratory to industry, or developed to developing countries, or from one application to another domain. Economists have analysed technology transfer on the basis of the properties of generic knowledge, focusing particularly on variables that relate to product design. A broader definition could be that technology transfer is the movement of knowledge, skill, organisation, values and capital from the point of generation to the site of adaptation and application.

The Small and Medium Enterprise (SME) sector is recognized for its contribution to employment, innovation and economic dynamism, and is considered as an engine of growth and an essential part of a healthy economy. Recognizing the distinctly positive impact of small enterprises on the economy, the governments of many industrially advanced countries have taken several policy initiatives for the growth and expansion of SMEs, and for improving their technological capability and market competitiveness.

SMEs in developing countries suffer from several inadequacies such as, technological backwardness, low production efficiency and poor quality of products. SMEs suffer from a number of disadvantages compared with larger firms in responding to and adapting new technological advances. They are often not aware of recent scientific advancements in the absence of a proper information dissemination system. Further, there are no motivational inputs and incentives for them for innovation and technological upgradation. They lack suitable qualified specialists and are largely unable to provide in-house R&D support required for the development of new products and processes and adoption of new technologies. Indigenous SMEs in developing countries are also facing intensifying competition in their local markets due to globalization, increasing liberalization, and the entry of multinationals through foreign investment.

It is, therefore, imperative for developing countries to take appropriate policy measures for building the capabilities of SMEs to develop, acquire and adapt new technologies; enhance efficiency of production and reduce production costs; and undertake total quality control, quality assurance and good manufacturing practices.

There is now increasing awareness in developing countries that while seeking technology from industrially advanced nations it will also be prudent, especially from a technology capacity building perspective, to cooperate among themselves to meet their technology needs. One option would be for the countries in the developing world that are advanced in science and technology (S&T) to cooperate with those countries that are less endowed in terms of S&T research and education. This approach, popularly known as South-South cooperation can contribute significantly in facilitating transfer of technology and in establishing links between research and industry for technological innovation.

This book is an extended and updated version of the papers presented at the International Conference on 'South-South Cooperation for Technology Transfer and Development of Small and Medium Enterprises (SMEs)' which was organized in Negombo, Sri Lanka during 18-22 August 2008 by the Centre for Science and Technology of the Non-Aligned and Other Developing Countries (NAM S&T Centre) in association with the National Science Foundation (NSF), Sri Lanka, and the National Science and Technology Commission (NASTEC), Sri Lanka. The conference brought together various stakeholders, *viz.* experts, policy makers, researchers, technologists and entrepreneurs to share knowledge, approaches and experiences.. The book also includes a few specially invited papers. Thus, the focus of the book is to share perspectives and experiences of developing countries on the transfer of technology and promotion of SMEs. The contents are divided into three parts. Section 1: 'Technology Transfer in the Global Context'; Section 2: 'Technology Transfer and SMEs in Developing Countries'; and Section 3: 'The Role of Institutions'.

Section 1 includes nine papers, which provide general perspectives on technology transfer, with particular reference to SMEs.

The first paper by K. Ramanathan draws attention to the problems commonly faced by SMEs in planning and implementing technology transfer (TT) and argues that a holistic approach needs to be adopted by SMEs in planning and implementing TT projects. The paper presents an overview of technology transfer and some important existing models of technology transfer. The strengths and weaknesses of these models are pointed out and an eclectic model titled, "The Life Cycle Approach for Planning and Implementing a Technology Transfer Project" is presented. It is envisaged that this model could help SMEs adopt a holistic approach in addressing many of the common problems that they face in TT projects.

In his paper, H.E. Juan Alfredo Pinto Saavedra, Ambassador of Colombia to India, emphasizes that South-South cooperation (SSC) must accelerate its transformation in the light of the paradigm change, and also explains the economic tendencies and international dynamics of micro, small and medium enterprises (MSMEs).

The UNDP's Special Unit for South-South Cooperation (SU-SSC) launched a project on South-South-Gate in 2006 which was conceptualised and developed on the basis of the lessons learnt in organizing the Africa Asia Business Forums (AABF). The paper of Mr. Denis Nkala, analyses the AABFs results, and explains the SS-Gate operational systems and lessons learnt from the project.

A. Balachandran, in his paper, underlines the role of business incubators (BIs) as an important support mechanism to accelerate the successful development of start-up and fledgling companies by providing necessary resources and services to the entrepreneur and facilitate technology transfer and development. Vijaya Kumar uses the case of Sri Lanka to highlight the challenges before developing countries in ensuring that technology transfer leads to technological learning and spin-offs, especially from foreign direct investment (FDI) and stresses that successful creation of intellectual property in the field of technology and transfer of this technology will require the development of effective scientific, educational, legal, financial, commercial and industrial infrastructure. Ananda Jayawardane points out that the development of the SME sector in partnership with large scale manufacturers is an essential and integral part of the development strategy of a country and that effective Public Private Partnerships (PPP) has become an important initiative in achieving this goal. He delineates the constraints and challenges in adopting this approach and emphasizes that a concerted effort by all the stakeholders is needed to more profitably exploit PPP for enhanced economic development

Tajammul Hussain in his paper presents a general overview on the genesis, need, and potential of South-South Cooperation (SSC); and the opportunities and challenges for SSC in SME development. In their respective research papers, Keith Jacobs *et al.* present the results of a study on the effect of information channels and networks on knowledge generation and technology transfer in SMEs, and Roselyn Gakure, M. Wachira presents the findings on the factors affecting utilization of information technology in organizations.

Section 2 of the book has twelve contributions, which describe, along with case studies, the current status of SMEs in different countries of the South and highlight important schemes and programmes in various countries to support the technology transfer and innovation management capacity of SMEs.

M. Bandyopadhyay, in his paper, presents a comprehensive overview of various policy initiatives, schemes and programmes aimed at adoption, development and transfer of technologies for capability building, and improving technological competitiveness of SMEs in India. He points out that successful policy measures taken by India to support the SME sector in their endeavour to become competitive could hold valuable lessons for other countries in the South and suggests that it would be of mutual benefit for other developing countries to collaborate with India in improving the technological competitiveness of their SMEs through various programmes. Another paper from India, by Rama Swami Bansal and V. P. Kharbanda dwells upon some case studies on how enterprises in India are facing the challenge of globalization and increased international competition, and particularly, how SMEs are moving towards clusters and are associating with R&D institutions to transfer technology that could help them to sustain their international competitiveness.

Some authors have elaborated upon the various technology transfer and business development services that are being provided to SMEs in their respective countries.. Anteng Setia Ningsih presents the programmes of the 'Business Technology Centre' in Indonesia, and Lakshman Wijeyewardena and Wasantha Senanayaka explain the role of business development service providers for the promotion of technology transfer in Sri Lanka.

Muditha P. Liyanagedara, I.Y. K. Jayaweera and Ranil D. Gunaratne present the results of a survey carried out in Sri Lanka to identify the factors that determine the destiny of the technologies developed and transferred by public sector R&D institutions. They identify the attributes of technologies that were successfully transferred and propose measures needed to ensure the successful commercialization and adoption of public sector R&D outputs. Bashir Rajab Kagere, in his paper reviews the role of 'Jua Kalis', the small scale enterprises comprising semi-skilled artisans in the informal sector, in Uganda's economy and suggests that its integration in the S&T system of the country could be beneficial to the growth of SMEs.

There are six other papers in this section, one each from Egypt (by Janet Ibrahim Youseef); Malawi (by Jonas F. Mwatseteza *et al.*); Nepal (by Dilip Subba); Pakistan (by Nighat Afza); Uganda (by Adam M Sebbit) and Vietnam (by Ngo Thi Loan), which may be regarded as country papers that describe the current status of the development and transfer of technology to SMEs in the respective countries.

Section 3 has four papers that highlight the role played by S&T institutions in the development of SMEs and transfer of technology to industry. These are, i) by Arun Seetharaman, *et al.* which presents the role, thrust areas, and important highlights of technology transfer activities of the International Advanced Research Centre for Powder Metallurgy and New Materials (ARCI), Hyderabad, India; ii) by Matheo L. Raphael on the functions and activities of the Tanzania Commission for Science and Technology (COSTECH); iii) by D. P. Kothari on the experiences of the Vellore Institute of Technology (VIT), Vellore, India; and iv) by Abdulilah Mohammed Saleh on the role of the Central Organization for Standardization and Quality Control (COSQC), Iraq.

Often, it is assumed that technology transfer is a relatively predictable process whereby buyers of technology acquire, assimilate, and then improve the purchased technology, often with assistance from government policies. This approach tends to oversimplify the magnitude of the problem faced by SMEs in planning and implementing TT projects. It is hoped that this volume will serve as useful reference material that draws attention to the problems faced by SMEs in transferring technology, possible approaches that can be taken to address these problems, and the various national initiatives that are being taken by countries of the South to strengthen the technology transfer capacity of SMEs so that they can sustain their competitiveness and grow. Experts, policy makers, researchers, academicians, and chambers of commerce in developing countries may find this publication useful in evaluating practices adopted in several developing nations and in initiating South-South cooperation programs for the development of SMEs in their countries

K. Ramanathan

Keith Jacobs

M. Bandyopadhyay

Introduction

'Technology Transfer' is a process of utilising and marketing the existing knowledge, facilities and capabilities to fulfil public and private needs and has crucial relevance for the society because a newly generated technology would lose its worth unless passed on to a user who can apply the same to create a tangible benefit. Technology Transfer involves a range of formal and informal collaborations between technology developers and technology seekers. In the prevailing technology driven global era, many governmental bodies, academic and R&D institutions and companies have created specialised Technology Transfer bodies that primarily dedicate themselves to identifying researches with potential commercial interest and evolving strategies to exploiting the same. Recent years have also seen a marked increase in technology transfer intermediaries, which are mostly multidisciplinary in nature and include economists, engineers, lawyers, marketers and scientists who work on behalf of the research institutions, governments and even large multinationals.

Small and Medium Enterprises (SMEs) are one of the most vital segments of a nation that significantly contributes to the gross domestic product (GDP), creates employment and earns foreign currency through export. This sector mobilises scarce resources and caters to the needs of fragmented domestic market particularly in the emerging economies. As studies indicate, SMEs contribute nearly 40-60 per cent of the total output or value addition to a nation's economy. Their importance is even more reinforced if we think of the huge unemployment in the developing countries. Promotion of SMEs is deemed as a prominent approach to sustainable development, alleviation of poverty and raising the standard of living. Globalisation has distorted the terrain for SMEs. Currently the governments of the industrialised nations too have accredited the importance of SMEs in industrial economic systems. Due to the limited resources and relative inability to absorb the costs and risks associated with domestic technology development, SMEs most often utilise the process of technology transfer to take advantage of technology and innovation.

From a modern optimistic viewpoint, SMEs can be easily assisted and facilitated to grow, multiply and replicate into a sufficient mass across industries and sectors leading to proper proliferation through applications of modern technology. In this process the coordinating unit of an R&D organisation for technology transfer process greatly helps in the strategic implementation of technologies by facilitating the sharing of information, national and international experiences and technical know-how, dissemination of information on best practices, data collection and networking.

In order to share the knowledge, approaches and experiences concerning the issues and prospects on the development of SMEs and the transfer of technology to improve technological competitiveness of SMEs for their national economic development, the Centre for Science and Technology of the Non-Aligned and Other Developing Countries (NAM S&T Centre) in association with the National Science Foundation (NSF) and National Science and Technology Commission (NASTEC) of Sri Lanka organised an international conference on 'South-South Cooperation for Technology Transfer and Development of Small and Medium Enterprises (SMEs)' at Negombo, Sri Lanka during 18-22 August 2008. This conference brought together 72 stakeholders comprising senior experts, policy makers, researchers, technologists and entrepreneurs from 17 countries–Belarus, Colombia, Egypt; India, Indonesia, Iraq, Kenya, Malawi, Nepal, Nigeria, Pakistan, South Africa, Tanzania, Uganda, Vietnam and Zambia and the host country, Sri Lanka–and also the representatives of a few inter-governmental organisations, *viz.*, UNESCAP-APCTT, New Delhi; UNDP Regional Centre in Bangkok, Commission on Science and Technology for Sustainable Development in the South (COMSATS), Islamabad; and NAM S&T Centre, New Delhi, who extensively brainstormed and exchanged knowledge and experiences on SMEs and issues related to technology transfer in the developing world.

The Conference was conducted in 11 technical sessions broadly categorised under the themes 'Technology transfer in the global context', 'Technology transfer from around the world: Case studies', 'Implementation of technology transfer: Laws and policies and market development', 'Implementation of technology transfer: Partnership development and IPR issues', 'Strengths and barriers for effective technology development and transfer', 'Technology assimilation and networking' and 'Entrepreneurship development and quality assurance of the product standardization for effective technology transfer'. An exclusive session was devoted to panel discussion on 'Strengths and barriers for technology development and transfer: Experience from Sri Lanka and way forward', which was led by Dr. A.T. Ariyaratne, Founder–President, Sarvodaya Shramadana Movement, Moratuwa, Sri Lanka.

The present publication titled 'Technology Transfer and Small and Medium Enterprises in Developing Countries' is based on the presentations made by the participants of the Negombo Conference and also includes some papers contributed by other eminent specialists. The articles and review papers in the publication have been classified under three sections, namely, 'Technology Transfer in the Global Context'; 'Technology Transfer and SMEs in the Developing Countries' and 'The Role of Institutions'. A set of recommendations that were adopted as 'Colombo

Declaration' at the concluding session of the Conference is placed as an Appendix at the end of the book.

I gratefully acknowledge the dynamic involvement and untiring efforts of Dr. K. Ramanathan, Head of the UNESCAP Asian and Pacific Centre for Transfer of Technology (APCTT), Prof. Keith Jacobs, Assistant Dean of Cape Peninsula University of Technology of South Africa and Mr. M. Bandyopadhyay, Senior Expert and Administrative Officer, NAM S&T Centre (formerly Adviser, International Cooperation, Department of Science and Technology, Government of India) for technical editing of this publication. I am indebted to Dr. A. T. Ariyaratne, President of the Sarvodaya Movement, Moratuwa, Sri Lanka for sparing his valuable time to attend the Conference and delivering his address and also writing a 'Foreword' for this book. Last, but not the least, the valuable services provided by the entire team of the NAM S&T Centre under the guidance of Mr. M. Bandyopadhyay and Dr. V.P. Kharbanda, particularly Mrs. Manjari Manisha, Mr. Gaurav Gaur and Mr. Pankaj Buttan in compiling the presented papers and giving a shape to this volume are deeply appreciated.

I hope that this publication will serve as a valuable reference material on the role of technology transfer in enhancing the competitiveness of small and medium enterprises (SMEs) for the entire scientific community and policy makers in developing countries.

Arun P. Kulshreshtha

Director, NAM S&T Centre

Contents

Section III
The Role of Institutions

– Section I –

Technology Transfer in the Global Context

Chapter 1

An Overview of Technology Transfer and Technology Transfer Models

K. Ramanathan
Head
UNESCAP Asian and Pacific Centre for Transfer of Technology (APCTT)
Qutab Institutional Area, New Delhi – 110 016, India
E-mail: kramanathan@apctt.org

ABSTRACT

In today's business setting, interest in the profitable exploitation of a firm's technological assets, through technology transfer, has intensified. Factors that have facilitated international technology transfer include globalisation of business, liberalisation of the economic regimes of many countries, and the impetus given to the protection of intellectual property after the formation of the World Trade Organization (WTO). These factors have collectively resulted in commercial transfer of technology becoming an important element of the international business setting. Experience over the decades has shown that the technology transfer process can be problematic and transferees often lack the skills to manage it effectively. While the literature is rich in terms of the coverage of the areas of concern it is sparse when it comes to possible approaches that can be taken to remedy these problems. This paper proposes a "Life Cycle Approach for Planning and Implementing a Technology Transfer Project" that could help address some of the important process management problems faced by managers of technology transfer projects and avoid common errors that they tend to make. The paper commences with an overview of technology transfer and some important existing models of technology transfer. The strengths and weaknesses of these models are pointed out and an eclectic model titled, "The Life Cycle Approach for Planning and Implementing a Technology Transfer Project" is presented. It is envisaged that this model could help in addressing many of the common problems that are faced by transferees of technology. The paper concludes with an initiative that the UNESCAP Asian and Pacific Centre for Transfer of Technology (APCTT) is currently developing, based on the Life-Cycle approach, to help technology transfer capacity building in the Asia-Pacific region.

Keywords: *Technology transfer, Small and medium enterprises, Technology transfer models, Life-cycle approach, Technology transfer problems, Technology transfer capacity building.*

1. INTRODUCTION

Firms have many ways of exploiting their technological assets for profitability and growth. While internal exploitation of technological assets, through designing, developing, manufacturing, and selling products and processes continues to be important, interest in their external exploitation through technology transfer has intensified in recent years. This may be attributed mainly to the globalisation of business, liberalisation of many developing economies, and greater emphasis on the protection of intellectual property after the formation of the World Trade Organisation (WTO). Indeed, today, the transfer of manufacturing technology has become an important part of the international business strategy of firms.

However, the importance of technology transfer from a development perspective is nothing new. More than three decades back, Mansfield (1975) pointed out that, "One of the fundamental processes that influence the economic performance of nations and firms is technology transfer. Economists have long recognized that the transfer of technology is at the heart of the process of economic growth, and that the progress of both developed and developing countries depends on the extent and efficiency of such transfer. In recent years economists have also come to realize (or rediscover) the important effects of international technology transfer on the size and patterns of world trade."

Technology transfer (TT) is an area of interest not just to business, economists, and technologists but also to other disciplines such as anthropology and sociology (Zhao and Reisman, 1992). While anthropologists emphasize the impact of TT on changes in patterns of culture and society, sociologists are more concerned with its role as a vehicle to develop the capacity of individuals and societies to cope with modernization and related changes accompany it. For economists, as argued by Mansfield (1975), the focus is on economic growth and achievement of economic goals. However, from the perspective of business and technologists the main focus of TT is to improve the competitive advantage of firms through the enhancement of customer value (Ramanathan, 2001). It is envisaged that, through the improvement of competitive advantage, a firm and its partners collaborating in the TT will gain financial and other strategic benefits.

Mayer and Blaas (2002) point out that, SMEs have begun to utilize technology transfer as a strategic means of meeting challenges posed by the globalisation of business. Due to their small size and skill resource constraints, they cannot carry out internal R&D to generate their own technologies but still need a flow of new technology to be able to compete. This need has created a new niche-market for technology transfer (Morrissey and Almonacid 2005). The importance of technology transfer, from an economic and competitiveness perspective, has also stimulated university–industry technology transfer. This is evident from the emergence of technology transfer offices in most research offices and universities (Siegel *et al.* 2004). Ramanathan (2000) shows that in today's international business setting, depending on the attributes of the technology, its intended use, and the motivations of the transferee and transferor, a wide range of TT modalities are available. The focus need not merely be on the purchase of plant and equipment or licensing.

However, planning and managing a technology transfer project, especially an international technology transfer (ITT) project, is not easy. Based on the experience gained over three decades, Godkin (1988) provided a comprehensive list of problem areas associated with technology transfer. Many of these problems still persist and with rapidly changing technological and business trends new problems have emerged. The productive entities that have been most affected by these problems are small and medium enterprises (SMEs). While large organizations may be able to gain access to the resources needed to overcome these problems this is usually not the case with SMEs. Evidence exists to show that governments, international agencies, and non-governmental organizations (NGOs) have all attempted to alleviate these problems by introducing various supportive measures. Yet, many of these measures make the tacit assumption that technology transfer (TT) is a relatively predictable process whereby buyers of technology (transferees) acquire, assimilate, and then improve the purchased technology, often with assistance from government policies (Cusumano and Elenkov, 1994). This approach tends to oversimplify the magnitude of the problem faced by SMEs in planning and implementing TT projects.

The main objective of this paper is to draw attention to the problems commonly faced by SMEs in planning and implementing technology transfer and argue that a holistic approach needs to be adopted by SMEs to successfully bring in the technologies they need, through TT projects, so that they can compete effectively and grow in today's global business setting. The rest of this paper is divided into four parts. The next part, which is the second of this paper, presents an overview of technology transfer and suggests that in today's business setting it may have to be viewed with fresh perspectives. This is followed by a review of common TT models that have been developed over the past few decades. Their strengths and weaknesses are described briefly. This is followed by the fourth part of this paper that argues that firms may consider adopting what may be called the "Life Cycle Approach for Planning and Implementing a Technology Transfer Project" to overcome the difficulties and common problems that they face in transferring technology. The fifth and concluding part then presents an initiative being planned by the Asian and Pacific Centre for Transfer of Technology of the United Nations – Economic and Social Commission for the Asia and Pacific (UNESCAP–APCTT) to help SMEs overcome these problems.

2. AN OVERVIEW OF TECHNOLOGY TRANSFER: DEFINITION AND MODES

Interest in technology transfer goes back to over six decades. During the colonial era, technology transfer by colonial powers to production entities in their colonies was mainly in the primary sector such as mining, plantation and agriculture (Ramanathan 1989). Those transfers were aimed at the development of methods and techniques in order to obtain the maximum output in export industries such as mining and plantation agriculture and the development of infrastructure for such industries. After independence, in the late 1940s and early 1950s, many governments in the newly independent countries showed great interest in acquiring technology for import-substitution and often relied heavily on firms from their former colonial rulers to gain access to the technologies needed (Bar-Zakay, 1971; Ramanathan 1989). In this era, transfer of technology by multinational companies (MNCs) became common, and public international bodies and not-for-profit organisations also became involved in such activities with the aim of improving living conditions in the recipient country by producing goods to be sold in the local market (Robinson, 1988). However, the transfer of technology of MNCs, through foreign direct investment and licensing, became controversial, and considerable literature emerged that was highly critical of their motives and mode of operations (Robinson 1988; Ramanathan 1989; Takii 2004). In fact, technology transfer became so controversial that attempts were made by UNCTAD to develop a "code of conduct" for technology transfer. In recent times great changes have taken place in the global setting. The demise of the former Soviet Union, the emergence of many new CIS nations, the opening up of China and India, the transition of many countries from centrally planned to market-oriented economies, and the privatisation or breaking up of large state-owned enterprises in many developing nations have provided opportunities for international technology transfer to take place on an unprecedented scale (Radosevic 1995, Sadowski, 2001).

Definitions of Technology Transfer

Even though technology transfer is not a new business phenomenon, the considerable literature on technology transfer that has emerged over the years agrees that defining technology transfer is difficult due to the complexity of the technology transfer process (Robinson 1988; Spivey *et al.* 1997). The definitions depend on how the user defines technology and in what context (Chen 1996; Bozeman 2000).

The term technology transfer can be defined as the process of movement of technology from one entity to another (Souder *et al.* 1990; Ramanathan 1994). The transfer may be said to be successful if the receiving entity, the transferee, can effectively utilise the technology transferred and eventually assimilate it (Ramanathan, 1994). The movement may involve physical assets, know-how, and technical knowledge (Bozeman, 2000). Technology transfer in some situations may be confined to relocating and exchanging of personnel (Osman-Gani, 1999) or the movement of a specific set of capabilities (Lundquist 2003). Technology transfer has also been used to refer to movements of technology from the laboratory to industry, developed to developing countries, or from one application to another domain (Philips, 2002). In a very restrictive sense, where technology is considered as information, technology transfer is sometimes defined as the application of information into use (Gibson and Rogers 1994). In a similar vein economists such as Arrow (1969) and Dosi (1988) have analysed technology transfer on the basis of the properties of generic knowledge, focusing particularly on variables that relate to product design. Mittleman and Pasha

(1997) have attempted a broader definition where they state that technology transfer is the movement of knowledge, skill, organisation, values and capital from the point of generation to the site of adaptation and application.

It may be useful to examine the distinction between technology transfer and technology diffusion. Sociologists such as Rogers and Shoemaker (1971) and Rogers (2003) have defined technology transfer in the context of diffusion of innovations. This has led to confusion where many researchers, and even practitioners, refer to the terms technology transfer and technology diffusion interchangeably. The literature on technology diffusion, in general, suggests that the term refers to the spreading, often passively within a specific technological population, of technological knowledge related to a specific innovation of interest to that population. Technology transfer, on the other hand, is a proactive process to disseminate or acquire knowledge, experience and related artefacts (Hameri 1996). Furthermore, it is intentional and goal-oriented but not a free process (Autio and Laamanen 1995). Transfer also presupposes agreement and therefore involves agreement, unlike diffusion (Ramanathan 1991; Hameri 1996).

The work of Hayami and Ruttan (1971) and Mansfield (1975) provide some of the earliest insights on the modes of technology transfer which are of relevance even today. Mansfield (1975) classified technology transfer into vertical and horizontal technology transfer. Vertical transfer refers to transfer of technology from basic research to applied research to development and then to production respectively and horizontal technology transfer refers to the movement and use of technology used in one place, organisation, or context to another place, organisation, or context. Souder (1987) refers to the former as internal technology transfer and the latter as external technology transfer. Souder further elaborates upon vertical technology transfer as a managerial process of passing a technology from one phase of its life cycle to another. This elaboration is valuable because it serves to reinforce the fact that it may be possible to horizontally transfer technology at any stage of the technology life cycle. Hayami and Ruttan (1971) and Mansfield (1975) refer to "material, design, and capacity transfer." Material transfer refers to the transfer of a new material or product while design transfer corresponds to the transfer of designs and blueprints that can facilitate the manufacturing of the material or product by the transferee. Capacity transfer involves the transfer of know-why and know-how to adapt, and modify the material or product to suit various requirements. While Hayami and Ruttan focused on agricultural technology transfer, Mansfield emphasised manufacturing technology.

Modes and Mechanisms of Technology Transfer

The above technology transfer concepts were put in perspective by Amsden (1989) and Habibie (1990). Amsden (1989) argued that while in developed countries the technology/product cycle took the route, {Research to Development to Design to Production} whereas in technologically less advanced developing countries, it tends to take the route, {Production to Design to Development to Research}.

According to Amsden, learners do not innovate and must compete initially on the basis of low wages, state support, high quality and productivity. The route that must thus be pursued should be based on transfer, absorption, and adaptation of existing technology. This viewpoint fits in with the material, design, and capacity transfer progression. Habibie (1990), often referred to as the architect of the Indonesian aircraft industry, states that, "technology receivers must be prepared to implement manufacturing plans on a step-by-step basis, with the ultimate objective of eventually matching the added-value percentage obtained by the technology transferring firm." He refers to such an approach as "progressive manufacturing" and popularised the slogan, "begin at the end and end at the beginning" implying that a transferee firm should start with production and move backwards to research as also pointed out by Amsden.

Steenhuis (2000) has combined these ideas and developed the concept of "the technology building." The technology building has two wings; the innovation wing consisting of the research, development, production, and distribution stages of the transferor; and the exnovation wing that consists of the distribution, production, development, and research stages of the transferee. The innovation and exnovation wings refer to the technology development stages of the transferor and transferee respectively in accordance with the Amsden and Habibie models of technology development. Steenhuis points out that transfer of technology can take place between the stages of both wings of the technology building in a variety of combinations. The terms innovation and exnovation, as used by Steenhuis, while useful, may cause confusion to practitioners since the term innovation is used in many different contexts. Thus, in this chapter the technology development stages of the transferor and transferee will be referred to as "technology generation" and "technology assimilation" respectively.

To avoid looking at technology transfer in a restrictive manner it may thus be useful to view technology transfer possibilities between the "generation" and "assimilation" processes of the transferor and transferee (Ramanathan, 2000). This is shown schematically in Figure 1.1.

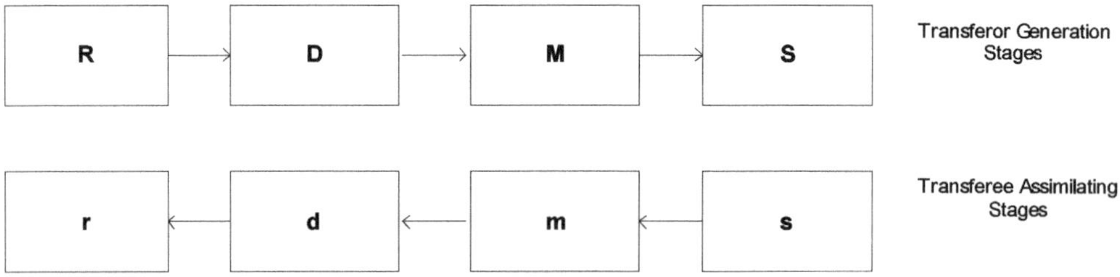

Figure 1.1: The Technology Development Chains of the Transferor and Transferee

Using the "technology development chain" concept outlined in Figure 1.1 above, Ramanathan (2000) points out that the simplest form of technology transfer could be said to take place when an owner of technology (the transferor) transfers the technology needed by a business partner (the transferee) to sell and service a product produced by the owner. This may be depicted as an [S: s] mode of transfer. The representation within parentheses implies that a product at the end of the "generation" stage is simply being sold and serviced by the transferee. The technology likely to be transferred here is that needed by the transferee to sell, repair, and provide other elements of after-sales service to customers buying the product. The objective of the transfer is to effectively maximise the sales of the product in the region managed by the business partner. Another possible variation is [M: s] if the transferee is the sole distributor of the product made by the transferor. These two types of technology transfer arrangements with a predominantly sales focus may be referred to as a "sales intensive mode" of technology transfer.

Based on similar considerations of business objectives, Ramanathan (2001) provides a classification of possible modes and possible transfer mechanisms that may be used. These are summarized in Table 1.1.

Table 1.1: A Possible Taxonomy of Technology Transfer Modes

Transfer Mode	*Possible Transfer Mechanisms*
Sales Intensive	
[S: s] or	Sales and service agreement either
[M: s]	as an agent or sole distributor
Manufacturing Intensive	
[M: m,S] or	Subcontracting arrangements, original
[M: m,s] or	manufacturing arrangements (OEM),
[D:m,S] or	production licensing, and joint ventures
[D:m,s]	
Development Intensive	
[R:d,M,S]or	Original design manufacturing (ODM),
[R:d,m,S] or	production licensing, joint ventures
[R;d,m,s]	
Research Intensive	
[R:r,D,M,S]	Joint R&D and production, university – industry
[R:r,d,M,S]	licensing, Government R&D institute –
[R;r,d,m,S]	industry licensing
[R: r,d,m,s]	

Source: Adapted from Ramanathan (2001).

The term "mode" is used to refer to the transfer links between the phases of the technology development chains of the transferor and transferee while the term "mechanism" is used to describe popular business arrangements that are deployed to transfer technology. The classification does not include the transfer of technology by multinationals to their wholly-owned subsidiaries operating in other locations. The classification proposes that technology transfer arrangements be examined under four main groups namely: sales intensive; manufacturing intensive; development intensive; and research intensive. Each of these categories involve different strategic issues from a business perspective

Based on the above, the following conclusions may be drawn:

- Commercial technology transfer may be defined as a mutually agreed upon, intentional, goal-oriented, and proactive process by which technology flows from an entity that owns the technology (the transferor) to an entity seeking the technology (the transferee). The transfer involves cost and expenditure that is negotiated and agreed upon by the transferee and transferor. The transfer may be said to be successful if the transferee can successfully utilise the technology for business gains and eventually assimilate it.
- Technology transfer can be vertical or horizontal technology transfer. Vertical transfer refers to transfer of technology from basic research to applied research, development, and production respectively and horizontal technology transfer refers to the movement and use of technology used in one place, organisation, or context to another place, organisation, or context.
- In today's globalised and liberalised business setting, many technology transfer modes could be deployed depending on how the technology development chains of the transferor and transferee are linked. Technology transfer can commence from a simple level to a much more comprehensive one with time. The mode chosen would depend on the corporate strategies of the transferor and transferee and the technological capability of the transferee.

The next part will examine some of the popular models of technology transferor that have been developed over the years to help transferees and transferors of technology understand the technology transfer process better.

3. POPULAR TECHNOLOGY TRANSFER MODELS

Since the early 1970s, considering the difficulties and complexities faced by managers of technology transfer projects, researchers, consultants, and practitioners of technology transfer have been proposing models of technology transfer that could facilitate the effective planning and implementation of technology transfer projects. Both qualitative and quantitative models have been proposed. Jagoda (2007) points out that, "Qualitative models often have as their objective the delineation of activities involved in managing TT and the elicitation of factors and issues that can influence the success and/or effectiveness of TT. Quantitative models, on the other hand, aim at quantifying parameters of significance in TT and analysing them with a view towards minimising goal incompatibility between the transferors and transferees of technology." In this paper, emphasis will be on the qualitative models. The mathematics involved in the quantitative models will not be elaborated upon and only their major findings will be presented.

A Brief Overview of Some Qualitative TT Models

(a) The Bar-Zakay Model

Bar-Zakay (1971) developed a rather comprehensive TT model based on a project management approach. He divided the TT process into the Search, Adaptation, Implementation, and Maintenance stages. He depicted the activities, milestones, and decision points (go or no-go) in each of these stages as shown in Figure 1.2. The upper half of the figure delineates the activities and requirements of the transferor (referred to as the "donor" by Bar-Zakay) and the lower half that of the transferee or the "recipient." The activities to be carried out are specified in detail in this model and the importance of both the transferor and transferee acquiring skills to undertake technological forecasting, long-range planning, and gathering of project-related intelligence is emphasised. The model uses the term "donor" for the transferor giving the impression that the owner of technology is giving away a valuable asset out of altruistic reasons. This is clearly not the case and the use of such terms must be avoided.

The Bar-Zakay model also suffers from another disadvantage. Jagoda (2007) points out that, "The model has limited relevance today since many of the activities, terms, and ideas expressed reflected the setting of the late 1960s to early 1970s, when buyers of technology were mainly passive recipients who depended greatly on aid programs for the purchase of technology. It was also an era when government controls were instrumental in determining the rate, direction, and scope of technology flows."

The lessons that can be learnt from the Bar-Zakay model are the following:

- There is a need for a comprehensive examination of the entire TT process from "search" right through to "post-implementation" activities.
- A process approach must be adopted in planning and implementing TT projects
- It is important to have milestones and decision points so that activities can be strengthened, mistakes corrected, or even the project terminated at any point in time.

(*b*) The Behrman and Wallender Model

Behrman and Wallender (1976) have proposed a seven stage-process for international technology transfer that may be more relevant to multinational corporations. The seven stages are:

- Manufacturing proposal and planning to arrive at decisions regarding location and preparing a business case including good resource assessments.
- Deciding the product design technologies to be transferred.
- Specifying details of the plant to be designed to produce the product and other aspects related to construction and infrastructure development.
- Plant construction and production start-up.
- Adapting the process and product if needed and strengthening production systems to suit local conditions.
- Improving the product technology transferred using local skills.
- Providing external support to strengthen the relationship between the transferor and transferee.

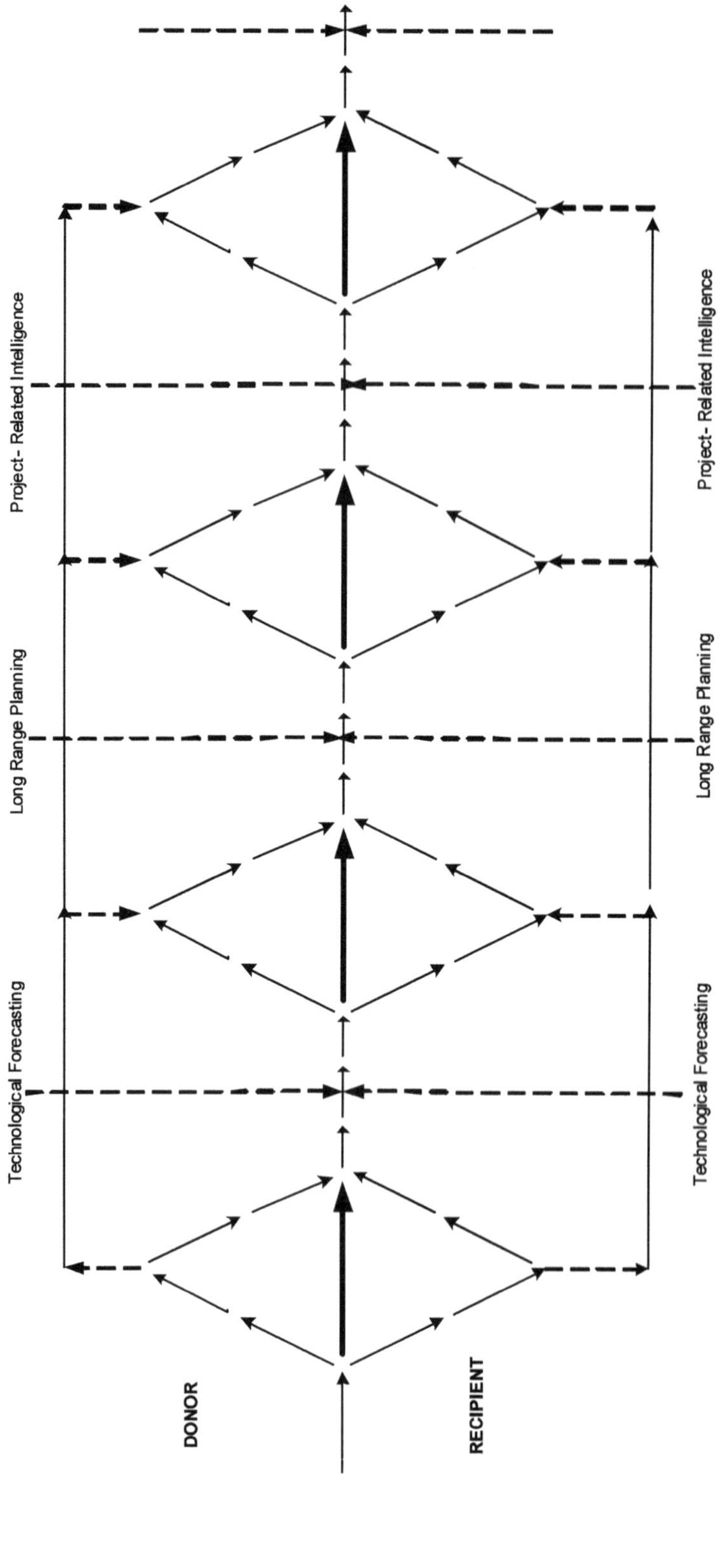

Figure 1.2: The Bar-Zakay Model of Technology Transfer
***Source*: Jagoda (2007)**

One of the weaknesses of this model is that, during the first three stages, the transferor develops the technology transfer project with minimal involvement of the transferee thereby reinforcing dependency. However, in the fifth and sixth stages there is considerable scope for the transferee to assimilate and improve both product and process technology. This serves to emphasise the fact that technology transfer does not stop with commencement of production and unless there is a mechanism to foster assimilation the project cannot be considered to have delivered.

The lessons that can be learnt from this model are the following:

- ☆ There is a need for the transferee to be involved right from the beginning in the planning and implementation of a TT project.
- ☆ A technology transfer project does not end with commencement of production.
- ☆ Unless explicit measures are in place to ensure assimilation of the transferred technology, the technology transfer cannot be said to have been successful.

(c) The Dahlman and Westphal Model

Dahlman and Westphal (1981) carried out considerable work in the Republic of Korea and, based on their experience in rapidly industrialising countries during the 1980s, in the Far East, have proposed a nine stage process model as follows:

- ☆ Carry out pre-investment feasibility to gather information and carry out a techno-economic analysis to establish project viability.
- ☆ Carry out a preliminary identification of technologies needed, based on the feasibility study.
- ☆ Carry out basic engineering studies that involve the preparation of process flow diagrams, layouts, materials and energy balances and other design specifications of the plant and machinery and the core technology to be transferred.
- ☆ Carry out a detailed engineering study that involve the preparation of a detailed civil engineering plan for the facility, including construction and installation specifications and identification of the peripheral technology needed to make the transfer effective.
- ☆ Carry out the selection of suppliers for equipment and subcontracting services to assemble the plant and machinery and plan for the co-ordination of the work among various parties
- ☆ Prepare and execute a training and education plan, in consultation with the suppliers of technology, for the workers who would be employed in the technology transfer project.
- ☆ Construct the plant.
- ☆ Commence operations.
- ☆ Develop trouble-shooting skills and put in place arrangements to solve design and operational problems as they arise, especially during the early years of operation.

This model may be regarded as an improvement of the Behrman and Wallender model with great emphasis on transferee involvement at all stages of the TT project. Its major weakness is that it assumes that the transferee will have access to high-level engineering skills. This may not be true in many developing countries. It also pays very little attention to negotiation and post-implementation assimilation initiatives.

The important lessons that this model presents include the following:

- ✰ A TT project is best studied using a sequential process perspective.
- ✰ Any TT project should not be commenced without a careful feasibility study since such projects often require heavy resource commitments.
- ✰ The transferee should be involved in the planning right from the beginning.
- ✰ It is important for transferees to develop sound engineering and project management skills without which the TT process cannot be managed effectively.

(*d*) The Schlie, Radnor, and Wad Model

Schlie *et al.* (1987) propose a simple and generic model that delineates seven elements that can influence the planning, implementation, and eventual success of any TT project. These seven elements are given below.

- ✰ The transferor, which is the entity selling the technology to the recipient.
- ✰ The transferee, which is the entity buying the technology.
- ✰ The technology that is being transferred.
- ✰ The transfer mechanism that has been chosen to transfer the chosen technology.
- ✰ The transferor environment which is the immediate set of conditions, in which the transferor is operating. Attributes of the transferor environment that can influence the effectiveness of the transfer process include, among others, economic status, business orientation (inward versus outward), stability, attitude and commitment to the transfer project, and operating policies.
- ✰ The transferee environment which is the immediate set of conditions under which the transferee is operating. Attributes of the transferee environment that can influence the absorptive capacity of the transferee include physical and organisational infrastructure, skills availability, attitude and commitment to the transfer project, technological status, business orientation (inward versus outward), economic status, and stability.
- ✰ The greater environment which is that surrounding both the transferor and the transferee. There may be layers of this environment that are sub-regional, regional, and global. Even if the immediate operating environments of the transferor and the transferee are favourable to the technology transfer, if the layers of the greater environment are not supportive, then cross-border and international technology transfer could be adversely affected. Factors in the greater environment such as political

relationships between countries, exchange rates, investment climates, trade negotiations, balance of trade, relative technological levels, and the status of intellectual property protection regimes could have a great influence on the success of a TT project.

The seven elements of this model are valid even in today's business setting. The way that they manifest themselves can however change with time. The weakness of this model is that it offers no guidelines as to what a transferee should do.

The valuable lessons that emerge from this model are as follows:

- ☆ The many changes that have taken place and are taking place in the global business setting today have made it imperative for managers of technology to gain good insights into the transferee environment, transferor environment, and the greater environment when planning and implementing a TT project.
- ☆ The choice of the technology transfer mechanism should be based on a sophisticated understanding of the other six elements.

(*e*) The Chantramonklasri Model

The Dahlman and Westphal Model has been further improved by Chantramonklasri (1990) who proposes a five phase model as shown in Figure 1.3.

The five phases of this model are as follows:

- ☆ Carrying out a pre-investment and feasibility study
- ☆ Developing engineering specifications and design based on the feasibility study
- ☆ Commence capital goods production based on the engineering specifications and designs that have been developed.
- ☆ Commissioning and start-up including comprehensive of the workforce
- ☆ Commence commercial production

While the first two phases of this model are valid it is not clear whether the required capital goods can be produced within the transferee setting unless the transfer arrangement also includes the transfer of technology needed to manufacture these. While this may be valid in large, technologically advanced countries such as China and India, it may not be so in other smaller developing countries. As in the Dahlman and Westphal Model the negotiation and assimilation elements are missing. The lessons that may be learnt in this case are similar to those of the Dahlman and Westphal Model.

(*f*) Other Qualitative Models of Technology Transfer

There are several other models that have been developed. However, due to limitations of space these will only be described briefly.

Lee *et al.* (1988) have developed a longitudinal model of technology transfer based on a study of developing and rapidly industrialising countries. They point out

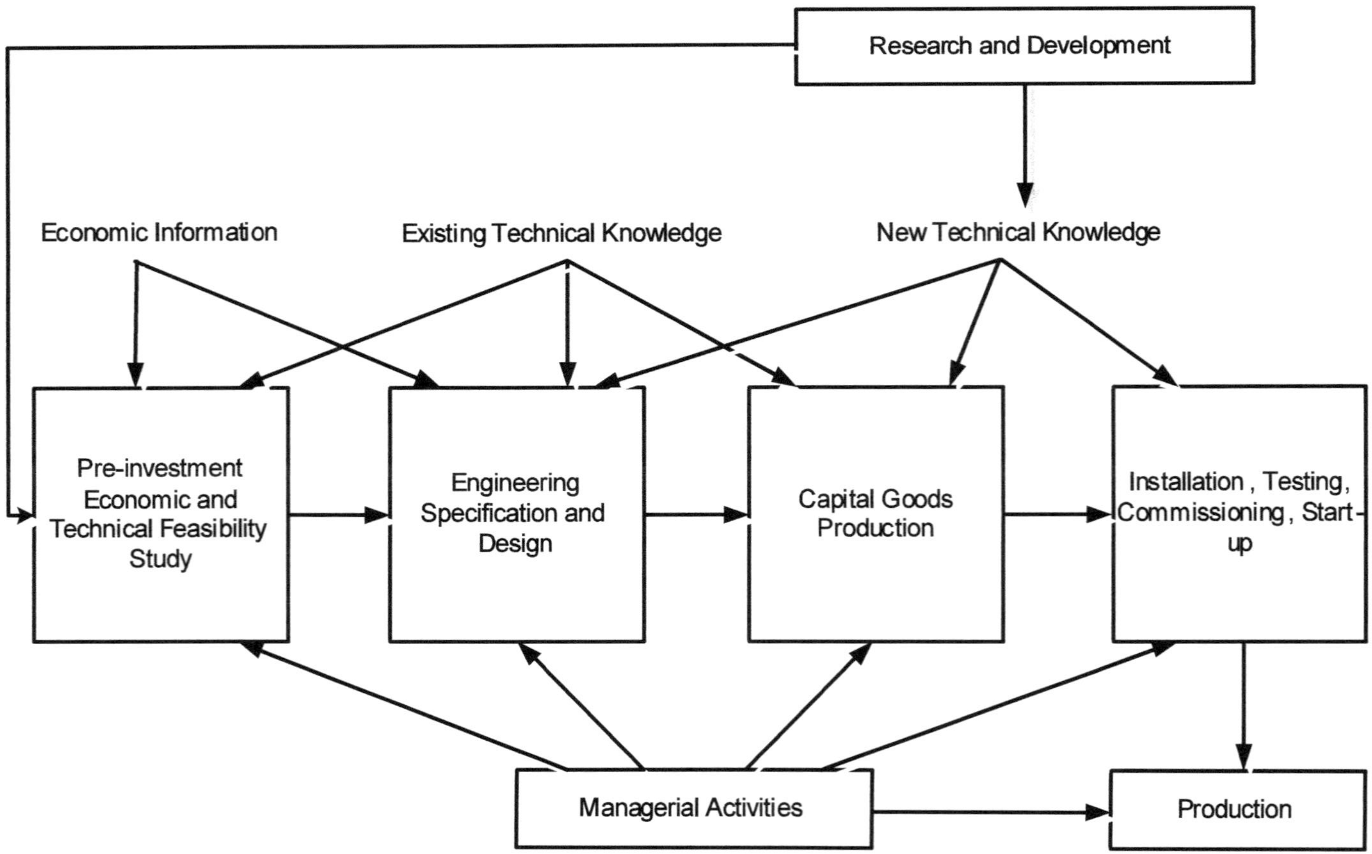

Figure 1.3: The Five-phase Model of International Technology Transfer
***Source*: Jagoda (2007)**

that a transferee firm needs to put in place strategies to be able to go through the stages of acquisition, assimilation, and eventual improvement. As the firm advances technologically, it needs to choose appropriate mechanisms of transfer, depending on the stage of the life cycle of the technology and their own technological capability profile. They also note that the mechanisms chosen by the transferor to transfer technology will depend on the relative newness of the technology, its strategic importance to the transferor firm, and the level of intellectual property protection needed.

Reddy and Zhao (1990), in a model similar to that of Schlie *et al.* (1987) state that any international technology transfer (ITT) project should examine three main components, which they refer to as the home-country component, host-country component, and transaction component. The home country is that of the transferor and the host country is that of the transferee. The home-country component involves an examination of issues such as home-country government policies on technology transfer (restrictions etc.), the role and strategy of transferring firms from a foreign direct investment point of view, the nature and importance of technology to be transferred, and the firm's global R&D investment strategies. The host-country component involves issues such as host-country government policies related to foreign investment and technology transfer, the relative suitability of the technology being considered for transfer, the technological capability of the transferee and the scope for upgrading, mechanisms of transfer being considered, and the scope for assimilation of the transferred technology. The transaction component consists of important business issues such as the pricing of technology, intellectual property protection, payment modalities, potential conflicts, and measures for ensuring effective transfer.

Keller and Chinta (1990) argue that effective technology transfer would be determined by the extent to which the transferor and transferee manage the barriers that impede transfer and strengthen initiatives that facilitate it. The facilitating initiatives refer to the willingness of the partners to adapt their respective strategic and operational postures to ensure a "win-win" outcome. The barriers could be political, legal, social, cultural, economic, and technological. They also stress the importance of selecting the correct mechanism to transfer the technology.

The UNIDO (1996) model, in what appears to be an endorsement of the Bar-Zakay approach, suggests that, in the manufacturing sector, once the need for a TT project is established, the steps of search, evaluation, negotiation, contract execution, and technology adaptation and absorption should be followed sequentially to ensure effectiveness.

Durrani *et al.* (1998) have proposed a generic model consisting of five steps:

- ✰ Establishing market-place requirements
- ✰ Identifying technology solutions
- ✰ Classifying the identified technology solutions
- ✰ Establishing sources from where the desired technology could be acquired
- ✰ Finalising the technology-acquisition decision

This model stops with the technology acquisition decision. Its major lesson is that it stresses the importance of establishing the need for a technology transfer project and the need for identifying multiple sources of technology for enabling a better choice of transferor.

Bozeman (2000) has proposed a contingent effectiveness model of technology transfer. While the emphasis is on technology transfer from universities and government laboratories to industry, the model is also relevant to inter-firm technology transfer. In this model, the key elements of the transfer process are:

- The transfer agent (the transferor)
- The transfer mechanism
- The transfer object (the content and form of the technology being transferred)
- The transfer recipient (the transferee)
- The demand environment (market and non-market factors vis-à-vis the need for the technology).

This model also stresses importance of establishing the need for a technology transfer project and the need for identifying multiple sources of technology for enabling a better choice of transferor. Six "out-the-door" measures are proposed. These are market impact, economic development, political benefits, opportunity costs, and development of scientific and human capital as a result of the transfer. The importance of impact assessment is a valuable lesson that this model imparts.

(g) A Brief Overview of Some Quantitative TT Models

The literature is sparse when it comes to quantitative models of technology transfer. Some of the more important models are described briefly. For the sake of brevity, the mathematics has been left out and the interested reader may wish to refer to the original publications.

Perhaps, the earliest quantitative model is due to Sharif and Haq (1980). This model proposes the concept of potential technological distance (PTD) between a transferor and transferee and argues that when the PTD is either too great or too small between the transferor and transferee, the effectiveness of the transfer is low. It suggests that when a transferee first looks for a potential transferor it is important to look for one with an "optimal" PTD. From a practical point of view, a potential transferor at the firm level may not be willing to easily divulge information that could enable an assessment of the PTD. The greatest value of the model is that it draws attention to the need for incorporating the concept of a PTD in deciding the transferor.

Raz *et al.* (1983) have presented a model of technological "catch-up" that shows how a technology leader, through technology transfer, can assist the rate of technological development of a technology follower. The model examines three phases of growth of a technology follower namely, the slow initial phase with high technological capability gap, the faster learning phase with the decreasing gap, and catch-up phase when the technological gap is very small or closed. They argues that this type of analysis would enable technology leaders to develop clear policies, based on considerations of competitiveness, security, and other related issues, when entering into technology transfer agreements.

Using an econometric model, Klein and Lim (1997) have studied the technology gap between the general machinery and electrical and electronic industries of Korea and Japan. Their findings suggest that technology transfer from leaders can play a critical role in upgrading the technological levels of follower firms. Their study also shows that the followers should supplement the transfer by independently putting in place measures to assimilate, modify, and localise the technology transferred from the leader. This model thus emphasises, based on empirical evidence, the need for post-implementation activities that facilitate assimilation and modification of the transferred technology. It also clearly delineates the need for a firm, as it grows technologically, to link its technology transfer activities with internal R&D.

It may be said that the main contribution of the quantitative models is their emphasis on the need for partners in technology transfer projects to develop skills to be able to use formal, analytical approaches that can generate needed information for better technology transfer planning.

An examination of the models of technology transfer shows that there are several valuable lessons that they convey. These are summarized below.

- ☆ It is important to expend comprehensive analytical effort in establishing the need for a technology transfer project prior to the commencement of a TT project.
- ☆ A TT project should not be commenced without a careful feasibility study since such projects often require heavy resource commitments.
- ☆ A process approach must be adopted in planning and implementing TT projects and to ensure effective technology transfer there is a need to comprehensively examine the entire process from "technology search" right through to "post-implementation" activities.
- ☆ The many changes that have taken place and are taking place in the global business setting today have made it imperative for managers of technology to gain good insights into the transferee environment, transferor environment, and the greater environment when planning and implementing a TT project.
- ☆ Multiple sources of technology must be identified to enable a good choice of transferor.
- ☆ The transferee must be involved right from the beginning in the planning and implementation of a TT project.
- ☆ It is important for transferees to develop sound engineering and project management skills without which the technology transfer process cannot be managed effectively.
- ☆ Partners in TT projects need to develop skills to be able to use formal, analytical approaches that can generate needed information for better technology transfer planning.

- It is important to have milestones and decision points so that activities can be strengthened, mistakes corrected, or even the project terminated at any point in time.
- The mechanisms chosen by a transferor to transfer technology will depend on the transferor and transferee setting, the technological capability of the transferee, the relative newness of the technology, its strategic importance to the transferor firm, and the level of intellectual property protection needed.
- As a transferee firm advances technologically, it needs to choose appropriate mechanisms of transfer, depending on the stage of the life cycle of the technology and its own technological capability profile.
- A technology transfer project does not end with commencement of production. Unless explicit measures are in place to ensure assimilation of the transferred technology the technology transfer cannot be said to have been successful.
- The success of a technology transfer project would be determined by the extent to which the transferor and transferee manage the barriers that impede transfer and strengthen initiatives that facilitate it.

However, what may also be noted is that there is no model that tries to capture all of these important considerations. An eclectic model that presents all this wisdom in a process-oriented approach would be very useful to managers of technology transfer project. Such a model must also have the capacity to address many of the problems faced by firms, especially small and medium enterprises (SMEs), when planning and implementing technology transfer. The next part will first present a summary of common problems faced by SMEs in planning and implementing technology transfer and then propose an eclectic, process model called, "the Life-cycle Approach for Planning and Implementing Technology Transfer" that tries to incorporate the wisdom of the models discussed. It is envisaged that the adoption of this process model will enable SMEs to manage the common problems they face in planning and implementing TT projects.

4. THE LIFE CYCLE APPROACH FOR PLANNING AND IMPLEMENTING A TECHNOLOGY TRANSFER PROJECT

This part commences with a brief presentation of common technology transfer problems faced by SMEs. This will be followed by the "Life Cycle Approach for Planning and Implementing a Technology Transfer Project."

Technology Transfer Problems Commonly Faced by SMEs

Based on the work of Jagoda (2007) and Ramanathan (2007), problems faced by SMEs in planning and managing technology transfer may be classified into three categories namely, technology transfer process issues, corporate capability issues, and operating environment and NIS issues.

The problems are summarized below.

(a) Technology Transfer Process Issues

Problems during the Technology Justification and Selection Stage

- ☆ Wrong selection of technology based on misjudgements when preparing a business case for a TT project
- ☆ The cost of buying, installing, operating, and maintaining the technology is too high
- ☆ The technology selected is too complex for easy understanding and assimilation of the transferee
- ☆ The technology needs considerable adaptation to suit local conditions
- ☆ Obsolescence of technology while the transfer is in progress

Problems during the Planning Stage

- ☆ Transferor (seller) underestimates the problems in transferring the technology to a developing country setting
- ☆ Transferor does not fully understand transferee needs
- ☆ Transferee managers are not involved in the planning which is carried out only by the transferor
- ☆ Too much attention is paid to the hardware to be purchased and not enough attention is paid to skills and information acquisition
- ☆ Overestimation of the technological capabilities of the transferee by the transferor thereby leading to unrealistic expectations on how well the transferee can meet target dates
- ☆ Poor market demand forecasting by the transferee of the outputs to be produced by using the transferred technology
- ☆ The objectives of the transferor and transferee are not compatible
- ☆ Mechanisms chosen for implementing the transfer are not appropriate

Problems during Negotiations

- ☆ Differences in negotiation approaches and strategies
- ☆ Lack of trust between the transferor and transferee
- ☆ Goal incompatibility during negotiations
- ☆ Inability to reach agreements on pricing, product, and marketing strategies
- ☆ Both parties try to achieve results in an unrealistically short period of time

Problems during Technology Transfer Implementation

- ☆ Shortage of experienced technology transfer managers
- ☆ Lack of trust in transferor developed systems by the transferee
- ☆ Inability to achieve quality targets
- ☆ Delay in obtaining supplementary materials, needed for quick implementation, from the local environment

- ☆ High cost and poor quality of locally available materials needed to implement the technology transferred
- ☆ Inadequate tracking of the technology during implementation
- ☆ Cost overrun due to poor implementation

(b) Corporate Capability Issues

Problems due to Inadequate Skills

- ☆ Inability of the transferee to attract the required skills due to financial and industrial restrictions
- ☆ Lack of experience of the transferee's workforce and absence of required skills at the industry level
- ☆ Lack of training of transferee personnel
- ☆ Absence of incentive systems at the transferee firm for learning and assimilating new technologies
- ☆ Language barriers that inhibit effective communication between transferor and transferee personnel and restrict effective transmission and assimilation of relevant information

Problems due to Ineffective Management

- ☆ Lack of visible and committed top management support for the project
- ☆ Lack of top management guidance to decide the type of the technology to be acquired, remuneration, incentives associated with the transfer, and the control of the flow of information.
- ☆ Differences in working methods and practices between the transferor and transferee managers
- ☆ Individual or organisational competition for the ownership of the technologies and the presence of the "not-invented-here" syndrome
- ☆ Failure of top management to identify transferee and transferor personnel who would work closely form project initiation through to full implementation

(c) Operating Environment and National Innovation System (NIS) Issues

- ☆ Shrinking of local markets due to adverse changes in the economic levels of the country
- ☆ Poor physical infrastructure
- ☆ Inadequate supportive institutional infrastructure to provide support in terms of finance, information, skill development, and technology brokering
- ☆ Inadequate mechanisms for intellectual property protection
- ☆ Lack of local suppliers who can deliver quality supplies and lack of policies to develop such suppliers
- ☆ High dependency on foreign suppliers and imports

- Lack of good education and training institutions to upgrade skills
- Ineffective legislation and incentives such as tax holidays, tariff adjustments, and industry parks to promote technology transfer
- Bureaucratic delays at various levels of government in obtaining approvals and clearances for finalizing technology transfer agreements
- Ineffective and sometimes excessive government intervention and regulation
- Foreign exchange restrictions
- Inability of new ventures to compete with former monopolies, often owned by government
- Uncertain tax environments

These problems continue to affect SMEs and even large firms in many developing nations. While a SME may not be able to handle problems related to the operating environment and the NIS, it should nevertheless guard against these while working with the relevant Business Associations and Chambers of Commerce to lobby government to rectify these.

Life Cycle Approach for Planning and Implementing a Technology Transfer Project

The "Life Cycle Approach for Planning and Implementing a Technology Transfer Project" is based on the stage-gate structure developed by Jagoda and Ramanathan (2005) for developing a systematic approach for planning and managing ITT. For the sake of convenience and expository ease, henceforth this model will be referred to as the TTLC (Technology Transfer Life Cycle) approach.

The TTLC approach takes a holistic view of a TT project from its "conception" right up to its "conclusion" and is based on the recognition of the fact that a life cycle of a TT project can be looked at from a process perspective as consisting of six major stages as follows.

- Identifying the technology needed and making a business case to obtain corporate approval
- Searching for possible technology sources and assessing offers
- Negotiating with short-listed suppliers and finalizing the deal
- Preparing a TT implementation plan
- Implementing and assimilating
- Assessing the impact of the TT project.

This life cycle has been developed based on the lessons learnt from the study of popular models of technology transfer that have been reviewed in the previous section of this paper. The major stages in the life cycle are shown schematically in Figure 1.4. It can be seen that, in this generic framework, each stage is associated with a gate. The stages are made up of prescribed tasks with cross-functional and simultaneous activities. The gate or controlling point is at the entrance to each stage. Using the

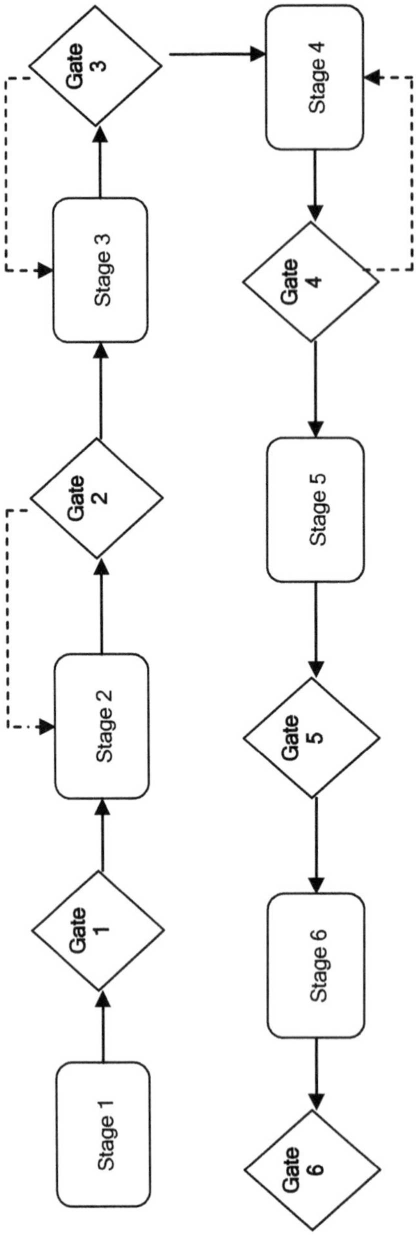

Stage 1: Identifying CVD enhancing technologies
Stage 2: Focused technology search
Stage 3: Negotiation
Stage 4: Preparing a TT project implementation plan
Stage 5: Implementing technology transfer
Stage 6: Technology transfer impact assessment

Gate 1: Confirming identified technologies
Gate 2: Technology and supplier selection
Gate 3: Finalising and approving the TT agreement
Gate 4: Approving the implementation plan
Gate 5: Implementation audit
Gate 6: Developing guidelines for a new project

Figure 1.4: The Life Cycle Approach for Planning and Implementing Technology Transfer

information generated at each stage, in-depth and critical evaluation is carried out at the gate that follows the stage. Based on the evaluation, a decision may be taken to go forward, kill the project, put it on hold, or recycle it. It is envisaged that, through this approach, proactive measures could be taken to avoid or minimize problems thereby enhancing the chances of successful TT. The main advantage of such an approach is that it could ensure that major activities are not carried out carelessly or even missed.

Stage 1: Identifying CVD Enhancing Technologies

All enterprises whether they are large firms or SMEs can compete effectively only on the basis of "customer value creation." Customer value may be defined as a function of quality, delivery, flexibility, convenience, and cost (Ramanathan, 2001). Quality represents how well a specific good or service meets customer expectations. Speed describes the time needed to design, produce, and deliver the good or service as characterized by determinants such as cycle time and speed to market. Flexibility reflects how easily and quickly the firm can modify goods or services to meet customer needs in terms of aspects such as options and extent of customisation possible. Creating convenience for the customer implies not only speed of service, but also self-service, process visibility, and easy to use, streamlined, consistent, and reliable customer service. Lastly, cost refers to all objective and subjective costs that the customer incurs to acquire, use, and dispose of the good or service and includes dimensions such as discounts, rebates, and incentives. Customer value is enhanced as quality, speed, flexibility, and convenience increases while cost decreases. These five determinants of customer value creation may be referred to as Core Value Determinants (CVDs) (Ramanathan, 2001). To ensure sustainable competitive advantage a firm must offer its customers a CVD profile that sets it apart from its competitors. Thus, in Stage 1 what is important is for the transferee firm to decide what technology or technologies it needs to create a unique CVD profile that will enhance its competitive edge vis-à-vis its competitors. The key activities that must be carried out at this preliminary stage of the technology transfer project are the following.

- ☆ An informal Technology Transfer Steering Committee (TTSC) is set up to study how competitors are using technology to enhance customer value and what technologies are available that could deliver even greater value.
- ☆ A list of technologies needed is developed and technology roadmaps are constructed to understand future trends of these identified technologies.
- ☆ Information for this is obtained through Internet searches, study of technical publications, exchange of communication with potential suppliers of technology, contacts with universities etc.
- ☆ A quick market assessment that examines market size, market potential, and likely market acceptance of the proposed initiatives is carried out, mainly through the use of marketing expertise and contacts with key users.
- ☆ A technical assessment is also carried out to estimate, the resources and capabilities needed to adopt the new technologies, time needed, costs involved, likely risks, and possible barriers (including policy, legal and regulatory aspects).

Gate 1: Confirming Identified Technologies

Gate 1 is a "critical but supportive" screen. The decision-makers, usually a top management team, develop a set of "must meet" criteria to review the proposal. The criteria could include:

- Strategic alignment
- Project feasibility in terms of technical and resource considerations
- Magnitude of opportunity
- Market attractiveness
- Sales force and customer reaction to the proposed technology
- Regulatory, legal, and policy factors

Financial returns are usually assessed at this gate using simple financial calculations such as payback period. The decision-makers will, at this gate, modify, confirm the composition of the TTSC which will then be in charge of the project.

Stage 2: Focused Technology Search

This is probably the most important stage where detailed investigation is carried out by the TTSC. It is here that a strong business case for the technology transfer is built. This includes specifying in detail the following:

- How the technology sought is expected to enhance customer value by influencing the CVDs
- What components of technology are needed (hardware, skills, information, and organizational arrangements)
- The extent to which the abilities to use the technology are available in-house and what gaps have to be bridged
- The resource commitments needed and the expected benefits
- Prioritised shortlisting of suppliers for the technology based on their business strategy, technological capabilities, experience in handling TT projects, past performance, and cross cultural expertise.
- Competitive analysis to assess the impact of the technology sought on competitiveness

Based on a consideration of these aspects, a business case is developed that includes clear technology specifications, discounted cash flow (DCF) analysis, project justification, and business plan. Development of this business case requires multidisciplinary interaction and cross-functional cooperation. If this stage is carried out poorly it could have adverse impacts at the remaining stages and cause serious difficulties.

Gate 2: Technology and Supplier Selection

This is the final gate prior to the formal negotiation and launch stage where the project can be killed before it enters a heavy spending phase. This gate gives the go-ahead for a "heavy spend." Gate 2 critically examines the analysis of Stage 2 and rechecks against the major criteria used in Gate 1. The following steps need to be followed very carefully at this gate.

- All suggestions with regard to technology choice, components of technology needed, capability gaps to be bridged, resource commitments needed, expected benefits, and supplier profile ratings are critically examined.
- The technology will be assessed very rigorously using techno-economic, socio, and politico-legal factors.
- The preferred supplier ranking will be reassessed rigorously based on strategic fit and process support capability and may be modified from the ranking proposed in Stage 2.
- The financial analysis (DCF) is rechecked very carefully here.
- The TTSC may have to revise the analysis in the light of the critical evaluations (as indicated in the figure) and submit the new analysis for further evaluation.

If the decision is a Go-decision then the TTSC is converted to a full technology transfer project team that is empowered, multifunctional, and headed by a leader with authority.

Stage 3: Negotiation

This is a critical stage where the TTSC now negotiates with the shortlisted suppliers. A critical issue in TT negotiation is the valuation of the technology to be transferred. The extent to which both parties can influence price depends on their respective bargaining power. The transferor's power arises out of the resources possessed such as ownership of a desired technology, brand name, reputation, management expertise, capital, and international market access. Transferee power often tends to have its roots in local knowledge and networks, access to local markets, raw materials and low cost labour, and political connections. To ensure effective negotiation, frequent contact and communication between both parties is imperative. The following activities need to be carried out at this stage.

- Agreeing upon a basis for the valuation of the technology and reaching agreement on issues related to payments and intellectual property protection – both short-term and strategic benefits have to be examined.
- Delineation of each party's contribution and responsibilities towards the TT project
- Discussion of issues and methods related to the transfer of codified and uncodified aspects of technology including training
- Creation of effective channels of communication between both parties including visits to each others facilities
- Consultation with government authorities to ensure concurrence with government policies and identification of possible barriers, likely policy changes and government support available.

- ☆ Finalizing the most appropriate mechanism(s) for transferring the technology components sought.
- ☆ Preparation of a detailed transfer agreement with emphasis on ensuring intellectual property protection
- ☆ Reaching agreement upon payment amounts, procedures, and time frames

Gate 3: Finalizing and Approving Agreement

This gate is operationalised once the negotiations have reached a satisfactory level and the parties express the desire to finalize the agreement through the drawing up of a legal agreement. This gate will critically evaluate the following:

- ☆ The comprehensiveness of the detailed transfer agreement
- ☆ The adequacy of intellectual property protection arrangements
- ☆ The appropriateness of the proposed mechanism(s) for technology transferring
- ☆ The suitability and affordability of the payment amounts, procedures, and time frames

Stage 4: Preparing a Technology Transfer Project Implementation Plan

At the beginning of this stage a transferor of technology would have been chosen and since the creation of a sound organisational infrastructure is critical to the implementation of TT, this stage focuses on making organisational arrangements to receive the technology. The main activities during this stage are the following:

- ☆ Identification of changes to be made to the organisational structure and work design based on an understanding of the transfer components
- ☆ Identification of changes to be made in the knowledge management system and policy regimes to accommodate the new technology
- ☆ Development of pragmatic training and education schedules for the workforce that matches with the components to be transferred
- ☆ Formulation of measures to build good relationships between the transfer personnel
- ☆ Formulation of a realistic TT project implementation plan that can form the basis of a working relationship between the transferor and transferee
- ☆ Milestones are specified to help strengthen project management and control.

Gate 4: Approving the Implementation Plan

At this gate, the following aspects will be carefully scrutinized:

- ☆ Whether agreement has been reached with the transferor with respect to the schedule
- ☆ Adequacy of the training arrangements
- ☆ Adequacy of the modification of the infrastructure

- ☆ Intellectual property protection measures
- ☆ Durations of critical activities
- ☆ Quality assurance procedures
- ☆ Payment schedules

If these are satisfactory then a go-ahead signal will be given. Otherwise revisions will be needed. At this gate an initial payment to the transferor, if specified in the agreement, will also be approved.

Stage 5: Implementing Technology Transfer

Technology transfer implementation requires good project management. Changes to product or process technology may sometimes be essential to the successful implementation of a TT project. Very often, firms in developing nations are confronted with finding suitable people at this stage and close cooperation with the transferor may be needed to locate required skills. Scheduling the timely arrival of allied materials, parts, and services is essential to ensure successful implementation of the project. Training programs will also have to be scheduled and conducted either in-house or at transferor approved locations. The major activities at this stage include the following:

- ☆ Identification of changes to be made to the product or process to suit local conditions and making the necessary adaptations.
- ☆ Recruitment and selection of personnel not already available within the organization and conducting training programs for existing staff.
- ☆ Development of improved remuneration plan to facilitate change management.
- ☆ Formulation of arrangements with ancillary suppliers of materials, parts and services based on a make vs. buy analysis
- ☆ Maintaining links with government authorities to keep track of policy changes
- ☆ Commissioning the transferred technology on or before schedule

Gate 5: Implementation Audit

At this gate the scheduled activities and the goals set for the TT project are evaluated. The focus should be on gaining an understanding of barriers to the successful implementation of TT. The audit may focus on the evaluation of project implementation with respect to critical factors such as:

- ☆ Commitment displayed
- ☆ Conflicts experienced
- ☆ Time frames
- ☆ Cost incurred
- ☆ Quality achieved
- ☆ Extent of learning and skill upgrading

- New knowledge generated
- Communication effectiveness

The compilation of a comprehensive audit report outlining the lessons learned and identifying critical success and failure factors is important at this gate so that future TT projects could benefit from these insights.

Stage 6: Technology Transfer Impact Assessment

Assessing the impact of a TT project is difficult because it is a complex process with multiple outcomes that could emerge throughout the life of a project. Also, the intangible benefits of a TT project are difficult to evaluate. However, a well structured impact assessment could be extremely beneficial and the impacts need to be assessed from customer, market, financial, technological, and organizational perspectives. The following activities are proposed for this last stage.

- Development of a "Balanced Scorecard (BSC)" approach to assess impacts.
- Identification of the variances (if applicable) between actual and expected outcomes and the formulation of organizational corrective measures.
- Examining the feasibility of improving the transferred technology.
- Identification of new or complementary technologies that could be transferred to consolidate the gains made.

Gate 6: Developing Guidelines for Post-Technology-Transfer Activities

At this gate important decisions have to be taken as to whether to continue to use the technology by improving it incrementally or go for another TT project. Successful TT projects can lead to strong and long partnerships between the transferor and the transferee and new projects could be initiated in a variety of ways. At this gate guidelines may be formulated, based on the experience gained at all the previous stages and gates for post-technology-transfer activities such as:

- A new technology transfer project
- Internal development
- A mix of both in partnership with the transferor.

These decisions can then be fed into the corporate planning process of the organization.

Summary Remarks on the TTLC Approach.

The TTLC approach is not purely conceptual. Its practical relevance, usefulness, and validity have been established through several case studies carried out by Jagoda (2007) in Australia and Sri Lanka. The main advantages of the TTLC approach are the following:

- The TTLC approach ensures that a TT project is considered holistically and incorporates much of the wisdom shared by various researchers and practitioners through their technology transfer models.

- The TTLC approach is structured to enable SMEs avoid many of the problems that they normally face when planning and implementing a TT project.
- It is a good way to incorporate cross-functional cooperation in planning and managing TT projects and also ensures that important activities are not forgotten or carried out carelessly.
- A single empowered team is responsible from start to finish. This avoids turf wars.
- All projects may not have to go through all the stages. Low risk projects may go quickly to the latter stages.
- The approach must not be seen as a bureaucratic system. It actually facilitates the development of a streamlined system with clear agreed upon, and visible road map.

Clearly the success of the approach will depend upon the skills possessed by the managers involved in the TT project to carry out the activities effectively at the stages and gates. Thus, organizations that are serious in competing in today's global business setting must develop such skills on a priority basis.

5. CONCLUDING REMARKS

This chapter commenced with an overview of technology transfer and then presented some important models of technology transfer that have been developed over the decades to help firms plan and manage technology transfer projects. The strengths and weaknesses in these models were pointed out and an eclectic, process model titled, "The Life Cycle Approach for Planning and Implementing a Technology Transfer Project" is presented. It is envisaged that this model could help in addressing many of the common problems that are currently faced by transferees of technology, especially SMEs. This concluding section will briefly present an initiative that the UNESCAP Asian and Pacific Centre for Transfer of Technology (APCTT) is currently developing, based on the Life-Cycle Approach, to help technology transfer capacity building in the Asia-Pacific region.

The discussions presented thus far in this paper show that, unless a critical mass of skills to plan and implement a TT project are created in a country, SMEs of that country will continue to face difficulties in managing TT projects. Either the SMEs will have to develop these skills within the organisation or they must have access to these skills through technology intermediaries, SME support institutions, or consultants. The challenge for governments is to develop innovative ways by which the required critical mass of skills can be developed quickly within a range of SME-oriented institutions and within the SMEs themselves.

To help member countries in the Asia-Pacific region develop this critical mass of skills, APCTT is currently in the process of seeking funds to commence a technology transfer capacity building programme in SMEs.

To ensure that this capacity building effort reaches a larger group of SMEs through a "multiplier effect" the APCTT programme would adopt a "training of trainers" strategy. From participating member countries, carefully chosen trainers from SME support institutions, R&D organizations, universities, business and industry associations, and chambers of commerce would be trained extensively on technology transfer fundamentals and the use of the TTLC approach. These trainers would then be expected to return to their countries and train SMEs in their own countries through business and industry associations and relevant government organizations. APCTT would provide the trainers with all the relevant training materials that could be translated into local languages, if needed, and also used for online training and education.

Under this technology transfer capacity-building "training of trainers programme" participants will be imparted skills in the following areas:

- Preparation of a business case, based on market assessments, realistic forecast demands of the final product, and reasonably accurate operating costs, to show how the proposed TT project can enhance competitiveness, profitability, and growth.
- Construction of technology roadmaps in priority areas to define future trends to avoid the pitfalls of buying outdated, inappropriate technology and ensure non-obsolescence of technology;
- Preparation of a detailed technology transfer agreement
- Negotiation in today's global business setting;
- Preparation of a detailed technology transfer implementation plan based on the decisions reached during negotiations; and
- Preparation of a scheme to assess the impact of a technology transfer project from market, financial, technological and organizational perspectives.

The training materials for this programme will be developed by APCTT. The training will be organized by APCTT in cooperation with ESCAP, other relevant international partner agencies, government agencies, chambers of commerce and business and industry associations.

It is envisaged that this capacity building programme will help SMEs to better utilise existing APCTT web-based initiatives such as the Technology4sme (www.technology4sme.net), Business-Asia (www.business-asia.net), the Asia-Pacific Traditional Medicine and Herbal Technology Network–APTMNET (www.apctt-tm.net), the Biotechnology Information Network for Asia – BINASIA (www.binasia.net), and the Asia-Pacific Technology Information Tracking and Unified Data Extraction (APTITUDE) search engine. To supplement these web-based initiatives, APCTT also comes out with publications such as the Asia Pacific Tech Monitor and the Value Added Technology Information Services (VATIS) in the areas of Ozone Layer Depletion Protection, Waste Technology, Biotechnology, Food Processing and Non-Conventional Energy.

There is considerable knowledge that exists today that can help to develop approaches to enable SMEs to effectively plan and implement TT projects. What is perhaps missing is a well-funded and potent mechanism to collect this knowledge, codify it, and disseminate it among the nations of the South which perhaps need it more than the technologically and economically advanced nations of the world. The proposed APCTT initiative could be a useful starting point.

DISCLAIMER

The views expressed in this Background Study are those of the author and do not necessarily reflect the views of the Secretariat of the United Nations Economic and Social Commission for Asia and the Pacific. The description and classification of countries and territories used, and the arrangements of the material, do not imply the expression of any opinion whatsoever on the part of the Secretariat concerning the legal status of any country, territory, city or area, of its authorities, concerning the delineation of its frontiers or boundaries, or regarding its economic system or degree of development. Designations such as *'developed'*, *'industrialised'* and *'developing'* are intended for convenience and do not necessarily express a judgement about the stage reached by a particular country or area in the development process. Mention of firm names, commercial products and/or technologies does not imply the endorsement of the United Nations Economic and Social Commission for Asia and the Pacific.

REFERENCES

Amsden, A.H., 1989. Asia's Next Giant: South Korea and Late Industrialisation. Oxford University Press, New York.

Arrow, K., 1969. Classificatory note on the production and transmission of technological knowledge. American Economic Review, Papers and Proceedings, May, pp. 244-250.

Autio, E. and Laamanen, T., 1995. Measurement and evaluation of technology transfer: Review of technology transfer mechanisms and indicators. International Journal of Technology Management, 10(7/8), pp. 643-664.

Bar-Zakay, S.N., 1971. A technology transfer model. Technological Forecasting and Social Change, 2, pp. 321-337.

Behrman J.N. and Wallender, H.W., 1976. Transfers of Manufacturing Technology within Multinational Enterprises. Ballinger Publishing Company, Cambridge, MA.

Bozeman, B., 2000. Technology transfer and public policy: A review of research and theory. Research Policy, 29, pp. 627-655.

Chantramonklasri, N., 1990. The development of technological and managerial capability in the developing countries. In: M. Chatterji, ed. Technology Transfer in the Developing Countries, the McMillan Press, London.

Chen, M., 1996. Managing International Technology Transfer. Thunderbird Series in International Management. International Thompson Press, London.

Cusumano, M.A. and Elenkov, D., 1994. Linking international technology transfer with strategy and management: A literature commentary. Research Policy, 23, pp. 195-215.

Dahlman, C.J. and Westphal. L.E., 1981. The managing of technological mastery in relation to transfer of technology. Annals of the American Academy of Political and Social Science, 458 (November), pp. 12-26.

Dosi, G., 1988. The nature of the innovation process. In G. Dosi, G., Ed. Technical Change and Economic Theory. Printer Publications, London.

Durrani, T. S., Forbes, S. M., Broadfoot, C. and Carrie, A. S. 1998. Managing the technology acquisition process. Technovation, 18 (8/9), pp. 523–528.

Gibson, D.V. and Rogers, E.M., 1994. R&D Collaboration on Trial: The Micro-electronics and Computer Technology Consortium. Harvard Business School Press, Boston.

Godkin, L., 1988. Problems and practicalities of technology transfer. International Journal of Technology Management, 3(5), pp. 587-603.

Habibie, B.J., 1990. Sophisticated technologies: Taking root in developing countries. International Journal of Technology Management, 10(1), pp. 489-497.

Hameri, A.P., 1996. Technology-transfer between basic research and industry. Technovation, 16(2), pp. 51-57.

Hayami, Y., and Ruttan, V.W., 1985. Agricultural Development: An International Perspective. The Johns Hopkins University Press, Baltimore.

Jagoda, K. I., 2007. A Stage-gate Model for Planning and Implementing International Technology Transfer. Doctoral Thesis. University of Western Sydney, Australia.

Jagoda, K. and Ramanathan, K., 2005. Critical Success and Failure Factors in Planning and Implementing International Technology Transfer: A Case Study from Sri Lanka, Refereed Proceedings (in CD-ROM) of the Portland International Conference on Management of Engineering and Technology–PICMET 05, Portland, Oregon, U.S.A, July 31-August 4.

Keller, R.T. and Chinta, R.R., 1990. International technology transfer: Strategies for success. The Executive, 4(2), pp. 33-43.

Klein, J.J. and Lim, Y. K., 1997. Econometric study on the technology gap between Korea and Japan: The case of the general machinery and electrical and electronic industries. Technological Forecasting and Social Change, 55(3), pp. 265-279.

Lee, J., Bae, Z. T., Choi, D. Y., 1988. Technology development process: A model for a developing country with a global perspective. R&D Management, 18 (3), 2pp. 35-250.

Lundquist, G., 2003. A rich vision of technology transfer technology value management. Journal of Technology Transfer, 28(3-4), pp. 284.

Mansfield, E., 1975. East-West technological transfer issues and problems, international technology transfer: Forms, resource requirements, and policies. American Economic Review, 65(2), pp. 372-376.

Mansfield, E., Romeo, A., Schwartz, M., Teece, D. Wagner, S., and Brach, P., 1982. Technology Transfer, Productivity and Economic Policy. Norton and Company, New York.

Mayer, S. and Blaas, W., 2002. Technology transfer: an opportunity for small open economies. Journal of Technology Transfer, 27(3), pp. 275-289.

Mittleman, J.H. and Pasha, M.K., 1997. Out from Underdevelopment Revisited: Changing Global Structures and the Remarking of the Third World. St. Martin's Press, New York.

Morrissey, M.T. and Almonacid, S., 2004. Rethinking technology transfer. Journal of Food Engineering, 67(1-2), pp. 135-145.

Osman-Gani, A.A.M. 1999. International technology transfer for competitive advantage: A conceptual analysis of the role of HRD. Competitiveness Review, 9, pp. 9.

Phillips, R.G., 2002. Technology business incubators: How effective as technology transfer mechanisms. Technology in Society, 24, pp. 299-316.

Radosevic, S. 1999. International Technology Transfer and Catch-up in Economic Development. Edward Elgar Publishing, Massachusetts.

Ramanathan, K. A Taxonomy of International Technology Transfer Modes, Proceedings of the Third International Conference on Operations and Quantitative Management, Sydney, December 17-20, 2000, pp 203-209.

Ramanathan, K., 1989. Evaluating the national science and technology base: A case study on Sri Lanka. Science and Public Policy, 15, 304-320.

Ramanathan, K., 1991. Technology assessment at the firm level, MOTIC Series #6, Bangkok: School of Management, Asian Institute of Technology.

Ramanathan, K. 1994. The polytrophic components of manufacturing technology. Technological Forecasting and Social Change, 46, 221-258.

Ramanathan, K., 2000. A Taxonomy of International Technology Transfer Modes, Proceedings of the Third International Conference on Operations and Quantitative Management, Sydney, 17-20 December.

Ramanathan, K., 2001, E-strategies for technological capability development, Proceedings of the Portland International Conference on Management and Technology, July 29-August 2, Portland, US.

Ramanathan, K., 2005. Competing through total technology management. Refereed Proceedings (in CD-ROM) of the Portland International Conference on Management of Engineering and Technology–PICMET 05, Portland, Oregon, U.S.A, July 31-August 4.

Ramanathan, K., 2007. The role of technology transfer services in technology capacity building and enhancing the competitiveness of SMEs. Mongolia National Workshop on "Subnational Innovation systems and Technology Capacity-building Policies to Enhance Competitiveness of SMEs." Organized by UN-ESCAP and ITMRC (Mongolia). Ulaanbaatar, Mongolia, 21-22 March.

Raz, B., Steinberg, G. and Ruina, A., 1983. A quantitative model of technology transfer and technological catch-up: The case of developed countries. Technological Forecasting and Social Change, 24, pp. 31-44.

Reddy, N.M. and Zhao, L., 1990. International technology transfer: A review. Research Policy, 19(4), pp. 285-307.

Robinson, R.D., 1991. International Technology Communication in the Context of Corporate Strategic Decision-making. In: R.D. ROBINSON, Ed. The International Communication of Technology: A Book of Readings. Taylor and Francis, London.

Rogers, E., 2003. Diffusion of Innovations. 5th Ed. The Free Press, New York.

Rogers, E.M. and Shoemaker, F.F., 1971. Communication of Innovations: A Cross Cultural Approach. Free Press, New York.

Sadowski, B.M., 2001. Towards market repositioning in Central and Eastern Europe: International Cooperative Ventures in Hungary, Poland and the Czech Republic. *Research Policy*, 30, pp. 711-724.

Schlie, T.M., Radnor A. and Wad, A., 1987. Indicators of International Technology Transfer. Centre for the Interdisciplinary Study of Science and Technology, North Western University, Evanston.

Sharif, M.N. and Haq, A.K.M.A. 1980. A time-level model of technology transfer. IEEE Transactions of Engineering Management, EM-27(2), pp. 49-58.

Siegel, D.S., Waldman, D.A., Atwater L.E. and Link, A.N., 2004. Toward a model of the effective transfer of scientific knowledge from academics to practitioners: Quantitative evidence from the commercialisation of university technologies. Journal of Engineering and Technology Management, 21, pp. 115-142.

Souder, W.E. 1987 Managing New Product Innovation, D.C. Heath, New York.

Souder, W.E., Nashar, A.S. and Padmanathan, V., 1990. A guide to the best technology transfer practices. Journal of Technology Transfer, 15(1-2),

Spivery, W.A., Munson, J.M., Nelson M.A. and Dietrich, G.B., 1997. Coordinating the technology transfer and transition of information technology: A phenomenological perspective, IEEE Transactions on Engineering Management, 44(4), pp. 359-366.

Steenhuis, H.J., 2000. International technology transfer: Building theory from a multiple case-study in the aircraft industry. Doctoral Thesis. University of Twente, the Netherlands.

Takii, S., 2004. Productivity differentials between local and foreign plants in Indonesian manufacturing. World Development, 32(11), pp. 1957-1969.

UNIDO, 1996, Manual on Technology Transfer Negotiations. UNIDO (ID/SER.O/18), Vienna.

Zhao, L.M. and Reisman, A., 1992. Toward meta research on technology transfer. IEEE Transactions on Engineering Management, 39(1), pp. 13-21.

Chapter 2

South-South Cooperation for Technology Transfer and Development of SMEs: How Can it be Possible?

Juan Alfredo Pinto Saavedra
Ambassador of Colombia to India
85, Poorvi Marg, Vasant Vihar, New Delhi – 110 057
E-mail: edelhi@minrelext.gov.co, edelhi@cancilleria.gov.co, embassyofcolombia@yahoo.co.in

1. SOUTH-SOUTH COOPERATION MUST ACCELERATE ITS TRANSFORMATION

Solidarity between the countries of the southern hemisphere and their integration, constitutes the elements of first order in the fight against universal inequality and favours the existence of an international order that is more just and consequently less conflictive. The main criticism against the South-South Cooperation is that it has difficulties showing material results and often becomes a mere exercise of rhetoric. Therefore, the said cooperation must move on from intentions to realities, from the intangible to the concrete, from declarations to actions and produce macro-economic impacts beyond the traditional pilot experiences.

2. CHANGE OF PARADIGMS IN SOUTH-SOUTH COOPERATION

In order to fabricate realities in South-South Cooperation (SSC) it is necessary to modify the paradigms upon which it operated during the second half of the 20th century, now acting upon the contemporary patrons of the competitive advantage. If the primitive form of the SSC, insisted in the supply of natural resources, in the geographical location, in the similarity of the offer, in the technological dependence and in the predominance on the flow of investments between the countries of northern hemisphere, it is necessary to adjust and adapt to such variables to the present times, considering the created competitive advantages, the relocation and specialization of the production, the much varied degree of development between countries of the southern hemisphere, the existence of powerful developing countries such as generators of technology and that have the capability of emitting a flow of investment from such nations towards the yet developing economies.

3. ECONOMIC TENDENCIES AND INTERNATIONAL DYNAMIC OF THE MSMEs

Beyond the traditional allusions to the micro, small scale and middle scale enterprises (MSMEs) as the key factors in the employment, the democratization of the property and political stability, the making of policies, plans and programs for the promotion of the MSMEs, imposes the recognition of the manifestations of strong incidence in the fulfilment of this business stratum at World level and particularly the developing countries, among others: the expansion of the informality in the micro scale business sector, concentration of the production in the global firms and transit from SMEs (Small and Medium Enterprises) towards MSEs (Micro and Small Enterprises), the declension of MSMEs in the manufacturing sector and its expansion in the areas of commerce and services, the policies oriented towards the base of pyramid till now more efficient in consumer habilitation than in business promotion. Also it is necessary to understand that being a huge growth in the services equally in both the countries with advanced economies as well as the southern countries, such a development does not have the same genesis and neither the same meaning since in the northern countries it is the consequence of the industrial development while in the southern hemisphere many a time it is product of sub development of the manufacturing sector. In the same manner, without denying the goodness of the international outsourcing and its input to the massive generation of employment it is essential to specify that quiet often it doesn't consist of another thing a career of low salaries and labour marginalization.

The associativity on the other hand, as the paradigm of a great frequency in the policies of promotion of SMEs of the last generation constitutes an appropriate but not universal response to the problems of the introduction of the MSMEs to the markets and the technological development.

One of the recurrent problems in the formulation of the policies for the development of the MSMEs is the idealization of the instruments and their exaltation to the condition of panaceas which can lead to unforgivable errors, for example a centre of technological development requires the existence of a critical mass of producers and specialists, if such a nucleus does not exist the centre will be eternally condemned to the dependence on the public subsidy and to manage technologies of the second last generation. The same would happen with what are called incubator of the technological base which have to be preceded by strong processes of human capital.

Upon examining more successful policies of productive transformation and more successful promotion we find as highlighted elements the combination of horizontal and vertical measures emphasis on the access to the factors, facilities in the incorporation of technology, formation of specific human talent for the MSMEs, impulse to the articulation without absolute dependence upon big companies, strong support in the modern logistics for the participation in the supply chain and the measures for the consolidation of business established in the local and regional environment.

4. NEW SCENARIOS FOR THE DEVELOPMENT OF THE MSMEs

The MSMEs are like pigeons: they exist in whole of the World, they are able to survive in the contexts of difficulties and they make their nests in any small place in order to reproduce, however, upon thinking systematically in their promotion it is necessary to discern where the territories that can be more fertile are. In a general sense without dispensing with the considerations referring to the grades of incorporation of technical progress it is evident more fertile terrains for the MSMEs are in the niche of the manufacture, the spaces of the term micro-marketing, gastronomy, the small scale hotel trade and the tourist services, the operation of the communitarian public services in small towns and villages, the techno entrepreneurships in IT, biotechnology and the compact and ultra advanced units of the farming production in the concept of intelligent re-primarization.

5. HOW TO MAKE TRANSFER OF TECHNOLOGY TO MSMEs POSSIBLE?

In the first term, the public agencies in charge of the scientific-technological development and the governing of education and the professional training must concentrate their major efforts and resources in supporting the incorporation of the technical progress in SMEs, to the improvement of its technological management, to its productive transformation, the promotion of techno entrepreneurship, to the support of the innovation in SMEs and to the promotion of the associativity with technological component, regional anchoring and international projection.

It is essential to carry out maximum efforts in canalizing the direct foreign investment from the powerful nations not aligned towards the emerging economies of Asia, Africa and Latin America. Also it is necessary to bring about the conditions so that the foreign investment articulates with SMEs on the national level.

The strengthening of the catalytic institutions for South-South Cooperation (SSC), such as the Centre for Science and Technology of the Non-aligned and other Developing Countries (NAM S&T Centre) with major support by the multilateral organisations, are the tasks that can not be postponed.

Promotion of the renamed brands of the global middle scale firms through the processes of commercial agency and productive articulation with SMEs at the national level is a vital strategy for the existence and growth of intermediate economic networks, within which the processes of technological transfer operate.

Finally the impulse to the programs of horizontal cooperation using productive strength and of the knowledge of the developing countries is pertinent and of extreme utility. For example, fruit processing in Thailand, bio-products in Cuba, flower-crop in Colombia, Jewellery in Sri Lanka, phosphate in Morocco and many more deserve to be used by other developing nations without any affecting their legitimate position in the global markets.

Chapter 3

Concept and Institutional Roles for the Implementation of the South-South Global Assets and Technology Exchanges System

Denis Nkala
Chief, Regional South-South Unit (Asia and Pacific),
United Nations Development Programme, UNDP Regional Centre,
Bangkok-10200, Thailand
E-mail: denis.nkala@undp.org

ABSTRACT

The SS-GATE was launched by the United Nations Development Programme's Special Unit for South-South Cooperation in 2006. The SS-GATE has three tracks, Track I has been developed while Track Ii and Track III are pending. Track I provides extensive matching facilitation for Small to Medium Enterprises to find the right partners for technology transfer and investment. The SMEs are supported by four networks. These are, Technonet Africa, Technonet Asia, The Islamic Chamber of Commerce and Industry (ICCCI) and the China-Africa Business Council (CABC).

These networks have 19 participating organizations (POs) in 16 countries. The software for Track I (MS-1) was developed by the Shanghai United Assets and Equity Exchange (SUAEE). The software can be adapted for application to many markets to facilitate matching between buyers and sellers at various geographical locations as well as different types of business activities. The immediate possible expansion of Track I for SME development exists in the global and national levels. The SS-GATE was developed on the basis of lessons learnt in organizing four Africa Asia Business Forums between 1999 and 2007.

Keywords: *South-South global assets, Technology exchange system, Small to Medium Enterprises.*

1. INTRODUCTION

The South-South GATE was conceptualized on the basis of the experiences that the United Nations Development Programme's Special Unit for South-South Cooperation (SU-SSC) has had in trying to bring together business people from Africa and Asia. Under the Africa Asia Business Forum (AABF), the Unit intended to stimulate business activity between Africa and Asia through facilitating contacts, creation of joint ventures and other corporate linkages. Another goal was to stimulate trade and Asian investment into Africa.

At the commencement of preparing the SS-GATE, the SU-SSC had already co-organized three Africa Asia Business Forums. AABF I took place in Kuala Lumpur, Malaysia in 1999, AABF II took place in Durban, South Africa in 2001 while AABF III took place in Dakar, Senegal in 2004.

The AABFs were conceptualized under the second Tokyo International Conference on African Development (TICAD II) where an agreement was made to expand linkages between Asia and Africa through South-South Cooperation. In fact, AABF IV, which took place in Dares Salaam, Tanzania in 2007, was organized by the TICAD Unit in the United Nations Development Programme's Regional Bureau for Africa (RBA).

A number of other organizations were partners to the SU-SSC during the organization of the AABFs. They have included multilateral organizations such as the United Nations Industrial Development Organization (UNIDO) as well as other major donor countries apart from the government of Japan.

In 2006, the Special Unit for South-South cooperation commenced the development of the SS-GATE. The results of the AABFs were a contributing factor for the conceptualization and development of the GATE.

Analysis of AABFs Results

The AABF constituted of national and regional workshops to select participating organizations. The matching took place at the actual venue. Significant results were achieved under the AABFs. Table 3.1 shows the results achieved under the four AABFs.

Table 3.1 shows that the number of memoranda of understanding (MoU) signed were increasing with each AABF except AABF III. The number of participants remained reasonably high and only fell below 200 in AABF IV. AABF IV was the most successful in terms of MoUs signed and their value. The AABFs showed that there was interest in doing business between African and Asian countries. However, the trend does show a drop in interest of Asian business people. AABF IV was mostly an African intraregional forum. While in AABF II 75 per cent of the MoUs signed were between Asian and African companies, in AABF IV most of the MoUs were signed between African participants.

Table 3.1: Review of Results Achieved Under AABFs

Indicator	*AABF I*	*AABF II*	*AABF III*	*AABF IV*
MoUs signed	23	97	55	118
Contracts Value US$m	24.5	80	48.4	156
No. of Participants	230	230	212	159
Asian Participants	120	72	26	30
African participants	110	158	121	129
Asian Countries	6	6	–	7
African Countries	23	17	–	17
Main Component	Trade	Trade	–	Trade
Main Sector	Agriculture	IT	–	Agriculture
Amount Spent US$ m	>1.2	2.228	>2.0	Not available

Notes: Results derived from AABF II Evaluation (I and II) and TICAD website for AABF IV.

The costs of the AABFs show that at least US$1 million was spent in each of the first three. Most of the costs went to project personnel, consultants and the actual workshop. Business people only paid for a return ticket from their homes to the Business Forum. Most of the expenses were shouldered by the sponsors.

Evaluations have shown that the AABFs had some weaknesses that needed to be addressed. In general, the AABFs organizers invited businesses from all sectors of the economy in the participating countries. The process entailed training of participants in negotiating business contracts, technology acquisition, doing business abroad, doing business in Asia, legal issues in doing business abroad and accounting issues. During the workshops, clinics that looked closely at some businesses were conducted. The consultancy teams also attempted to match Asian and African businesses with common interests to negotiate and sign agreements.

The 2002 evaluation of AABF II provides observations and recommendations that were instrumental in the design of SS–GATE. The evaluation concluded that there were too many sectors invited to the workshops and that future workshops required a focus on the priority areas for both Asian and African countries. An assessment of the MoUs during the evaluation concluded that 68 per cent seemed unsuccessful as the parties may either have stopped correspondence or one of them had given up altogether.

Consistent findings came out of the AABFs evaluations. Investments expected by African participants were limited. Most of the Asian participants were more interested in trade. Participants also concluded that the business forum should be organized on a continuous basis with an established information system and mechanism for follow-up support, including consultancy services and financial support.

On the positive, the evaluations indicated that the AABFs helped establish contact between African and Asian companies and provided opportunity for them to have a first hand knowledge and to explore business opportunities. However, at the same time, the AABFs only happened occasionally so that there was a gap in the facilitation process in between the successive AABFs. The AABF process only involved two regions although there is potential to include other regions such as Latin America and the Caribbean as well as the transitional economies of Eastern Europe. There was also some evidence that only a few of MoUs actually materialized into actual contracts. The costs of holding the forums were high although it could be argued that the modality choices selected to organize the AABFs ultimately resulted in the high costs and could be modified.

2. THE SOUTH-SOUTH GATE

The Special Unit for South-South Cooperation envisaged a system that could facilitate technology exchanges between developing countries continuously. The system needed to retain some of the business support functions that the AABFs had. In fact some o the companies matched in the SS-GATE could still participate in the AABF and conduct face to face negotiations. Since a lot of ground work would have been done, the signed MoUs would be more concrete and lead to real collaboration between the parties. The SS-GATE also needed to be efficient and self sustaining.

The Unit conceptualized the SS-GATE as a three track system. Track I is a technology and assets exchange which can facilitate business dealings in assets, equity, technology and access to financial resources by SMEs in developing countries. Track I facilitates project submission, partner matching and a system for support services to the SMEs. Track II and Track III, which are not yet developed are focused on support for infrastructure projects and underfunded projects respectively. The relevant track for this publication is Track I.

Development and Piloting of Track I

Track I has been developed since 2006. The software was developed and continues to be improved by the Shanghai United assets and Equity Exchange (SUAEE). The SUAEE is leading the piloting of the Track I and will operate the functional system in collaboration with other institutions.

The vision for the SS-GATE is "To provide a global and sustainable South-South transaction platform that facilitates market-driven and transparent exchanges of technology, assets, services, and financial resources among the private sector and civil society for inclusive growth of countries in the South" (SUAEE, 2008).

The piloting of Track I commenced at the end of 2007 and should have been completed in July 2009. The pilot period was extended to December 2009. The full fledged implementation is expected to commence in January 2010. The purpose of the pilot is to demonstrate the practicality of the SS-GATE. A working model of an online system has been created with a support system of institutions. The system will demonstrate the facilitation of matching, establish institutional mechanisms for SME support to ensure successful transactions and to ensure financial and institutional sustainability.

The piloting will take place in two phases. Phase I, which is ongoing, includes African and Asian Countries up to this point. The countries involved are China, Malaysia, Pakistan and Thailand in Asia; Nigeria and Tanzania in Africa, and Turkey. A number of networks have been identified to support the implementation of the SS-GATE. These are shown in Table 3.2.

Table 3.2: Institutions Involved in Track I Piloting

Implementing Organization	*Description of Organization*
Shanghai United Assets and Equity Exchange (SUAEE)	✰ Serves as a comprehensive platform for the exchange of assets and equity for SME's in China and around the world.
	✰ Delivers policy consultation, information dissemination, project promotion, investment guidance, mergers and acquisitions planning, project financing, and transaction management services.
Collaborating Networks	Participating organizations (P.O.s)
TECHNONET Africa	✰ Manufactures Association of Nigeria, Nigeria
	✰ Nigeria Association of Small scale Industrialists (NASSI), Nigeria
	✰ Small Industry Development Organization (SIDO),Tanzania.
TECHNONET Asia	✰ SIRIM Berhad, Malaysia,
	✰ Ministry of Industry, Department of Industrial Promotion, Thailand
Islamic Chamber of Commerce and Industry (ICCI)	✰ TBD Pakistan
	✰ Union of Chambers and Commodity Exchanges of Turkey (TOBB), Turkey
China-Africa Business Council (CABC)	China Africa Business Council. (CABC plays the dual role of network and Participating organization).

Note: From SUAEE with permission.

Table 3.2 shows the institutional arrangements for support to the SS-GATE. The networks are part of the management structure for SS-GATE. They were selected on the basis of a long term partnership with the Special Unit for South-South cooperation. Their function is to provide knowledge about the SS GATE and to work with a number of participating organizations affiliated to them. The second group of institutions are participating organizations (POs). The POs are the service providers to the SMEs. They are each allocated work stations to access SUAEE online. Their role is to help the SMEs process proposals. The way the SS-GATE functions is depicted in Figure 3.1.

Preparations to launch Phase II have been underway since July 2008. Saudi Arabia and Vietnam will be additional Asian countries while Cameroon, Ghana, Mozambique, Senegal and South Africa will join from Africa. Phase II software will

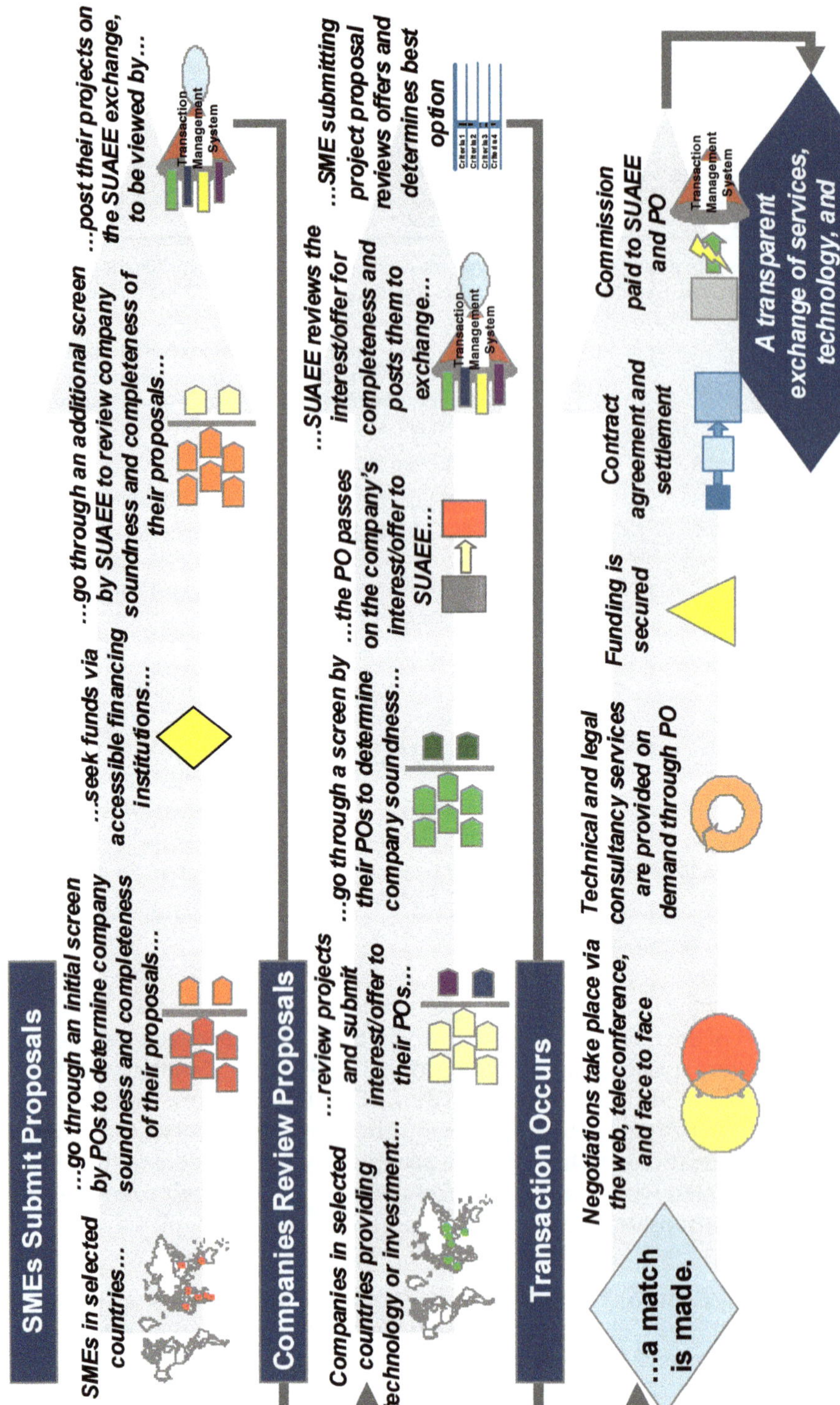

Note: Participating Organization (PO)= Country-level member organization of a SS-GATE collaborating network

Figure 3.1: SS-GATE Track I Pilot Business Model

be able to facilitate matching of proposals from anywhere in the world including intra-regional arrangements.

The SS GATE will be available soon to networks and POs. Negotiations are also taking place for the "franchising" of the SS-GATE. At regional level, Technonet Asia and Technonet Africa are exploring customizing the software for a SS–GATE Africa and SS GATE Asia respectively. The Islamic Chamber of Commerce and Industry (ICCI) intends to create SS-GATE ICCI in Arabic. Plans for expansion are also at country level. There is interest from the Standards and Industrial Research Institute of Malaysia (SIRIM) to create SS-GATE Malaysia in Bahasa, Malaysia. There is interest in Turkey as well to create SS Gate Turkey in Turkish.

3. OPERATIONAL SYSTEMS OF SS - GATE

The operational system that is being piloted is shown in the illustration below. The SUAEE have undertook a training of the participating organizations. The system in place will continue to undergo refinement as the lessons learnt from the pilot are used to adjust the operational systems.

The Transactions process proceeds in the following steps:

(1) Collecting project proposals (2) Screening by Pos (3) Screening by SS GATE Centre (4) Listing and broadcasting by SS GATE Centre (5)Receiving offers from providers (6) Screening offers by Pos (7) Screening by SS-GATE Centre (8) Feedback offers to owners (9) Selecting one of the best providers (10) Match identified (11) Negotiations (12) Supporting Services (13) Contracting.

Note: In the future, the automatic matching facilitation will become possible.

The illustrated process will be quick or slow depending the preparedness of the SME submitting or responding to the proposal. The participating organization may forward or return the submission depending on whether it meets the SS-GATE's and the PO's criteria for a successful proposal. Some of the criteria are the legality of the firm and sufficient reporting on its assets, management and financial situation. Depending on their experience with working with the SMEs, each PO may have its areas of emphasis for vetting the proposal. The full process of the SS-GATE is illustrated in Figure 3.1, the SS-GATE pilot business model.

Addressing the Weaknesses of AABFs.

The SS GATE will be available to interested SMEs regardless of time and location on a continuous basis. It will have a self sustaining basis, as the SMEs will pay minimal cost of the services provided. The transactions will be competitive and transparent therefore should provide the best value for the participants.

4. LESSONS LEARNT FROM THE PILOT

The lessons learnt from the pilot show that the SS-GATE should strengthen the participating organizations, especially the networks such as the Technonets and ICCI. The International Telecommunications Union, a UN Agency, will be used to train the networks.

Participants in the pilot training have insisted that the SS-GATE should collaborate with other projects to promote SMEs and other UN Agencies such as UNIDO, the United Nations Education and Science Organization (UNESCO) and the United Conference on Trade and Development (UNCTAD). For example, the SS-GATE will connect its system to UNIDO's so that the businesses matched under UNIDO system can finalize their negotiations on the SS=GATE. The SS-GATE will also provide information to businesses on opportunities, risks and access to financial resources.

5. CONCLUSION

The SS GATE targets SMEs in the developing countries. Through an efficient and self financing system it will enable SMEs around the world to enter into business arrangements. It is a complete system that facilitates a transaction from proposal to matching, negotiation and signing of contracts. The involvement of supporting institutions will enhance the integrity of the process and build confidence among SMEs.

Since the first draft of the article, 20 projects have been successfully implemented under the pilot, most of them among Chinese companies.

ACKNOWLEDGEMENTS

The author of the paper would like to express gratitude Mr. Tatsuo Fujimura, Director of the SS GATE Project Management, who, on behalf of SUAEE has presented work and provided useful direction as well as giving permission for the use of some illustrations prepared by the organizations.

REFERENCES

Shanghai United Assets and Equity Exchange. (2008). Track I operation guideline. Shanghai, China. SUAEE.

Special Unit for South-South Cooperation. (2002). Second Africa-Asia Business Forum Evaluation report, 2002. New: United Development Programme.

ABBREVIATIONS

AABF:	Asia Business Forum
ICCI:	Islamic Chamber of Commerce and Industry
ITU:	International Telecommunications Union
MoU:	Memorandum of Understanding
Pos:	Participating Organizations
RBA:	Regional Bureau for Africa
SIRIM:	Standard and Industrial Research Institute of Malaysia
SS-GATE:	South-South Global Assets and Technology Exchange
SU-SSC:	Special Unit for South-South Cooperation
SUAAE:	Shanghai United Assets and Equity Exchange

TICAD:	Tokyo Africa International Conference for African Development
UNDP:	United Nations Development Programme
UNESCO:	United Nations Education and Science Organization
UNIDO:	United Nations Industrial Development Organization
UNCTAD:	United Nations Conference on Trade and Development

Chapter 4

Role of Business Incubation Systems to facilitate Technology Transfer and Development

A. Balachandran
General Manager,
VIT-Technology Business Incubator (VITTBI),
VIT University, Vellore – 632 014, India
E-mail: abalachandran@vit.ac.in; vittbi@vit.ac.in; balac68@yahoo.com

ABSTRACT

This paper deals with a suggestive framework of using Business Incubators for the purpose of providing "Technology Transfer" support to start ups and Small and Medium Enterprises_(SMEs). This is in the context of South-South Cooperation, using several networks and linkages that are available in the Business Incubation arena.

Keywords: *Technology Transfer, Business Incubators, Small and Medium Enterprises (SMEs).*

1. INTRODUCTION

The Technology transfer issues pertaining to SMEs in developing nations are much different from that of USA and Europe. The challenges are mainly due to lack of an enabling ecosystem, which is needed for facilitating an active interaction among various actors, involved in the Technology transfer activities. The challenges could be as varied as 'limited interaction' between Technology providers and technology seekers, lack of awareness about emerging technologies and related opportunities, the after effect of globalization and liberalization and also the cultural and regional factors that affect the growth of SMEs.

The Business Incubators in India have started establishing stronger local, regional and global networks. Many of the Indian Incubators are sponsored by the Department of Science and Technology (DST) and are generally hosted by reputed academic institutions.

This paper attempts to portray a suggestive framework that explores the possibility of using the Business Incubation systems for facilitating Technology Transfer among SMEs under the South-South Cooperation (SSC) umbrella.

2. TECHNOLOGY TRANSFER: ISSUES AND CHALLENGES IN THE LIGHT OF LACK OF A FAVOURABLE ECOSYSTEM

The author had a couple of opportunities in the past to interact with Technology Transfer Officers, Technology Licensing Officers and members of Association of University Technology Managers in the United States of America and Europe. Though there is a general grievance that the interaction between Academia and Industry is difficult, the context is very different from a developing country's perspective. The lack of a favourable eco system in the developing country's context breaks down to many of the following issues:

2.1 Strong Industry-Academia Linkage

Universities in USA have a strong patenting culture that results in huge revenue generation not only to the University but also to the inventors. Billions of dollars are generated through these systems and this in turn benefits everyone in the process. The direct beneficiaries are the inventor (mostly academic experts), the university system, the industry and the economy in general.

It is widely believed that the "Bayh-Dole act of 1980" (National Knowledge Commission, 2008) has paved way for the success in University Technology Transfer and the consequent economic benefits. This act gave US universities, small businesses and non-profit organisations "intellectual property control of their inventions" and other intellectual property that resulted from such funding. Prior to the enactment of Bayh-Dole, the U.S. government had accumulated 30,000 patents. Only approximately 5 per cent of those patents were commercially licensed. Hence, the provisions of the act provided an incentive for universities to protect their innovations and, therefore, for industry to make high-risk investments resulting in products made from those innovations. This is evident from the royalty income generated by the top 20 US Universities in the year 2004. The New York University topped the list with USD 109 million and California Institute of Technology occupying the 20th position with an income of USD 9.86 million.

In the same lines, the National Knowledge commission in its recommendation to the Prime Minster of India in 2007 has stated that:

> *"Our consultation with diverse stakeholders has revealed that there is a need to provide incentives to increase innovations, collaborations, licensing and commercialization. It is therefore recommended to enact a legislation that creates a uniform legal framework for the government funded research and gives universities and research institutions ownership and patent rights".*

This must have resulted in the introduction of a bill on "The Public Funded R&D (Protection, Utilization and Regulation of Intellectual Property) Bill, 2007 drafted by the Central Government of India. While this is expected to be the Indian version of "Bayh-Dole", one should be aware of the basic reason of success of "Technology-Transfer" in the USA. There has been a strong linkage between Industry and Academia even before the Bayh-Dole act as evidenced from the fact that initiatives like Stanford Research Park (earlier known as Stanford Industrial Park) and Silicon Valley were started as early as 1951 and the act only provided motivation for accelerating the growth. Hence, the basic issue is to build on a strong linkage between Academia and Industry.

Excepting few institutions like Indian Institute of Science (IISc), Indian Institutes of Technology (IITs), Council of Scientific and Industrial Research (CSIR) labs and very reputed private institutions, the Industry- Academia interaction needs a huge thrust and momentum to spread the culture across the country.

2.2. Nascent Stage of Patenting Culture

The number of patents filed is an indicator of the innovation and technology potential of any country. According to WIPO statistics (WIPO Patent Report, 2008) in the year 2005, India was at the 11th position with around 20,000 patents in the Top 20 countries list of Patent filings. Japan leads the list with over 350,000 patents, USA in the second position with little over 200,000 patents and China in the 3rd position with around 100,000 patents. In the year 2007, the Indian scenario hasn't changed much with around 35,000 according to Indian Patent office. Around 80 per cent of the patent filed in this period is understood to have been filed by foreign companies. In the same period, the number of patents filed in China has risen to over 2, 45,000.

Another interesting dimension to this is the number of researchers per million of population. Number of persons doing research and development in Scandinavian countries is 7,000 per million of population and 4,700 per million of population in the US. In India, there are 156 researchers per million of population. Hence, there are basic issues relating to increasing the focus on research initiatives and also to promote research careers aggressively.

2.3 Lack of Government Funding

2.3.1. On R&D

Government Expenditure on R&D (GERD) as percentage of GDP" (Brandon Shackelford, 2006) in India is only 0.8 per cent as compared to China's 1.23 per cent. Developed countries have R&D expenditure of up to 3 per cent of GDP. Israel and Sweden are the only countries with over 4 per cent GDP spending on R&D. Unlike in developed countries; most of the R&D expenditure is still incurred by the Government

While the points mentioned above highlight some of the issues and challenges at the macro level, there are challenges at the micro level also.

2.3.2. On Technology Development and Transfer

It is estimated that there are over 1.4 million SMEs in India and around 30 per cent of them are in the manufacturing space. It is also vital that technological inputs to these firms (including ICT intervention) are essential to ensure their continued relevance and competitiveness. Though there have been many schemes in Ministry of Micro Small and Medium Enterprises, Ministry of Textiles, Ministry of Science and Technology for supporting SMEs, the intervention quantum is still in adequate and might take considerable time for majority of the industries to realize the benefits.

2.4 Lack of Awareness about the Emerging Technologies and Opportunities

- Widespread lack of awareness about emerging technologies and derived opportunities there in–as a means for creating a competitive edge in the trade and technology market for value addition.
- Indian SME needs more information, orientation and facilities for protecting their IPR, understanding the underlying issues in technology transfer.
- There is an immense need to educate the SMEs for Technology adaptation, upgradation and in turn to become knowledge driven industries.
- Improving awareness on the importance of networking with Technology providers
- Harnessing the potential of ICT as an enabler in bridging the gaps and to propel the sensitization and facilitation activities.

2.5 Effect of Globalization

SMEs are generally known for their labour intensive activities and also for use of local resources. However, these factors were also responsible for certain factors that amounted to lack of competitiveness in the light of globalization. Using run of the mill technology, inadequacy in meeting the quality requirements in packaging, finishing, entry of foreign players in the domestic market and inability to adopt or source new technology are the challenges. While the Ministry of Micro, Small and Medium Enterprises (MSME) is taking a number of initiatives to overcome these problems and to make the SMEs competitive, the percolation to the length and breadth will take some time.

2.6 Cultural and Regional Factors

Traditionally SMEs were created by first generation entrepreneurs as a means of self-employment. Due to cultural issues, the growth of the SMEs is limited to the ability of the individual to manage the affairs, in most cases. Unlike the SME contemporaries in the developed nations, there is less compulsion on accelerating a planned growth and expansion. As the day-to-day management of the business itself consumes considerable amount the promoter's time, it is seldom possible to concentrate on newer avenues. This is also due to the perceived losses due to delegation, lack of availability of skilled personnel for building a second line and lack of trust etc.

3. ROLE OF BUSINESS INCUBATORS

3.1 Business Incubators

The National Association for Business Incubation (NBIA), USA defines Business Incubation in the following manner:

> *"Business incubation is a business support process that accelerates the successful development of start-up and fledgling companies by providing entrepreneurs with an array of targeted resources and services. These services are usually developed or orchestrated by incubator management and offered both in the business incubator and through its network of contacts. A business incubator's main goal is to produce successful firms that will leave the program financially viable and freestanding. These incubator graduates have the potential to create jobs, revitalize neighbourhoods, commercialize new technologies, and strengthen local and national economies."*

There are around 5000 business incubators in the world and the incubation industry is slightly more than 2 decades old. It has come a long way from the " Real Estate and Shared resources model" in the first generation to the current "in house debt/equity financing and success sharing" 3rd generation model. More than 3/4th of the business incubators are located in North America, Asia and Western Europe.

3.2 Status of Business Incubators in India

The National Science and Technology Entrepreneurship Development Board (NSTEDB), Department of Science and Technology (DST), Government of India has been in the forefront in setting up Business Incubation Centres to encourage development of start-up companies and to generate employment among S&T graduates. The NSTEDB has already catalyzed and supported several Science and Technology Entrepreneurs Parks (STEPs), Technology Business Incubators (TBIs), in addition to many Innovation and Entrepreneurship Development Cells (IEDCs), in different parts of the country. NSTEDB has a plan to set up many more such institutions in the coming years. There are a number of other institutional mechanisms existing and promoted by other Departments of the Government of India in developing business incubation in emerging areas of science and technology such as the Software Technology Parks of India (STPIs).

IEDC's primary activities include organizing Entrepreneurship Awareness Camps (EAC), Entrepreneurship Development Programmes (EDP), Faculty Development Programmes (FDP) and Skill Development Programmes (SDP) in the college/institution for the benefit of S&T persons.

STEP's offer facilities such as nursery sheds, testing and calibration facilities, precision tool room/central workshop, prototype development, business facilitation, computing, data bank, library and documentation, communication, seminar hall/ conference room, common facilities such as phone, telex, fax, photocopying etc. It offers services like testing and calibration, consultancy. Training, technical support services, business facilitation services, database and documentation services, quality assurance services and common utility services.

TBIs have modern work space, communication facilities, computing facilities, vital equipment needed in identified area library and information centre, training and conference facilities. TBIs also are expected to undertake activities like providing specialised services to existing SMEs in the region, facilitating technology commercialization, consultancy, training including short courses, technology related IPR issues, legal and quality assurance services, marketing, assistance in obtaining and other clearances, common facilities, assistance in preparation of business plans, technology shows/technology clinics/trade fairs.

3.3 Number of Business Incubation Environments in India

There are 40 Business Incubators and Science and Technology parks functioning in the country that are supported by the DST. Another 16 are under various stages of establishment. There are around 40 "Software Technology Parks (STPIs)" that act as Business Incubation systems under the Ministry of Communication and Information Technology. While the STPIs primarily focus on the I.T. businesses, the incubators supported by the DST focus on different technologies like Agriculture, Auto components, Biotechnology, Consumer Durable, Design, Electronics, High technology, Information Technology, Manufacturing, Medical Electronics and others. Many incubators work on a "Mixed Technology" model. DST is also planning to increase the rate of creation of the business incubators to meet the emerging requirements.

Besides its Cluster development programme, Ministry of MSME is planning another 100 incubation systems in the next four years and some of the existing Business Incubators under the DST program will be further strengthened under this MSME initiative.

The following is an extract from the report on "Entrepreneurship" (National Knowledge Commission 2008) released by the National Knowledge Commission:

Quantity

> *There is a need to massively increase the number of incubators in the country. USA has about 1400 incubators while China and Korea have about 800 and 400 respectively. According to 'The Indian STEPS (Science and Technology Entrepreneurs' Parks) and Business Incubation Association (ISBA), the apex professional body supporting business incubation', India has only about 100 incubators. Scaling up the numbers also means widening the reach beyond centres of excellence in metropolitan cities (i.e. beyond IITs and the top-grade engineering and business schools), going beyond alumni entrepreneurs and exploring areas beyond high tech. Further, there is a need to widen the incubation horizon beyond the idea-stage to accommodate issues relating to scaling up. There are initiatives already being proposed. For example, there are plans in Gujarat to set up a number of incubation centres with possible venture capital infusion in certain selected regions. Expansion of incubation centres on a globally competitive scale will necessitate much greater involvement of private players as well as PPPs to supplement the current governmental initiatives.*

Hence, it is evident that there will be a conscious effort to scale up the incubation system in India not only in numbers but also in making them available across India.

3.4 Business Incubator Networks

The DST supported Business Incubators in India are parts of one or more of the following networks:

3.4.1. Indian Science and Technology Parks and Business Incubators Network (ISBA)

The Indian STEP and Business Incubator Association (ISBA) was set up in 2004 as a registered professional body to promote business incubation in India.

ISBA aims to promote business incubation activities in the country through exchange of information, sharing of experience, and other networking assistance among Indian Business Incubators, Science and Technology Entrepreneurs Parks (STEPs) and other related organizations engaged in the promotion of start-up enterprises.

ISBA organizes Annual Conferences to facilitate networking of Business Incubation managers with various other stakeholders in the system and to identify opportunities in promoting technology business incubation and co-operation among the incubators and other related organizations.

3.4.2. Asian Association of Business Incubators

The Asian Association of Business Incubation (AABI) promotes business incubation activities by facilitating information exchanges among Asian Incubators, incubator clients and related organizations. AABI comprises of associations that organize incubation resources and facilities, organizations operating incubation programs, and more, located in Beijing (China), Shanghai (China), Hong Kong(China), Chinese Taipei, Japan, Korea, India, Kazakhstan, Kyrgyzstan, Pakistan, Uzbekistan, Singapore, Malaysia, Philippine, Australia and New Zealand,totally covers 16 nations and regions. Since ISBA is part of the AABI network, Indian incubators have linkages with all the incubator associations in Asia.

3.4.3. infoDev network

*info*Dev is a global development financing program among international development agencies, coordinated and served by an expert Secretariat housed in the Global ICT Department (GICT) of the World Bank, one of its key donors and founders. It acts as a neutral convener of dialogue, and as a coordinator of joint action among bilateral and multilateral donors–supporting global sharing of information on ICT for development (ICT4D), and helping to reduce duplication of efforts and investments. *info*Dev also forms partnerships with public and private-sector organizations who are innovators in the field of ICT4D. Priorities and strategies for *info*Dev are guided by a governance framework, approved by *info*Dev Donors in June 2005.

3.4.3.1. infoDev Incubators Initiative

Developing countries are increasingly embracing innovation and entrepreneurship as a key to their social and economic development, and have implemented national business incubator initiatives as one tool for supporting this agenda. Business incubators can help entrepreneurs to leverage ICT in a global marketplace where information and networks are important competitive assets for spurring innovation.

Since 2002, *info*Dev's Business Incubator Initiative has provided financial and technical assistance to more than 70 incubators in over 50 developing countries worldwide and has helped establish five regional networks of incubators

There are six Indian incubators in this network. infoDev so far organized three major events with the partnership of DST in India. An Asia regional workshop on Business Incubation was held in Bangalore during October 2003, a Global forum on Business Incubation was held in Delhi during October 2004 and 2nd Global forum was held in Hyderabad during February 2006.

Apart from this the Asian regional workshop is held in different Asian cities every year. infoDev also has established an online incubator support centre iDisc (www.idisc.net) which provides opportunity for online networking. All these initiatives are resulting in intensive networking activities among like-minded incubators and the benefits are increasingly visible.

The author himself had an opportunity to execute an infoDev project titled "Developing a business ecosystem for augmenting incubation" in VIT -Technology Business Incubator (VIT-TBI). Apart from own capacity building activities, the project also had a networking and knowledge dissemination component.

VIT-TBI established networks with incubators in Thailand, Sri Lanka, Nepal, Mauritius, Malaysia, Philippines, Viet Nam, Australia, New Zealand and United States. This helped the network to share the good practices and also challenges. Under this project VIT-TBI collated "Technology Opportunity and emerging needs" from two major industries with Technology focus namely Bharat Heavy Electricals Limited and Honeywell Technology Solutions Labs. This information was also made available for the benefit of all the Indian incubators under the DST support and also the infoDev global network. The project also shared all the pertinent information through a project portal www.knowledgebiz.org.

More recently, another Business Incubator, PSG-Science and Technology Entrepreneurs Park (PSG STEP) has been awarded an infoDev project titled "Incubation Asia Pacific Network Project".

According to PSGSETP, the Incubation Asia Network development initiative shall mount a series of project platforms to add directional value to the Incubation Asia Pacific Network. These synergistic set of activities range from Promotion of the Incubation Asia Pacific Network to existing and potential incubation community members, Dissemination of incubation value packages, specific process based Quality Enhancements Drives aiming at standardization and replication of incubation processes, Facilitation and Leveraging member driven initiatives, Participation Interfaces and Advocacy Initiatives, broad basing incubator performance by bringing in more resources, innovation, inter linkages with allied processes, planned and focused Reach outs with potential partners and developing New models and conceptual inputs for incubation. This activity spectrum shall be multifaceted under pinning to launch the Incubation Asia Pacific into vibrant sustainability levels and modes of performance.

3.4.4. EU-India Project to Promote Cross-cultural Ventures

Another successful incubator "TREC-STEP, Trichy" has carried out a major initiative in building networks among European Business Incubators and Indian Business Incubators for promoting cross cultural ventures. The incubators under ISBA in India, Swedepark (an Association of Science Parks and Business Incubators in Sweden), ADT (Association of German Science Parks and Business Incubators in Germany) and UKBI (United Kingdom Business Incubators Association).

The program focused on creating cross cultural ventures through a structured creation of an information exchange platform. Initially the Indian Incubation Managers were sensitized on the necessity for making themselves ready for global alliances and partnerships. This resulted in exchange visits comprising of Incubation Managers, interested entrepreneurs and SMEs, Government officials. A logical outcome of this project also resulted in creation of a "Venture Showcase" (TREC-STEP, Trichy, 2006), an online SME marriage portal that will provide information on technology seekers and providers from the participating countries.

The venture showcase is a platform for showcasing technologies in incubators which are ready to enter into joint venturing/collaborations with incubates/SMEs in other countries, through business incubator networks. The Venture shows case displays details of companies having aspiration for partnerships with incubates supported by other incubators and provides concise information about technologies, markets etc required for initial match making. Depending upon the opportunities and requirements, TREC-STEP and partnering incubators provide soft landing services for the cross-cultural ventures. Along with TREC-STEP's partnering networks, mentoring support for cross-cultural ventures shall be available on request from incubates.

Under this initiative, the author has witnessed development of a number of significant technologies in the incubators in Europe. The major advantages of such early stage alignment between technology provider and seeker are "affordable licensing fee or royalties", "possibility of collaborative development to suit the local needs", "minimal chance of being exploited as the background checks are done by the incubators", "facilitation by the incubators in arranging partnerships.

Some of the interesting examples are highlighted below:

3.4.4.1. Healthier Fast-Food: A Sweden Company

MicVac is the name of the company, which has developed a new and unique method of keeping the nutrition and good taste of microwave fast-food when warming it. MicVac develops, markets and sells two methods for processing and packaging convenience food. Both solutions (for pasteurised and unpasteurised foods respectively) use a valve that offers several benefits for both the producer and the consumer.

The MicVac method is based on the principle of cooking and pasteurizing food using microwaves directly in the sealed consumer package. The film cover on the package has a valve that opens during cooking, releasing the overpressure in the form of steam and oxygen. After cooking, the valve closes and a natural vacuum is created in the package – another of the product's many advantages. The Wisselpac steam technology is used for unpasteurised products. The producer packs the product and applies a Wisselpac valve to the package. The food is cooked by steaming in the consumer's microwave oven.

3.4.4.2. Electronic Nose by a UK Company

Scensive Technologies Limited produces bio mimetic sensor instruments for research and commercial applications. Current products are based on the proven Bloodhound™ electronic nose, which uses chemical sensor array technology.

The Bloodhound™ has been manufactured for some years, producing truly ground breaking results in a variety of applications and research fields. Recently upgraded by Scensive Technologies and available for both general purpose and specific application use.

The purpose of highlighting the above two examples is to illustrate the possibilities of Technology transfer through the business incubator networks. The complete listing is available in the TRECSTEP's website and the link is http://www.trecstep.com/model-sen/showcase.htm.

3.4.5. Euro Office Services

TREC-STEP also offers access to the EurOffice Network, the international network of Business Incubators and Science Parks, which provides pragmatic business services that facilitate access to global markets. The EurOffice Services initiative promotes dynamic networking opportunities between companies located in different regions of the Network. Innovative start-ups and small companies can select from the range of EurOffice services those, which will best suit their needs. Services are provided both in India and abroad.

The TREC-STEP helps companies to select the right EurOffice-services in accordance with their needs. TREC-STEP's wide experience in providing services for incoming firms, including soft landing and business development services, networking, mentoring and access to financing services, guarantees our professionalism in services related to the EurOffice Network.

4. SOME OF THE TECHNOLOGY TRANSFER FACILITATION ACTIVITIES IN INDIA

4.1. Department of Science and Technology

Mission Reach program of TIFAC division, indigenous technology development support through Technology Development Board (TDB), several sector focused technology development initiatives through project funding in reputed academic institutions and sponsoring of Technology Business Incubators are some of the major initiatives of the Department of Science and Technology.

4.2. Technology Development Board

Technology development Board aims at accelerating the development and commercialization of indigenous technology or adapting imported technology to wider domestic application. The Board provides financial assistance in the form of equity soft loans or grants. The Board has signed 137 agreements–134 with commercial; enterprises and 3 with other agencies adding up to a total project cost of Rs. 20,438 millions. The board has committed to provide financial assistance of Rs. 6629 millions in these projects till 2005

4.3. Council of Scientific and Industrial Research Initiatives (CSIR)

Apart from spearheading patenting culture and technology transfer activities in all the labs, the recent New Millennium Technology Leadership Initiative of linking Academia, Industry and Government seemed to have resulted in 57 Technology commercialization activities. This program had a funding outlay of Rs.5000 million and periodic announcements are made to enable industries to take part.

4.4. Asian and Pacific Centre for Transfer of Technology (APCTT)

APCTT is a regional institution of the United Nations Economic and Social Commission for Asia and the Pacific (UNESCAP) servicing the Asia-Pacific region. It was established in 1977 with the objective of facilitating technology transfer in the Asia-Pacific region. The Centre is headquartered in New Delhi with host facilities provided by the Government of India.

APCTT provides "Technology Exchange" services both in online and in periodical forms. It also provides matchmaking services for a fee.

4.5. Indian Institute of Sciences (IISc) and Indian Institutes of Technology (IITs)

Society for Innovation and Development (SID) of IISc, The Foundation for Innovation and Technology Transfer –(FITT) of IIT Delhi, Industrial Research and Consultancy Centre (IRCC)of IIT Bombay, Industrial Consultancy and Sponsored Research (IC and SR) and TeNet group of IIT Madras, Sponsored Research and Industrial Consultancy (SRIC) of IIT Kharagpur are some of the successful initiatives for transfer of technology from these institutions.

4.6. Technology Export Development Organization (TEDO)

Technology Export Development Organization (TEDO) is a joint initiative of Confederation of Indian Industry (CII) and Council of Scientific and Industrial Research (CSIR). The main objective of TEDO is to promote Indian technologies (Technical Know how/Technology Intensive Products, Services and Projects). The services of TEDO comprise of Training, Company Assessment for Readiness for Exports (CARE), Market and Export information services, Advisory Services and Promotion.

4.7. FICCI Lockheed Martin Corporation and IC² Institute, University of Texas, Austin-India Innovation Growth Program

The two-year, nation-wide project has been created to enhance the growth and development of India's entrepreneurial economy. The aim of this new program is to accelerate innovative new Indian technologies into new markets in the United States and around the world. The India Innovation Growth Program is the only program of its kind, because of its focus on teaching and using world-class commercialization strategies. Though this is a time bound program, the model is an interesting one where there is an inbuilt capacity building for the innovator.

5. SUGGESTED FRAMEWORK FOR ENGAGING BUSINESS INCUBATORS TO FACILITATE TECHNOLOGY TRANSFER

The attempt made here is to suggest a structured approach of enabling Business Incubators as facilitating agencies for Technology Transfer. This implies that the established business incubation systems across India will leverage their position, complimentary role to the existing technology transfer mechanisms. This is essential, as the requirements of SMEs in India cannot be catered with the existing systems alone owing to the numbers and location factors.

This framework focuses on augmenting Technology transfer activities for SMEs by effectively using the Indian Business Incubators network. The major reasons and benefits are discussed here and it also addresses some of the challenges highlighted in Technology transfer.

5.1. Linking SMEs (Industry)–Academia

Majority of the business incubators in India have strong connections with academia (as many are part of reputed institutions) and this brings in tremendous value to the SMEs as they will be interacting with an agency that will be more sensitive to their needs. The bridging between institution and needy SMEs will happen in a more effective manner.

5.2. Enabling Technology Development and IPR Support

Most of the business incubators are supported by the Department of Science and Technology. DST is also planning to increase the number of business incubators and currently it is growing at a rate of 15 per cent. Due to the pioneering efforts of DST in the recent years, a strong working relationship is established with other Government agencies like the Council of Scientific and Industrial Research, Technology Information Forecasting and Assessment Council (TIFAC), Ministry of Micro, Small and Medium Enterprises and Department of Biotechnology.

DSIR has established Technopreneur Promotion Program (TePP) outreach centres in 20 places in India in which 13 are Business Incubators under DST. This TePP program provides grant funding to innovative ideas and innovators can obtain support up to Rs.6 million for developing prototypes.

While up to Rs.1.5 million funding is available to individual innovators, the second phase seamless scale up support provides up to Rs.4.5 millions. The innovators have the option of starting their ventures or commercializing the technology through licensing deals. The wherewithal required for protecting the resulting IPR and also to showcase technologies is covered by TePP program. Recently DSIR launched a thematic Technology Development and Demonstration Program (TDDP) for start-ups in the SME category. The support by way of soft loan will be to the tune of Rs.100 million and carries an interest of a mere 2 per cent per annum.

5.3. Facilitating Access to Government Funding for Technology Development

In the past 3 years, 10 Business Incubators were provided a grant of Rs.10 million each under a Technology Development Board sponsored seed fund initiative. Under this qualifying Technology ventures will be funded to the tune of Rs.2.5 million at the incubator level with minimal procedures. NSTEDB division of DST is scaling up this initiative by increasing the threshold to Rs.5 million with an individual incubator grant to Rs.20 million. Incubators are also frequently sensitized about various other emerging funding schemes and opportunities and hence are in a better position to handle such requirements relating to arranging finance for technology development.

All the Business incubators have good working relations with the local District Industries Centre and are in a position to spot Technology Centric SMEs easily. This association also enables incubators to strongly advocate the SME funding requirements with bankers and also for sensitizing the SMEs on the Credit Guarantee Trust Fund scheme to avail collateral free loans from banks. Not with standing this, MSME is also launching 100 business incubators with an outlay of Rs.700 million in the next four years. This opportunity also will help many of the business incubators to build their capacity to support SMEs.

As explained earlier, the Business Incubators' strong local, regional and global network will be great boon to provide support to both Technology seekers and technology providers. The soft landing arrangements with incubators abroad will also help SMEs to look for business expansion overseas.

5.4. Increasing the Awareness

The business incubators can also play a vital role in increasing the awareness levels of the SMEs to know the emerging trends in Technology and business opportunities. This is very much possible by aligning with the local industry associations, clusters and develop a need gap metrics. This will also make the Host academic institutions to include SMEs in their conferences and workshops. The Business Incubators also can leverage their networking strength to trigger a channel of information flow to the connected SMEs. Already many Business Incubators are performing this task by way of Newsletters, Innovation funding programs, advertisements in the media and also through exhibitions. What is needed is a more dedicated approach targeting SMEs and consistency in the approach.

5.5. Effect of Globalization and Cultural and Regional Issues

It may be difficult for the Business Incubators to offer complete solution to these issues under this framework. However, since most of the Business Incubator service delivery will happen at the personal level, a case to case analysis of individual short comings and appropriate actions to overcome these deficiencies is possible.

6. CONCLUSIONS

While the above advantages seem to project a positive outlook, there are some challenges that need to be addressed to make this proposed model operational. The current core activity of Business Incubators is creation of successful start up ventures. To enable them to undertake this additional responsibility, a nationwide Business Incubator capacity building program is essential. A joint activity by DST and MMSME will help business incubators to provide this technology transfer support in addition to their core programs. The arrangement shall result in Technology Transfer facilitation cells in the business incubators. After creation of these cells, a nationwide promotion of this new initiative to reach out to SMEs either directly or through their associations has to be carried out.

This attempt of bringing in Business Incubators to support the SMEs Technology transfer issues will provide a new dimension to the business incubation movement and will also lead to evolving growth accelerator programs as a major value benefit to SMEs. The SMEs in need will get a personalized and holistic support in their entrepreneurial journey.

ACKNOWLEDGEMENTS

The author is thankful to Management of VIT-Technology Business Incubator, Mr. H. K. Mittal, Advisor and Head-NSTEDB, DST, New Delhi and colleagues in the ISBA for their support and cooperation to write this paper.

REFERENCES

Brandon Shackelford, 2006, U.S. R&D CONTINUES TO REBOUND IN 2004, an article from Info Brief, National Science Foundation, USA, http://www.nsf.gov/statistics/infbrief/nsf06306/nsf06306.pdf (accessed on 20th July, 2008)

Department of Scientific and Industrial Research, Technology Development and Demonstration Program (TDDP- Start ups), http://dsir.nic.in/(accessed on 18th July, 2008)

infoDev, 2006, "The infoDev Global Network of Business Incubators", http://www.infodev.org/en/Publication.6.html (accessed on 18th July, 2008)

Jennifer A. Henderson, J.D., M.P.H., 2002, Academia, Industry, and the Bayh-Dole Act: An Implied Duty to Commercialize, a report by Regulatory Affairs Program, Centre for Integration of Medicine and Innovative Technology

National Knowledge Commission 2008, A report on "Entrepreneurship in India".

National Science and Technology Entrepreneurship Development Board, 2008, A list of "Supported Organisations" http://www.nstedb.com/(accessed on 15th July, 2008)

Technology Development Board, 2006, A report on "Enabling Commercialization"

TREC-STEP, Trichy, 2006, a report on EU-India Venture Showcase, http://www.trecstep.com/model-sen/showcase.htm (accessed on 20th July, 2008)

WIPO PATENT REPORT Statistics on Worldwide Patent Activities, http://www.wipo.int/export/sites/www/freepublications/en/patents/931/wipo_pub_931.pdf (accessed on 18th July 2008)

Chapter 5

Technology Transfer: Challenges to Developing Countries like Sri Lanka

Vijaya Kumar

Chairman, Industrial Technology Institute, Colombo, Sri Lanka
Emeritus Professor, University of Peradeniya, Peradeniya, Sri Lanka
E-mail: vkumar@pdn.ac.lk

ABSTRACT

The transformation of Korea and Chinese Taipei from agricultural to industrial economies through science and technology has made them development models for developing countries. One can question the relevance of this model in today's post World Trade Organization world. South East Asia with its liberalised economy of 40 years has done better in industrialisation than South Asia which had an import substitution-based strategy until recently. Sri Lanka liberalised its economy more than a decade before India and Pakistan, but has been less successful than Sri Lanka in attracting high technology industries. Sri Lanka's National Innovation System is weak and the country has not done well in exploiting S&T for national development, mainly due to budgetary and human resource constraints and the low priority given during the budgeting process. Sri Lanka has therefore not been very successful in the domestic production of technology, most innovations being of low technology. The private sector has been weak in utilizing technology licensing for development. Although Sri Lanka was able to attract some FDI, there is little evidence of technological learning and spin-offs from FDI. Sri Lanka however has some success stories. Among them are pre-liberalization resource-intensive enterprises which developed high demand products for the export market, resource based technology intensive enterprises, state enterprises which formed strategic partnerships through privatization and labour intensive firms of the post liberalization era which have grown into world leaders. If Sri Lanka is to promote innovation and technology transfer, the rapid expansion of S&T education, especially at post-graduate level would be crucial.

Successful technology transfer will also require the development of a scientific, educational, legal, financial, commercial and industrial infrastructure that could facilitate the commercialization of research.

Keywords: *Technology transfer, National Innovation, Sri Lanka, Success stories.*

1. INTRODUCTION

Technological changes have played an important role in economic growth from the days of the Industrial Revolution, the initial domination of which by the United Kingdom has been attributed to the mechanisms available at that time for the transfer of useful knowledge and incentives available for entrepreneurs (Mokyr, 2002). Of the various types of economic growth, economists believe that growth generated by technological growth and the changes in institutions due to increases in human knowledge or Schumpeterian growth contributes most to the sustained growth of a nation. The creation of knowledge comes about from the research and development (R&D) efforts of individuals and firms and the technological changes they engender. Technological change is known to be the source of sustained increases in per-capita output.

Real world per-capita GDP has been estimated to have grown by an average of only 5 per cent for each century during the 15th and 16th century, an average of 17.5 per cent during the 17th and 18th century and by 248 per cent in the 19th century and 863 per cent in the 20th century (DeLong, 2000). The rapid increase since the 19th century has been ascribed to the Industrial Revolution which began in the late eighteenth century and the subsequent acceleration in technological development. International incomes too were not very different until the early nineteenth century when rapid increases in per-capita output in the U.K, Europe and the U.S. led to large differentials in the incomes of these countries compared with the rest of the world. The disparity grew even more sharply during the latter half of the 20th century. We see a wide variation in the per-capita income of countries today. At the upper income levels, we have Norway and Luxembourg with per-capita gross national incomes (GNI) of over US $ 75,000 and most developed countries with incomes over $ 30,000 in 2007 while at the lower levels many developing countries in Sub-Saharan Africa have per-capita incomes as low as $ 100 to $ 400.*

1.1 The NIE Model

Industrial development through the exploitation of technology has been the strategy used by the so-called Newly Industrialized Economies (NIE), Korea and Chinese Taipei. Their successful transformation from agricultural economies to industrial economies through massive investment in science and technology has made them the development models for many developing countries.

---

* Per-capita gross national incomes (World Bank Atlas method) are from World Development Indicators database, World Bank (http://siteresources.worldbank.org/DATASTATISTICS/Resources/GNIPC.pdf).

Although developing countries are aiming to emulate this model, one can question its relevance in today's post World Trade Organization (WTO) world.

In the early days of their development Korea and Chinese Taipei used an import substitution strategy and built up large industries using state intervention and financing with their products protected through tariffs and subsidies. Foreign Direct Investment was permitted only in selected areas and that too with preferred partners. The autocratic rule which existed at that time permitted labour control and a low wage strike free environment (Lall, 1994). However none of these is possible today with the WTO discouraging non-tariff barriers and subsidies, pressing for lower tariffs on imports and requiring that all countries be offered most favoured treatment. Furthermore, changes in the intellectual property rights regime after TRIPS preclude the use of reverse engineering and the non-recognition of product patents. Developing countries hoping to duplicate the achievements of these countries would find this particular route to development to be hard and challenging.

2. INDUSTRIALIZATION IN ASIA AND SOUTH EAST ASIA

Sri Lanka is considered a lower middle income country with a per capita GNI of US $ 1540 in 2007. Although India is much more industrialized than Sri Lanka, India also lower middle income has a per-capita GNI of US $ 950 while Pakistan is considered low income with a per capita GNI of US $ 870.

Growth rates during 2005-2007 in India, Pakistan and Sri Lanka have been not very different respectively being 9.3, 7.2 and 6.9 per cent. The percentage of GDP derived from industry is considered an important indicator of the extent of industrialization of the country and all three countries are similar in this respect with about 28 per cent derived from industry. The percentage of high technology exports (HTE) in the total exports of manufactured goods of a country has also been considered an index of industrialization. Sri Lanka has an HTE of less than 1 per cent compared with 2 per cent for Pakistan and 5 per cent for India.

If we compare the situation in the South East Asian countries, we find that Singapore is a high income country with a per-capita income of $ 32,470, Malaysia upper middle income ($ 6,540) country and Thailand a low middle income ($3,400) country. The growth rates of Singapore, Malaysia and Thailand during 2005-2007 were 7.7, 5.8 and 4.8 per cent and the percentage of GDP derived from industry in 2006 32 per cent, 45 per cent and 47 per cent respectively. High technology exports formed a large proportion of manufactured exports with the HTE of Thailand being 27 per cent and those of Malaysia and Singapore around 55 per cent. South East Asia with its globally liberalised economy of 40 years would be expected to do better in industrialisation than South Asia which had a restricted economy and an import substitution based strategy until the 90s. Sri Lanka in 1980 was the first South Asian country to liberalise its economy. Sri Lanka liberalised its economy (although not its capital account unlike the ASEAN countries) in 1980 more than a decade before India and Pakistan but it failed to attract high technology industries compared with other South Asian countries which opened up their economies much later (De Silva and Amaradasa, 2001).

3. TECHNOLOGY FOR DEVELOPMENT

A developing country planning to exploit technology for development can produce its technology domestically, license technology from owners of such technology or attract foreign direct investment accompanied by technology in technology based industries. Domestic production of technology which is the strategy used by the developed countries and by NIEs like Korea and Chinese Taipei requires a robust national innovation system including a strong human resource base in science and technology and a high expenditure on research and development, particularly by the private sector. The licensing of technology successfully used by Malaysia and Singapore requires the building of partnerships which is facilitated by a strong economy and a vibrant private sector.

The attraction of foreign direct investment, particularly greenfield FDI requires an enabling environment in terms of a stable economy, a reliable infrastructure and an attractive tax regime and is facilitated by the availability of a large domestic market. FDI, an important strategy in industrialization which can foster technology learning and facilitate domestic spin-off industries, has contributed to the rapid expansion of industrial and high technology exports in South East Asia. Many developing countries have been able to attract FDI through privatization exercises.

4. NATIONAL INNOVATION SYSTEM

Since the 1980s, the concept of the National Innovation System (NIS) has become popular and economists have tried to relate the NIS of a country with the technological changes which are believed to be necessary to ensure its long-term economic development (Freeman, 1982; Lundvall, 2007). It is generally understood a robust NIS requires a strong human resource base in science and technology facilitated by higher expenditure on tertiary education in science and engineering. Gross expenditure on research and development (GERD) and business (or private sector) expenditure on R&D (BERD) as a proportion of GDP are also considered to be important criteria in determining the strength of a country's innovation system. Ready accessibility to capital required for the setting up of new manufacturing industries and to incubators for commercializing research, the capacity a country possesses in undertaking research in emerging technologies, and the existence of an enabling environment for the attraction of capital or foreign direct investment (FDI) in terms of regulatory framework, intellectual property rights regime and tax incentives are considered to significantly contribute to the strength of a country's NIS. The success of a country in exploiting its NIS could be evaluated from the number of scientists and researchers active in the country as a proportion of its population, the FDI it has been able to attract, the number of quality publications and patents it produces and the royalties and technology licensing fees received and paid for by it.

4.1 South and South East Asia

Sri Lanka does not seem to have done particularly well in these areas. A comparison of human resources in Science and Technology showed that the percentage of students of ages between 17-24 in Sri Lankan Universities was similar to that of Pakistan and stood at a mere 4 per cent compared with 10 per cent for India,

28 per cent for Malaysia, 35 per cent for Thailand and 44 per cent for Singapore. The low figures for Sri Lanka probably reflect the higher age at which students enter Universities in Sri Lanka, 21 years for a first attempt entrant which means many of the senior University students in Sri Lanka do not fall into the 17-24 age category. The bigger impact however is the scarcity of University places in Sri Lanka as no private Universities are permitted to function unlike in other countries. The percentage of GDP spent on Education in Sri Lanka (2.1 per cent) is even smaller than that of Pakistan (2.3 per cent) while the percentage is much higher for India (3.7 per cent), Singapore (3.9 per cent), Thailand (4.2 per cent) and Malaysia (7.4 per cent). Pakistan has since initiated a programme to increase investment in higher education by 1500 per cent. However these statistics do not provide any information on the quality of education which can impact on a nation's ability to exploit S&T for development.

The number of researchers in R&D in Sri Lanka per million population (208) is also much less than Thailand (330), Malaysia (430) or Singapore (7000) although higher than Pakistan (120) and India (157). GERD/GDP has been for Sri Lanka extremely low at 0.14 per cent, compared with Thailand (0.24 per cent), Pakistan (0.6 per cent), Malaysia (0.7 per cent), India (0.9 per cent) and Singapore (2.2 per cent) while BERD/GDP is 0.01 per cent for Sri Lanka, 0.02 per cent for Pakistan, 0.09 per cent for Thailand, 0.1 per cent for India, 0.5 per cent for Malaysia and 1.3 per cent for Singapore.

In terms of annual publications in ISI journals per million population, Sri Lanka averages 14, compared with 11 for Thailand and 21 for Malaysia. US Patents granted to citizens during 2002-2006 were 5 for Sri Lanka compared with 40 for Thailand, 58 for Malaysia and 366 for India. Receipts from Royalties and licence fees were $ 12 million for Thailand and none for Sri Lanka and Malaysia. Foreign Direct Investment as a percentage of GDP was 1.2 per cent for Sri Lanka and 0.3 per cent for Thailand, most of the FDI in Sri Lanka being in the telecommunication sector.

4.2 Sri Lanka

Sri Lanka's National Innovation System is therefore weak and therefore it is not surprising that she have not performed well in science and technology development and in the exploitation of S&T for national development. In many developing countries S&T has been accorded a central position in the national planning process and its importance in achieving socio-economic objectives recognized.

However, Sri Lanka has neither attempted to formulate a science and technology strategy for development nor put into place the institutional mechanisms to implement such policies. This could be for a variety of reasons including budgetary and human resource constraints. Although developing countries like Sri Lanka have separate budgets for science and technology, science and technology development generally suffer from funding constraints as a low priority is given to science and technology when limitations in funding require rationalization of funds. Sri Lanka has to invest heavily in science and technology infrastructure before it could benefit from research and development and technology transfer, largely because it had failed to invest sufficiently in building up infrastructure in the past. As such decision makers are skeptical about the immediate benefits the nation could derive from science and technology and give low priority to science and technology. The escalation of the war in the North and East of the country has led to an increase of military expenditure to almost 4 per cent of GDP this year, placing a strain on the economy and thereby on any expansion of the S&T budget.

It is not surprising that Sri Lanka has not been very successful in the domestic production of technology, most innovations being of low technology. Unfortunately there has been no attempt to promote defence related research although the country has been engaged in a protracted civil war.

Defence research is believed to have helped provide countries like Japan and Korea with the capacity to attain rapid post-war technological growth. The private sector in Sri Lanka has also not been vibrant enough to utilize technology licensing for its development. Sri Lanka has however been able to attract FDI through privatization strategies and also some Greenfield FDI mostly in the telecom sector. FDI has not resulted in much technological learning or spin-offs.

5. SUCCESS STORIES

Although the performance indicators suggest that Sri Lanka is doing badly, there have been many success stories which are worth further study. Among them are resource intensive enterprises begun before economic liberalization which have developed high demand products for the export market like the Hayleys group of companies, Haycarb and Dipped Products, resource based technology intensive enterprises like Loadstar, state enterprises which became strong players in heavy engineering after privatization by forming strategic partnerships with foreign firms like the technology intensive Colombo Dockyards and labour intensive firms set up in the post liberalization era which have grown into world leaders.

5.1 Hayleys Group

Haycarb belonging to the Hayleys Group was incorporated in 1973. It is the world's largest coconut shell based activated carbon manufacturer manufacturing over 16 per cent of global production and an annual turnover in excess of US $ 30 million. It has established manufacturing units in Thailand and the Philippines and subsidiaries in Thailand, Australia and the U.K. Their products have moved up the value chain to serve high value niche applications such as vehicle cabin air filters, military and industrial gas masks and medical applications. Its subsidiary Puritas provides environmental solutions to local companies and assistance in the setting up of biomass based power plants. Dipped Products, another Hayley company is a protective hand wear producer and supplies 5 per cent of the world demand for rubber based domestic, industrial and medical gloves in 270 versions belonging to 60 different categories. Its annual turnover of over US $ 70 million and it has expanded its manufacturing into Thailand. The factors contributing to their success included the fact that they were based on Sri Lankan agricultural resources and the favourable import substitution policies which were in place before economic liberalisation. Technology was initially purchased but there has been little production of domestic technology.

5.2 Loadstar

Loadstar which is responsible for 20 per cent of the world solid tyre production is an example of an enterprise that is both resource intensive and technology intensive. Consequent to an increase in the export duty of raw rubber introduced in the early 1980s, the Belgian solid tyre manufacturer Bergounan set up a joint venture with a leading local rubber company to manufacture solid tyres in Sri Lanka, Bergounan (Lanka) Ltd. Key Belgian personnel left the company to float a new Belgian company, Solideal which in 1985 began a new joint venture, Loadstar, with a leading Sri Lankan light engineering family owned business. Presently, Solideal has a 60 per cent share and Jinasena's a 40 per cent share. The company with nine plants, a turnover of over $ 300 million in 2006 and a staff of 7000 is responsible for 2.5 per cent of Sri Lanka's total exports. It has established new rubber plantations but imports 15 per cent of its raw rubber requirements having expanded into pneumatic tyres, industrial wheels and rubberised tracks as well.

An R&D centre with 20 researchers has been established in Ghent, Belgium while a small research unit is functioning in Sri Lanka. A Swiss patent has been granted to the Belgian partner. Loadstar is an example of successful technology transfer with Sri Lankan staff technologically learning from the Belgian company. Although it does not appear to have led to the establishment of any spin-off companies several other foreign companies have established solid rubber manufacturing plants in Sri Lanka, probably exploiting the availability of trained staff.

5.3 Colombo Dockyard

Colombo Dockyard which began in 1974 as a state enterprise engaged in carrying out repairs on ships at the Port of Colombo was converted into a public limited liability company in 1983 and entered into a partnership with Onomichi Dockyard, a medium size shipbuilder from Japan who purchased 50 per cent of its shares in 1993. From 1977 it began supplying barges, tug boats and in 1982, an offshore patrol vessel for the Sri Lankan Navy. In 1983 it exported barges to Myanmar and in 1987 began shipbuilding for export to Maldives. A major restructuring took place after forming the partnership with Onomichi in 1993 and the firm has moved into the new areas of heavy engineering and offshore engineering.

Technological learning from this partnership has led to improvement in its Quality Management Systems and the internationalization of its activities obtaining ship building orders from India and the Middle East. Its annual turnover has exceeded US $ 70 million and its greatest limitation is the lack of space in Colombo Port for expanding its activities. The partnership with Onomichi contributed to technological learning while Onomichi, a medium size shipbuilder in Japan made use of the opportunity to expand into and access export markets in South Asia.

5.4 MAS Holdings and Brandix

The garment exporter, MAS Holdings has over 20 years developed a comprehensive supply chain for intimate apparel and sportswear in the country. The company which is owned by a family with roots in the textile trade has achieved technological learning through strategic partnerships with market leaders in the field particularly in the past few years. It has even developed branding and retail skills and successfully introduced its own brand of lingerie in India and has earned world renown for its involvement in the development of innovative swimsuits used by Olympic swimmers. The company employs 45,000 people in 28 factories in 5 countries and has an annual turnover of USD 750 million. A similar family owned garment exporter, Brandix has a turnover of USD 320 million and has embarked on a Brandix City in Visakhapatnam, India. Both companies have gained from the Multi-Fibre Agreement (MFA) which initially gave them guaranteed markets under a quota system. They seem to have also exploited the Free Trade Agreement with India to expand into India.

6. CONCLUSIONS

It may be concluded that on the basis of the accepted attributes of a NIS, Sri Lanka would not be expected to successfully exploit S&T for economic growth. However, the correlation between S&T investment and economic growth has not been unambiguously shown to be true for developing countries, especially those with poor infrastructure. The success stories suggest that in spite of low investment in S&T, firms have been able to innovate and contribute to economic growth. Most of them originated in the pre-liberalisation era and many were based on the use of Sri Lankan natural resources. Labour intensive post-liberalization companies like MAS Holdings and Brandix were initially protected by the Multi-Fibre Agreement. Technological learning from FDI has been a feature of a small number of companies including the resource intensive Loadstar and the privatized Colombo Dockyards. However if Sri Lanka is to continue to promote such innovation and building new competencies in technology transfer, the rapid expansion of S&T education at university level, especially in post-graduate training to increase the nation's capacity to absorb technology would be crucial. Successful creation of intellectual property in the field of technology and transfer of this technology will also require the development of the scientific, educational, legal, financial, commercial and industrial infrastructure that would facilitate the commercialization of research.

ACKNOWLEDGEMENTS

The author is grateful to the Rockefeller Foundation for awarding him a travel grant and a Residency at Bellagio Center, Italy, during which period this work was carried out.

REFERENCES

DeLong, J.B., 2000. Cornucopia: The Pace of Economic Growth in the twentieth Century NBER Working Paper 7602. http://www.j-bradford-delong.net/TCEH/2000/TCEH_2.html (accessed 5 Oct. 2007)

De Silva, M.A.T. and Amaradasa, R.M.W., 2000. Science and Technology in Industry, Proceedings First Biennial Conference on Science and Technology in National Development, National Science and Technology Commission, Sri Lanka, pp.107-125.

Freeman, C., 1982. The Economics of Industrial Innovation, 2nd edn, MIT Press, Cambridge, MA, USA. pp.250.

Intarakamnerd, P., Chairatana, P. and Tangchitpiboon, T., 2002. National Innovation System in Less Successful Developing Countries: The Case of Thailand. Research Policy, 8, pp. 1-13.

Lall, S., 1994. Technological Capabilities in Salomon, J.-J., Sagasti, F.R. and Sachs-Jeantet, C. (ed.), The Uncertain Quest: Science, Technology and Development, UNU Press, Tokyo, Japan, pp. 264-301.

Lundvall, B. A., 2007. National Innovation Systems – Analytical Concept and Development Tool. Industry and Innovation. 14 (1), pp. 95-119.

Mokyr, J., 2002. The Gifts of Athena: Historical Origins of the Knowledge Economy. Princeton University Press, Princeton, pp. 376.

Chapter 6

Public-Private Sector Partnership

A.K.W. *Jayawardane*
Professor of Civil Engineering and Dean/Faculty of Engineering, University of Moratuwa, Sri Lanka
E-mail: dean-eng@mrt.ac.lk; akwi@civil.mrt.ac.lk

ABSTRACT

The Small and Medium Enterprise (SME) Sector in Sri Lanka, as in many developing countries, is lacking in competitiveness to face the challenges and exploit market opportunities. The development of the SME sector in partnership with large scale manufacturers is an essential and integral part of the development strategy of a country. Effective Public Private Partnerships (PPP) has become an important initiative in achieving this goal. While government R&D institutions play a major role in PPP for technology development, university – industry partnerships are more innovative and wide ranging. These may include simple mechanisms such as regular and formal university industry interactions; carrying out industry- funded undergraduate and postgraduate research projects and research degrees addressing specific industry issues; regular technical clinics for industry; and individual, group or inter-departmental or international multidisciplinary R&D activities. They also include more complex interactions such as industry-funded R&D laboratories in universities; multi-party collaborations with university-industry cluster-relevant ministries; endowed research chairs; and formalised mechanisms for matching industry needs with university seeds (need-seed matching). Despite these success stories, PPP face many issues and challenges. These include–lack of funding from industry specially the SME sector, lack of capacity in the industry to absorb new technologies, the pressure to provide solutions in double quick time, lack of emphasis on technological development in industry, lack of commitment by some industrialists, and lack of incentives or recognition for their enhanced commitment. A concerted effort by all the stakeholders therefore is needed to more profitably exploit PPP for enhanced economic development.

Keywords: *Universities, R&D Institutions, Partnerships, SME sector, Technology development, Technology transfer.*

1. INTRODUCTION

The Small and Medium Enterprise (SME) sector in Sri Lanka, as in many developing countries, is lacking in competitiveness to face the challenges and exploit opportunities in the global market.

This sector is mostly confined to their local markets and sometimes can not compete even in national markets. However, the development of the SME sector, in partnership with large- scale manufacturers, is an essential and integral part of the development strategy of a country, especially through growth of the industrial sector. Effective PPP is popularly seen as an essential component in achieving this goal. Depending on the industry sector and the strengths of relevant public and private sector institutions, their complementary capabilities need to be harnessed. Such partnerships can be effectively used to identify strategic industries, formulate favourable policies, address industry common issues, match the expertise with technology needs, create widespread employment opportunities, and use raw materials more productively–leading to wealth creation and improved standards of living of society. This paper discusses PPP with specific reference to University-industry partnerships addressing (*a*) its need; (*b*) mechanisms for public private partnerships; (*c*) the case of University of Moratuwa, (*d*) uses and challenges; and (*e*) the way forward in PPP.

2. PUBLIC PRIVATE PARTNERSHIP – THE NEED

The need for Public Private Partnerships (PPP) is simply to harness the complementary strengths of the two sectors for mutual benefit, industry growth and economic development. It can be considered just like joint ventures formed for business success between two partners, with multiple partners or with an industry cluster.

Public institutions in these partnerships can be numerous including government policy-making bodies, ministries, provincial and local government authorities, the Treasury, research and development (R&D) institutions, science and technology (S&T) institutions and universities. Private sector partners can include service organisations, manufacturing organisations, educational organisations, professional bodies, trade associations, non-governmental organisations, industry cluster organisations, community organizations, and even individual entrepreneurs.

Understanding the complementary strengths of these sectors would be vital and strategic for forging partnerships for mutual benefit. These strengths can be broadly identified as shown in Table 6.1.

Table 6.1: Complementary Strengths of Public and Private Institutions

Strengths of Public Partners	*Strengths of Private Partners*
Usually not profit oriented	Highly motivated with the need
Work for national interest	Less bureaucracy
Medium to long term perspective	Liberal financial regulations
Stable institutions	Availability of funds
Concentration of experience	Aim for competitiveness, survival and growth
Connections with the government	

These needs and strengths can be effectively harnessed to achieve a wide range of objectives such as enhancing output of production; solving an industry problem; facilitating technology transfer and diffusion; quality and productivity enhancement; creation of new industries; enhancing competitiveness; as an educational partner in upgrading curriculum and teaching; and as an idea generator.

3. PROSSIBLE MECHANISMS FOR PUBLIC PRIVATE PARTNERSHIPS

Mechanisms for PPP can be broadly classified into market/financial oriented mechanisms or non- market/non financial oriented mechanisms, as shown in Table 6.2, which can be effectively used for university-industry collaborations.

Table 6.2: Market and Non-market Oriented Mechanisms for Public Private Partnerships

Market Oriented Mechanisms	*Non- Market Oriented Mechanisms*
R&D collaborations	Representations in government bodies
R&D consultancies	Industry interaction boards/committees
Trouble shooting consultancies	Exchange of employment
Technology development and transfer consultancies	Industrial placements
Fee levying education and training	Industrial exhibitions
Joint ventures	Technical clinics
IPR assistance	R&D competitions
Endowed R&D positions	Alumni associations and societies

4. CASE STUDY–THE UNIVERSITY OF MORATUWA, SRI LANKA

The University of Moratuwa in Sri Lanka is one of the 15 government universities and is popularly regarded as the best and most prestigious in Sri Lanka. The University has only three faculties–the Faculty of Engineering, the Faculty of Architecture, and the Faculty of Information Technology and hence it is a purely technical university. The University maintains close interaction with industry and produces highly employable graduates. The University provides a conducive environment for PPP, emphasizing R&D and innovation and entrepreneurship, some of which can be considered as pioneering initiatives in a Sri Lankan University. Some of these initiatives are described below.

4.1 Engineering Design Centre

The Engineering Design Centre (EDC) was established in the Faculty of Engineering in 2000 with seed funding from the Asian Development Bank Science and Technology Personnel Development Project. EDC is now a self-financed Centre managed by a Board of Management headed by the Dean of the Faculty and functions under the Director of the Centre. The objective of the Centre is to promote the importance of engineering design, carrying out engineering designs for the industry, trouble shooting, and promoting technology development and transfer. This centre provides much needed knowledge on engineering design to under-graduate students, promotes innovation and entrepreneurship and outreaches to meet industry needs in these areas. Some R&D activities carried out by the EDC and development initiatives are listed below.

(*a*) Development of a Tea Colour Separator at one eighth of the cost of an imported machine for the Tea Planters Association under a tripartite agreement with the Ministry of Industries, EDC and the Tea Planters Association. At present the technology is being transferred to a local manufacturer for commercial production.

(*b*) Design of a Rotary Transformer Testing Machine for FDK Lanka, one of the electrical component manufacturers in Sri Lanka.

(*c*) Coordinating a Japanese Government funded project with Toyohashi University in Japan aimed at enhancing and formalising university-industry interaction. The initiative is called Seed – Need Matching as the objective is to formally identify the needs of industry and match it with the seeds (expertise) in the University for more focused and applied R&D activities. It also involved the establishment of an intellectual property (IP) and Technology Licensing Office at the University for the benefit of the industry. Currently, a database is being prepared identifying the needs of industry and three "need – seed" matching projects are underway.

(*d*) EDC has established a Rapid Prototyping Facility at the University of Moratuwa under a tripartite collaboration with the Ministry of Industries, National Science Foundation (under the Ministry of Science and Technology) and the EDC. This facility is aimed at satisfying a much needed requirement among local manufacturers for accelerated product design and development with reduced cost. While this project benefits the electrical, electronic, ceramics, gem and jewellery, die and mould, chemical, and plastic sectors it also provides added facilities for university R&D.

4.2 University Industry Interaction Cell

This is another outreaching arm of the University functioning within the University administrative framework. It functions under a Director guided by the Board of Management headed by the Deputy Vice-Chancellor and can be considered as the internal commercial arm of the University.

The Cell acts as the agent between the University and industry to facilitate industry focused R&D projects, conduct Continuing Professional Development (CPD) courses for industry, and provide expertise in organising conferences/seminars and workshops, all of which are mostly self-funded. This takes out the burden of organising the CPD activities and other outreach activities by individual department of the University.

4.3 Uni-Consultancy Services–UNIC

This is the external commercial arm of the University functioning as a limited liability association outside the university framework. Its objective is to quickly respond to industry needs as the University's internal mechanisms have limitations in terms of administrative and financial autonomy. This is a self-funded project that is self sustaining and is the one of its kind in Sri Lanka. Income from this initiative is shared with the University. This mechanism has met with success in providing CPD courses overseas in partnership with reputed international universities.

4.4 Industry Sponsored Laboratories

Due to its close collaboration with industry and prompt response to industry needs, both the private sector and other ministries have come forward to help the University in developing and promoting new initiatives. Some of these include the following.

(*a*) Dialog-University of Moratuwa (UoM) Mobile Communications Laboratory

This is a tripartite agreement between Dialog Telecom – the premier mobile communication company in Sri Lanka, the University of Moratuwa (UoM) and UNI Consultancy Services (UNIC). This is a success story where Dialog Telecom has been providing total financial assistance to establish the R&D Laboratory at the UoM and funding for research and associated project staff. UoM provides laboratory space and expertise for R&D and UNIC provides funds and human resources management. The Laboratory functions under a Director and reports to the Board of Management which consists of both Dialog staff and UoM staff. Several new technologies have been developed, some of which have received national and international awards.

(*b*) Zone 24x7-UoM Laboratory

Zone 24x7 is a collaborative venture between a firm from the United States of America and a Sri Lankan company. The Sri Lankan partner has entered into an agreement with the UoM very similar to that described in (*a*) above. The objective of this laboratory, housed in the University, is to develop middleware technologies for the electronics sector with the idea of commercialising it both in Sri Lanka and the United States

(*c*) Cargills-UOM Food Processing Laboratory

Cargills is a large grocery chain in Sri Lanka and this collaboration with the UoM provides food processing technologies to meet the requirements of Cargills.

(*d*) Die and Mould Facilitation and Development Centre

This is a tripartite agreement with the Ministry of Industries, Die and Mould Manufacturers Association and the UOM. The Ministry provided funding to establish this centre at the UoM. The UoM provides laboratory space, expertise, and maintenance of the facility. The Association provides necessary advice while utilising the facilities. The Centre is managed by a Board of Management consisting of all three parties.

(e) Advanced Computational Laboratory

This is a donation to the UoM to establish an Advanced Computational Laboratory in the Civil Engineering Department by a major civil engineering contractor and the funds are to be disbursed over a period of five years.

In addition to above, there are several other mechanisms where various types of collaborations, such as funding for facilities and free licence to use software are provided for mutual benefit.

4.5 Collaborations for Industry Consultation

The UoM, especially the Faculty of Engineering, has institutionalised mechanisms called Industry Consultative Boards where industry feedback for university academic programs is taken at regular intervals while also serving as a forum for initiating other collaborative projects.

(a) Department Industry Consultative Boards (DICB)

This is a Board with 50-50 industry -academia participation that meets once in every 3-6 months to deliberate on matters of mutual benefit. They address curriculum revision, industrial training for under-graduates, research, development, visiting lecturers and consultants from the industry, industry placement for lecturers and so forth.

(b) Faculty Industry Consultative Boards (FICB)

This is very similar to the DICB meeting at regular interval once in 6-12 months taking up issues discussed at DICBs which are common to the Faculty.

4.6 Industry Endowed Positions

Industry endowed academic and research positions are rare in the developing world although this mechanism is productively used in the developed world. The UoM has just developed a common policy with minimum endowment as three times the salary of a Professor for posts of Endowed Professor/Senior Fellow. Negotiations are in progress for two positions in the areas of Entrepreneurship and Sustainability and Environment. These positions can be very productively used to conduct high-end state-of-the-art research or to introduce new specialties and develop academic programs in emerging disciplines.

4.7 Other Industry Collaborations

Possible industry collaborations are so numerous and wide ranging. Several other collaborations include:

(a) Industry Technical Clinics

This involves provision of free technical advice to industry, especially SMEs, where they visit UoM and chat with expert staff to obtain necessary assistance while visiting the University facilities. This type of interaction significantly enhances industry confidence in University expertise and facilitates friendlier collaboration based on trust.

(b) Student Mentoring

Industry and employers often criticise universities that they do not produce graduates with appropriate skills and attitudes and that fresh graduates lack leadership, communication and team skills and confidence thereby emphasizing that they lack much valued "soft skills". It is now recognized that universities alone cannot produce readily employable graduates with the requisite soft skills and that industry should be an active partner in building these skills. Here individual experts, within and outside the discipline, undertake to mentor undergraduates in their working environment and this initiative has produced gratifying results.

(c) Industry Open Days in the University

On these Open Days, various departments of the UoM display their student and staff development projects for industry, school children, and potential collaborative partners.

(d) Invited Visits to the University

These involve visits by industry experts to talk to students and staff to enhance their awareness of industry-related issues and seek possible assistance.

(e) Community Projects

These projects are initiated and implemented by students and staff to provide opportunities to students to demonstrate leadership and social responsibility by serving society.

All the aforementioned mechanisms can enable universities to identify industry problems and potential research for under-graduate and post-graduate projects and for which industry is willing to provide necessary funding.

5. ISSUES AND CHALLENGES

Despite successes there are several issues and challenges we need to address for more productive PPP. Given below are some of those applicable in Sri Lankan context-most of which can be used in other countries too.

(a) Lack of Industry Funding

This is a common problem especially in developing countries where industries find it difficult to invest in R&D and to support similar initiatives. One main reason for this is lack of competitiveness and short-term orientation. This is especially true in the SME sector. There is a need to jointly explore funding opportunities as demonstrated in the Die and Mould facilitation centre mentioned above.

(b) Large Technology Gap and Lack of Absorptive Capacity

This again is more prominent in the SME sector. Entrepreneurs in this sector do have innovative ideas but very often have insufficient theoretical background and technical capacities to absorb new technologies and systems.

(c) The "Quick Fix" Mentality of Industry

Very often industry looks for quick fixes and solutions which can sometimes mismatch with the objectives of public institutions – final outcome versus quality R&D. Unrealistic deadlines often discourages researchers in public institutions.

(d) Lack of Emphasis on Technological Development

This happens when organizations do not have a long term vision and there is lack of commitment from top management for technology-based development.

(e) Lack of Direction by the Government

Although government initiatives in PPP have vastly improved, there still exist lot of room for making the process more effective. Weaknesses include policy shortcomings, lack of strategic directions for thrust industries, lack of support for thrust industries/players, insufficient funding support, lack of outreach, and lack of incentives for public institutions.

6. THE WAY FORWARD

The above challenges call for careful interventions by both parties and the government. Following are some recommendations identified as appropriate having analysed the present situation of PPP in the Sri Lankan context.

1. Develop and Adopt a Visionary Science and Technology (S&T) Policy

Although we have been addressing this issue for a long time, a concrete document developed with inputs form all the stakeholders has not come out. The National Science and Technology Commission (NASTEC) and the Ministry of Science and Technology have initiated action on this. It is hoped that very soon Sri Lanka will have a proper and visionary S&T Policy.

2. Effective Performance Assessment of S&T Institutions/S&T Auditing

It is very often argued that S&T institutions are not productive and they do not do what is expected. The need to have an effective mechanism to carry out S&T auditing has been identified. NASTEC is working on this and very soon it is expected to have a performance assessment mechanism for S&T institutions.

3. Provide more R&D Funding Opportunities and Mechanisms

Investment for R&D is generally poor in developing countries and in Sri Lanka the investment was less that 0.2 per cent of the GDP. Despite some initiatives by the Government Ministries especially the Ministry of S&T, investment in R&D needs significant improvement in both public and private sectors.

4. More Effective and Coordinated "Need-Seed" Matching

Despite a few initiatives, a significant proportion of R&D conducted by public institutions do not focus on industry issues and hence do not contribute much in terms of industry development. Thus, a more focused and structured Need Seed matching has to take place in every public sector research institution.

5. More Effective Incentives for R&D Institutions and Achievers in Research and Development

R&D institutions, especially those in developing countries, receive step motherly treatment due to lack of funding and other urgent priorities for allocation of limited funding. Remuneration for researchers is low compared to industry practitioners and hence these institutions are unable to attract high quality researchers in general. Thus, more incentives and rewards for better achievers need to be introduced.

6. More Effective Intellectual Property Rights (IPR) Policies

Most public sector institutions do not have their own IPR policy which is essential for successful technology transfer and diffusion. The national Science Foundation (NSF) in Sri Lanka is developing a generic policy in this regard while some individual institutions are also doing the same. It is expected that this aspect will have necessary follow-up and legal backing to ensure that effective IPR polices are adopted by the institutions.

7. Compelling Technology Transfer and Technology Diffusion

Some technologies that have been developed and which have commercial potential are not effectively transferred or diffused to industry due to various reasons thereby no getting the full benefit of the investment that has gone into its development. A legal mechanism should be introduced to compel the institutions to transfer/ diffuse technologies developed such as the Bayh Dole Act (COGR, 1999) in the USA. The NSF and the Ministry of S&T have been taking action on this and it is expected that the situation will improve in the future.

8. Significant Change in the Attitude of People

A national drive should be introduced for promoting indigenous technology development, technology transfer, and promoting innovation and entrepreneurship. All stakeholders should subscribe to this movement with the government making a strong commitment towards technological leapfrogging for strengthening national competitiveness and raising the standard of living of its people.

REFERENCES

COGR – Council on Governmental Relations, 1999, The Bayh Dole Act – A Guide to the Law and Implementing Regulations, October.

Chapter 7

South-South Cooperation for Development of Small and Medium Enterprises (SMEs)

Tajammul Hussain
Director General (International Affairs),
Commission on Science and Technology for Sustainable Development in the South
(COMSATS), Islamabad, Pakistan
E-mail: hussain_tajammul@hotmail.com, husseint@comsats.net.pk

ABSTRACT

This paper emphasizes the importance of south-south collaboration for technology transfer and development of Small and Medium Enterprises (SMEs).

In a world where the accumulation of scientific knowledge and its technological applications are accelerating at a fast pace, it is a well-known fact that the developing world contributes only meagrely to modern science and technology due to lack of resources and skills. No developing country on its own has the capacity to shape the processes that can inspire the development of a global economy.

The developing countries are characterized by imperfections in planning, administration and management of its resources. All this has led to weak economies in the countries of the South over the past decades and centuries. It is also a region that has failed to arrive at a form of governance, where plurality prevails in various forms. Therefore developing countries are facing the severe burden of misery, poverty and destitution.

Small and Medium Enterprises (SMEs) play an important role in the economic development of the developing as well as developed countries. But the ratio of SMEs to all businesses is six times smaller in developing countries as compared to the industrialized countries.

It is, therefore, imperative for the countries of the South lacking in basic science and technology and Research and Development infrastructure to join together and try to help each other in the transfer of technology and development of SMEs. That is most practicable, as these countries can readily understand each other's difficulties and constraints due to being in the same economic state.

Keywords: *South-South Cooperation, Small and Medium Enterprises (SMEs), Technology transfer.*

1. INTRODUCTION

The concept of South-South cooperation, which encompasses both technical and economic cooperation among developing countries, may be said to have formally emerged as a universal principle during the 1970s, both as a strategy in support of the development efforts of the developing countries and as an instrument for promoting collective self-reliance in order to ensure their effective participation in the international economic system. This cooperation can contribute to the achievement of the internationally agreed development goals, including the Millennium Development Goals (MDGs). South-South cooperation is all about the tremendous force of solidarity, with which we can overcome even the biggest developmental challenges.

It is a fact that the South-South cooperation is needed today more than ever. No single country, even the most advanced among developing countries, has much hope of individually reaching the desired growth and development and influencing outcomes of international agenda. But, collectively, these countries can play a more effective role in achieving development objectives and improving their international standing.

2. INTERNATIONAL COOPERATION FOR SHARING KNOWLEDGE AND TECHNOLOGY

The notion of international cooperation came to existence with the UN Charter, which pledges to "employ international machinery for the promotion of the economic and social development of all people".

Technology co-operation enables the sharing of risks, rewards and progress of technology development and enables co-ordination of priorities. Mutual recognition of the value contributed by a country's investments in new technologies and innovation could usefully be built into an international cooperation.

International cooperation can help create an environment that enables developing countries to act and empowers people vulnerable to its adverse effects, building the resilience needed to prevent increased risk from translating into greater vulnerability.

The human development case for urgent international action is rooted in the ethical, social and economic implications of our ecological interdependence. The developed countries are required to support developing countries that are particularly vulnerable to the adverse effects of climate change, building their adaptive capacity and providing financial assistance.

International cooperation on adaptation can be thought of as an insurance mechanism for the world's poor.

International Cooperation can focus on:

- ✰ Sharing of knowledge and information between developed and developing countries;
- ✰ Coordinating R&D priorities in different national programmes;
- ✰ Pooling risk and reward for major investments in R&D, including demonstration projects.

Such an International Cooperation can be among South- South countries or North-South countries.

3. SOUTH-SOUTH COOPERATION

3.1 Genesis of South-South Cooperation (SSC)

Since a long time the scope of international cooperation was thought to be only limited between the industrialized world and the developing world because of the obvious fact that the industrialized countries were economically well off and had an easier access to advanced technologies and practices which could be shared with the developing countries for their development.

The 'North-South cooperation' and technology transfer from the North to South was considered as the bridge to fill the technological gap between the developed and the developing countries. For instance, the Vienna Convention (1979) on Science and Technology (S&T) for Development undertook to establish funds for S&T projects in developing countries. This was followed by international agreements with provisions for technology transfer, among which the prominent ones are the UN Conventions on Biodiversity, Climate Change and Combating Desertification. In 2005, the UN Commission on S&T for Development hinted on the responsibility of the developed countries to transfer technology to the South.

Over the years it was realized that within the group of developing countries also there were marked differences. This made people to think of cooperation among the developing countries to create a pool of expertise and technological capacities to supplement what they individually lack, thus giving rise to the concept of 'South-South Cooperation' (SSC). The term 'South' has often been used in the realm of international interaction to entail the fact that all the developed nations of the world (barring Australia and New Zealand) lie to the North of the developing nations.

3.2 Need for South-South Cooperation: Shared Goals and Challenges of the South

The countries in the South generally share certain commonalities such as similar developmental experience and are also faced with common challenges, such as high population pressure, poverty, hunger, disease, environmental deterioration, etc. In view of this, South-South Cooperation is clearly becoming more relevant.

Further, the technology is often limited in terms of its applicability to a specific ecosystem dynamics. For example, the technology applicable in the temperate zone is different from that required in the tropics and also the market scenario in response to the demand in the tropics is different from the one that exists in other developed countries.

3.3 South-South Cooperation: More Promising and Appropriate?

In addition to the above-mentioned factors, several geo-political developments emphasize the need for South-South cooperation. Some of these factors may be:

- The focus of research activities in the industrialized countries is gradually moving from the public to the private sector. The private sector aims at maintaining their technological edge over the developing countries, and hence is averse to sharing their technology. This is an impediment for the scope of potential international research collaborations with developing countries, which depend highly on the public-sector.
- The interest of many developed countries in helping developing countries seems to be receding, e.g. the monetary support for the flagship Consultative Group on International Agricultural Research (for promotion of sustainable agricultural development) has been diminishing over time.
- On the other hand, South-South trade has been rising at a rate of 11 per cent annually over the past decade. For example, the trade between Africa and Asia has risen from about US$ 6 billion to around $ 18 billion in the past decade. Also the average annual growth rate between 1980 and 2002 in exports of goods and services was higher in developing countries than developed countries. This gives an insight to the thought that SSC has been delivering visible beneficial results on the economic front (UN, 2004).
- The LDCs (Least Developed Countries) could find economic and sustainable solutions to address their needs and problems, by sharing and learning from the experiences of other developing country counterparts.

SSC can also boast of being based on abundant networks (social, natural, institutional, etc.) that support globalisation. Key players in this arena of networking in the South are the East African Community, the African Union, the Caribbean Community, and SAARC (UN, 2004).

3.4 The Potential of South-South Cooperation

Cooperation among the developing countries has begun to be favourably regarded in various parts of the world. Such cooperation among developing countries has considerable potential, which can be seen in the following contexts:

- The developing countries have many things in common, and their capacities and resources are often complementary and exist at different but relatively close levels of development. This would facilitate the transfer of technology and exchanges of experience and other resources among these countries;

- Today several developing countries have diversified their economies and have a large production capacity for goods and services;
- The information society revolution has opened new opportunities for exchanges of information among countries of the South and has made such exchanges easier;
- The accelerating trend towards globalization and trade liberalization has created new opportunities for exchanges of goods, services and information among developing countries;
- Today several developing countries have human resources properly trained in institutions of high technical quality;
- In addition to the governments, new participants in international cooperation are continually emerging in developing countries; for example, participants from civil society and the private sector, which play a growing role in South-South cooperation;
- South-South cooperation is seen as a necessary and fundamental component of international cooperation for development. South-South cooperation complements North-South cooperation and fits perfectly into triangular cooperation arrangements where the know-how and technology of one or several developing countries are combined with financial support from one or several developed countries to provide assistance and to transfer technology and know-how to one or several other developing countries.

4. SMALL AND MEDIUM ENTERPRISES DEVELOPMENT

4.1 Introduction to Small and Medium Enterprises (SMEs)

Small and medium enterprises (SMEs) or small and medium businesses (SMBs) or variations thereof are companies whose headcount or turnover falls below certain limits.

SMEs have been the backbone of growth of an economy in driving industrial development. Due to their sheer numbers, size and nature of operations, this segment of the economy has been recognized as promoting endogenous sources of growth and strengthening the infrastructure for accelerated economic expansion and development.

The potential of SMEs to promote domestic-led growth in new and existing industries and to strengthen the resilience of the economy in a competitive and challenging environment are inarguable. Economic growth in developed countries, such as Japan, Korea, Taiwan and many others, were significantly generated by SME activities. The percentage contribution of SMEs to GDP/total value added range from 50 per cent in Korea, 55.3 per cent in Japan, 57.0 per cent in Germany, 60 per cent in China compared to 47.3 per cent attained by Malaysia.

SMEs are generally divided into two broad categories:

1. Manufacturing, manufacturing-related services and agro-based industries -are enterprises with full time employees not exceeding 150.

2. Services, primary agriculture and information and communication technology are enterprises with full-time employees not exceeding 50.

4.2 The Need to Develop SMEs

On several social and economic grounds, small and medium-sized enterprises (SMEs) are of overwhelming importance in most Asian countries, including those in the Association of South-East Asian Nations (ASEAN). Typically, the SME sector accounts for more than 90 per cent of firms outside the agricultural sector in East and South-East Asia, and in Japan as well. It is also the biggest source of domestic employment, providing a livelihood for over three quarters of the region's workforce, especially women and the young. The relative share of SMEs in total output and exports is generally much smaller, i.e., one third or less. As such, the SME sector will remain the backbone of virtually every economy in this region and, for that matter, of the world in the foreseeable future.

Despite their economic importance, the SMEs suffer from a variety of structural and institutional weaknesses, which have constrained their ability to take full advantage of the rapidly advancing process of globalization and adjust to the economic liberalization measures. These weaknesses include the relatively narrow base of the sector and its focus on low value added products; the absence of an adequate information infrastructure, the technological obsolescence of much of the capital stock; the inadequacy of the existing support services for entrepreneurship development and promotion; and a relatively low level of integration in global value chains. Equally important, SMEs also fail to realize their full potential for creating productive employment because of the highly uneven dissemination of information resources. Therefore, there is an urgent need to enhance the international competitiveness of developing countries' SME sector and promoting the increased inflow of foreign investment and technology. One of the major bottlenecks in this context is the lack of knowledge at the end of SMEs.

SMEs and, by extension, all business firms have to manage growth and change in an environment where the pace, patterns and organization of production have evolved fundamentally since the late 1980s. Trade liberalization at the global and regional levels and the new information and communications technologies (ICT) have entwined to create rich opportunities as well as formidable challenges to all interdependent countries and enterprises.

4.3 Opportunities for South-South Cooperation in SME Development

The revolution in ICT has created unprecedented opportunities to narrow the knowledge-gap between the North and the South by providing equitable access to the world's stock of scientific knowledge to everyone, everywhere. Through electronic mail and the Internet, data can now be instantly transferred across vast distances, providing science-poor countries with the possibility of access to the latest scientific and technological information for addressing local and global problems. In fact, SMEs in the South with Internet facilities can now communicate easily with each other and their contemporaries in the North to form new virtual networks and global research teams.

However, many members of the world community cannot fully participate in and benefit from this information revolution. Unfavourable economic conditions and the high cost of wire-line infrastructure have made it difficult to provide these facilities to the people in the Third World, particularly those living in remote areas. For example, the total number of telephone lines in the 48 LDCs is 1 per cent of the number of lines in the USA alone, and only 1 per cent of the world's telephone lines are in Africa and about half of these are in South Africa alone.

On the other hand, rapid advances in wireless digital systems based on satellites or cellular transceivers can provide a much less expensive and permanent solution to communication problems in developing countries. Among many advantages of wireless telecommunication systems over wire-based systems are that they can be developed quickly and are not affected by natural hazards.

Several developing countries like Argentina, Brazil and China, are investing heavily in digital communication systems. Telephone networks in small countries like Botswana, Djibouti, Ghana, Maldives, Mauritius and Qatar, are now completely digital, having bypassed the older wire-based systems and leapfrogging to this new technology. This trend deserves to be emulated by other developing countries. The present information age will help SMEs exchange information on products, technologies, human resources, etc. freely, in overcoming obstacles of distance and borders.

Trade via the Internet has now become an intrinsic part of an increasingly large number of SMEs in the developed countries. About one half and one third of the medium- and small-sized enterprises in Europe maintain an e-mail contact address or a presence on the World Wide Web; but there are no comparable data in the case of SMEs in ASEAN or Asia.

There are now greater scope and more opportunities for inter-firm linkages for enhanced collective efficiency, technological and innovation capabilities, and hence competitiveness. In particular, the proliferation of complex networks of international production and cross-border supply chains have widened and deepened the potential and avenues for SME involvement. Furthermore, subcontracting and outsourcing relationships now cover processing and manufacturing activities and services of high value addition and technological sophistication, ranging from original equipment manufacturing, complete-package production, product design and engineering, and research and development (R&D) to various other high-end support services.

4.4 Challenges to South-South Cooperation in SME Development

One of the most critical challenges facing the developing world is how to bridge the huge gap between the North and the South in the production and utilization of scientific and technological knowledge. Measured in terms of publications, the science-rich North, representing 20 per cent of the humanity, contributes more than 90 per cent of the world's share of current scientific knowledge; while, the science-poor South, representing 80 per cent of the humanity, generates less than 10 per cent of this knowledge.

In terms of technological output measured by patents, the inequality is much greater. The South's 1995 share of patents, held by the two largest and most international patent systems in the USA and Europe, amounted to less than 1 per cent of the world's total.

What is more disturbing is that the North-South divide in scientific output and technological innovations is constantly widening. On the one hand, the North, with its huge investments in research and development (R&D), is rapidly advancing the frontier of scientific knowledge. On the other hand, developing countries are spending small proportions of their gross domestic product (GDP)–often less than 1 percent – on R&D. Put another way, the world's total R&D expenditure in 1994 was about US$470 billion; and only 10 per cent of that was attributed to the South. This makes it very difficult for the South to develop their capacity to catch-up.

Reducing these disparities will be a major challenge for South-South co-operation in the 21st century. Previous efforts by national governments and international development agencies to overcome poverty and stimulate growth in developing countries have not recorded much success.

Rapid globalization, driven by revolutionary advances in technology and information and communication systems and characterized by economic liberalization, free trade, and increased competition have often widened the gap between poor and rich nations. The impact of globalization on the composition of financial flows, for instance, has been dramatic. Overseas development aid, a major source of external funding for development projects in poor countries, slid from US$56.4 billion in 1990 to US$44.2 billion in 1996. At the same time, foreign direct investment and private financial flows soared from US$41.9 billion to US$256 billion. Such trends have benefited only a few developing countries with large economies. The primary challenge, then, is how South-South cooperation can help the majority of developing countries bridge the knowledge gap and effectively respond to and benefit from rapid globalization by enhancing their capacities in science, technology and knowledge. Such efforts will serve as cornerstones for the transition of the South to sustainable economic growth and development.

In developing countries SMEs typically find it hard to become internationally competitive and offer products and services of the right quality services and governments and non-profit agencies often need to provide assistance. The top challenge according to then MDGs *i.e.*, poverty reduction makes it hard for the developing countries to develop their SME sector. Poverty remains a key concern, and highlights the need for enhancing the participation of marginalized groups, particularly women, youth and minorities in the developing countries.

Due to poverty, most of the developing countries lack a strong education base. Creating a pool of qualified people with the knowledge and skills to contribute significantly to SME development, to be entrepreneurial, develop science and technology, deliver basic services, and to be enlightened leaders is more than necessary. None of the MDGs can be achieved satisfactorily without educating and training enough qualified people through all levels of education. Also the issue of illiteracy is made worse by the gender inequality issue.

Brain drains of the talented from the developing countries also create hindrance in SME development. Technological, entrepreneurial and professional experience from the Diasporas should be mobilized to contribute towards the development of SMEs in their home countries.

The almost unlimited opportunities for gainful growth through trade at the demand side are counterbalanced by formidable challenges on the supply side. Competition has become increasingly fierce among the global and regional economies and enterprises, SMEs included. There are also many more producers competing for both existing and new markets and market segments for goods, services, finance and other wealth-creating technologies and knowledge. The competitive strength of China was notable in this regard, even before the country became a member of World Trade Organization in December 2001. Indeed, market penetration and displacing pressures from China have been keenly felt by producers across Asia, particularly those suppliers (including SMEs) at lower stages of technology sophistication and relying on high import content.

Consumer preferences and market standards have become more sophisticated and exacting. Competitive advantage is now determined by several non-price parameters such as quality, health and safety, social equity in employment and production, and ecological compatibility of products and processes. Furthermore, market demand is constantly changing, a trend facilitated not least by the rapid advances in ICT, bio-engineering and new materials sciences. In consequence, there are more frequent introductions of new products and processes, faster and more innovative design changes, shorter product cycles and smaller output batches, higher quality and greater mass customization, more just-in-time sourcing and greater punctuality in delivery.

4.5 Policy Implications and Options for SME Development

The new development context requires a change in both perception and practices, in other words, a new or different mindset in the promotion of SME development. As is apparent from the preceding discussion, competitiveness is increasingly man-made; furthermore, it can be leveraged by factors other than location and natural resource endowments. One lever is through the maintenance of ongoing access to the available store of global information and knowledge, including market standards, marketing opportunities and innovative technologies. Another is embodied in the large gains in collective efficiency and flexibility through participation (whether or not at arm's length) in clusters of firms, or in networks of inter-linkages backward with suppliers, laterally with other producers and providers, and forward with users and consumers. Yet leverage relates to the firm's own capabilities for ongoing learning and improvements in efficiency and flexibility; indeed, business enterprises (both large and small) must become and remain learning organizations under the new development paradigm.

5. SCENARIO OF SME DEVELOPMENT IN PAKISTAN

Amidst the rapidly growing economies in South Asia, Pakistan became the slowest-growing economy in this region during the last decade. The Vision-2030 says that by 2030 Pakistan will be the world's fifth most populous country (around 230 million people), with an economy whose absolute size will be among top two dozen countries, on the basis of only a 6 percent sustained growth, and around 12th on purchasing power parity (PPP). Its people in under-25 age group will have an average of 10 years of education, while tertiary enrolments will be nearly 20 percent of the 17-23 age cohorts. It will also be a major regional hub for industry, education, services and arts. The pronounced slowdown in growth in the manufacturing, retail and wholesale sectors, in particular, is largely explained by the fact that their growth rate of gross fixed capital formation (GFCF) has halved. The trend slowdown in the GFCF growth rate is equally apparent in small and large scale enterprises. Even more worrying is the lack of structural change, measured by the inertia in the structure of value-added and the size distribution of firms in the SME sector during the last two decades.

The Pakistan Economic Survey 2001-02 reported: "the foundation of industrialization could not be established without an efficient network of SMEs".

Credit rationing constraints SME growth and investment in Pakistan by increasing both the cost and risk of growth and investment. SME's demand for formal sector finance is inhibited by the high cost of credit and lease financing. The high cost of credit is, among other factors, a consequence of the high cost of funds that arises because of the high-risk free returns offered under national saving schemes in 2002. Regulatory costs are heavily biased against SMEs that have entered their expansion phase; this sets a strong incentive for firms to remain small.

Evidence shows that poor infrastructure, particularly in the power sector, increases the cost of growth for firms of all sizes. Poor power provision entails high costs, poor quality of service, lack of reliability, and corruption in obtaining supplies. This situation significantly distorts the growth potential of SMEs in Pakistan.

Another factor hindering the SME development in Pakistan is low level of skill, training, and education among workers and management, which raises the cost of firm growth. Furthermore, the human capital constraint is exacerbated by the lack of investment in training by firms, in particular, by SMEs. High market transaction costs and inefficient formal contract enforcement inhibit the development of SME clusters and subcontracting networks, imposing high inventory costs and, perhaps, forcing SMEs into suboptimal diversification.

Lowering the financial, judicial, fiscal and regulatory, and infrastructure and human resource constraints would have a significant positive impact on the growth of SMEs in Pakistan. These constraints can be reduced with the help of other developing countries in the South by experience, knowledge and technology sharing.

6. CONCLUSION

The countries of the South must intensify co-operative efforts to enhance its indigenous capacity to generate, manage and utilize science and technology in ways that address their own basic needs. For this to take place, regional and inter-regional efforts must be vigorously pursued. The ultimate goal of these efforts should be to develop collaborative programmes in capacity building for policy formation and building human capital for the development of SMEs.

In the final analysis, South-South cooperation on issues related to technology cannot be isolated from the wider trends of globalization. Such cooperation should be designed as a strategic approach to leverage technical knowledge from wherever it is located, but not as an exclusive political device that could further isolate developing countries. In other words, South-South cooperation should start with basic units such as regional integration and then be extended to other developing countries as well as the global community.

It should also be stressed that many regions of the developing world have the same economic attributes as much of the developed world. For example, concerns about the 'digital divide' are just as serious in parts of the United States or Europe as they are in Africa, Asia or Latin America. These regions constitute another opportunity to extend international partnerships beyond national borders and allow for the establishment of genuine global technological communities. In the final analysis, it is this ability to leverage global resources that gives strategic meaning to South-South cooperation. Everything else is purely rhetorical.

A well-developed SME base is capable of providing immense opportunities for SMEs to become catalyst in the economy. Cooperation among the countries of the South should take place to establish a more comprehensive approach towards SME development such as increasing their access to financing, providing greater access to business facilities locally and abroad, enabling the business infrastructure. The way forward for SMEs is therefore is to enhance human capital development, be adaptive in the use of information technology and improve their operations and productive capabilities to move up the value chain in order to remain competitive. SMEs must be strengthened to be the next engine of growth.

REFERENCES

Mohamed H. A. Hassan, Challenges, opportunities and strategies: south-south cooperation in science and technology in the 21st century

http://tcdc.undp.org/CoopSouth/2000_june/02_Hassan.cc.pdf

http://en.wikipedia.org/wiki/Small_and_medium_enterprises

http://www.g77.org/marrakech/main.htm

http://www.smeda.org.pk/

Thitapha Wattanapruttipaisan, Promoting SME Development: Some Issues and Suggestions for Policy consideration

http://www.aseansec.org/FINAL_SMEs—BULLETIN-website.doc

<http://www.hm-treasury.gov.uk/d/Chapter_24_Promoting_Effective_International_Technology_Co-operation.pdf>

Chapter 8

The Effect of Information Channels and Networks on Knowledge Generation and Technology Transfer in Small and Medium Enterprises

Keith Jacobs[1*], Jasson Gryzagoridis[2] and Bingwen Yan[1]

[1]Cape Peninsula University of Technology, South Africa

[2]University of Cape Town, South Africa

**E-mail: jacobsk@cput.ac.za, jacobs.keith@gmail.com*

ABSTRACT

There is a widely held view that South Africa's economic growth is being retarded by insufficient support forthcoming from technology and innovation. If South African entrepreneurs were successful, this could have a positive impact on the South African economy, as more money would be spent on local goods and services. Important problems in Small and Medium Enterprises (SMEs) are perceived to be in the field of a lack of access to information and knowledge, lack of access to a networking environment, lack of access to markets, lack of access to consultancy and lack of decent training. In organizations, there are many obstacles to knowledge generation and technology transfer. Some of these obstacles are caused by mistrust of sharing information inside and outside of the company, ambiguous communication and a lack of a structure of incentive systems or internal competition. This largely stems from personal and organizational beliefs that knowledge should not be shared or transferred but stored and used when necessary by a few people only. The study mainly focuses on the engineering sector, where there is a need for research into how knowledge is created and subsequently disseminated in SMEs. The study attempted to collect data formulated from answers to interviews and questionnaires that occurred with staff and management, as well as data from observations collected from existing SMEs.

The companies are based in the Western Cape, Eastern Cape and Kwa-Zulu Natal in South Africa, and this paper looks at how information channels and networks can be established so that information and knowledge can be shared within SMEs in a non-threatening environment.

Keywords: *Knowledge generation, Technology transfer, Information sharing, Information channels and networks, Collaboration, SMEs (Small and Medium Enterprises).*

1. INTRODUCTION

In organizations, there are many obstacles to knowledge generation. Some of these obstacles are caused by mistrust of sharing information inside and outside of the company, ambiguous communication if there is communication at all and a lack of a structure of incentive systems or internal competition. This largely stems from personal and organizational beliefs that knowledge should not be shared but stored and used when necessary by a few people only. Another hindrance is unreflected action and routines (March, 1991). People in organizations are often trapped by implicit theories of behaviour (Argyris and Schon, 1993) that guide their decisions and behaviour.

Since these theories are unconscious, they are not open to new influences or change, and almost impossible to share with others. Nevertheless, they influence individual and group action. This study looks at how this behaviour can be changed in Small and Medium Enterprises (SMEs) so that information and knowledge can be shared in a non-threatening environment.

Theorists are proclaiming that knowledge is one major source of competitive advantage (Grant, 1991; Teece *et al.*, 1997). Learning must occur continuously in the face of environmental change and must therefore become an integral part of the organization's day-to-day activities (Brown and Duguid, 1991). Even with this understanding, SMEs find it difficult to share information and empower their people to make decisions and become part of the knowledge generation process. In order to be successful in innovation, collaboration is essential, although entrepreneurs are sceptic about collaborations or 'partnerships'. Some of them do have strong ties with academic institutions to get new information on technological development and some of them are involved in sector networks.

In view of the above, there is a need for research into how knowledge is created and disseminated in organizations, and how this knowledge is shared and can be converted into value adding activities within a firm.

2. BACKGROUND OF THE STUDY

In the last three decades several East Asian countries have achieved spectacular economic growth. Japan has emerged as the world's industrial superpower. Small to medium-sized companies called "Chu-Sho-Kigyo" make up more than 99 per cent of the Japanese industries and form the real foundation of the Japanese economy. Small to medium-sized enterprises employ over 80 per cent of the national workforce (Vosloo, 1994:96).

Small, Micro and Medium Enterprises (SMMEs) are perceived to play a fundamental role in job-creation, economic growth and equity in South Africa. In South Africa small and micro enterprises fall into one category called small enterprises. Therefore, in many cases, SMMEs are equivalent to SMEs. The country's first democratically elected minister of Trade and Industry **Mr. Trevor A. Manuel** said the following on the presentation of the White Paper on the ***National Strategy for the Development and Promotion of Small Business in South Africa***: *"Throughout the world one finds that SMMEs are playing a critical role in absorbing labour, penetrating new markets and generally expanding economies in creative and innovative ways"*.

In accordance with section 1 of the South African Small Business Act, small business is defined as a separate and distinct business entity, including co-operative enterprises and non-governmental organizations (NGO's), managed by one owner or more owners.

In general a medium enterprise may have a maximum of 200 employees, a maximum turnover US$ 5.4 million per annum and a total gross asset value of US$ 2.36 million. A small enterprise may have a maximum of 50 employees, turnover of maximum US$ 3.3 million and asset value of US$ 0.6 million.

The SME sector contributes significantly to South Africa's GDP (Gross Domestic Product) and employment and this explains the effort that is put into stimulating the development of SMEs in the different sectors. About 10 per cent of all SMEs in South Africa are based in the engineering sector.

During the period of apartheid until the early 90s, black entrepreneurs were discouraged to start up their own businesses. This explains the underdevelopment of SMEs in comparison with SMEs abroad. There is a widely held view that South Africa's economic growth is being retarded by insufficient support forthcoming from technology and innovation. If South African entrepreneurs were successful, this could have a positive impact on the South African economy, as more money would be spent on local goods and services. The main problems in SMEs that are perceived to be in the field of a lack of access to information and knowledge, lack of access to a networking environment, lack of access to markets, lack of access to consultancy and lack of decent training. It is also worth mentioning, in some instances, the lack of government protection of propriety rights.

This study was focused mainly on the engineering sector, where there is a need for research into how knowledge is created and subsequently disseminated in SMEs. Knowledge, of course is important for competitive initiatives such as improving customer satisfaction, developing new products and markets and providing faster response to customer needs. In addition, more research is needed to explore how action research in collaboration between industry and academia can contribute to knowledge production in both the organizational setting and in the academic institution.

In organizations, there are many obstacles to knowledge generation and technology transfer. Some of these obstacles are caused by mistrust of sharing information inside and outside of the company, ambiguous communication and a lack of a structure of incentive systems or internal competition.

This largely stems from personal and organizational beliefs that knowledge should not be shared or transferred but stored and used when necessary by a few people only.

3. PROBLEM STATEMENT

The engineering sector within the SMEs is playing a critical role globally in absorbing labour, penetrating new markets and generally expanding economies in creative and economic ways. In South Africa, nearly half of the GDP and employment comes from SMMEs that are flourishing all over the country.

Government studies indicate that up to 80 per cent of new businesses in South Africa fail within one year.

One of the reasons is that SMEs in South Africa have problems remaining regionally, nationally and globally competitive. The problems are not only of financial nature such as a lack of start-up financing or upgrading finances. Other problems include a lack of innovation, knowledge and skills to turn a good business idea into profitable one in a quality-oriented environment. A further problem is the ignorance of the importance of the development of new products in SMEs. The government estimates that the economy must achieve growth at a minimum of 6 per cent to offset unemployment, which is estimated to be about 30 per cent although unofficial sources put it as high as 38 per cent. In an effort to boost economic growth and spur job creation, the government is working to encourage small and medium enterprise development.

It is because of the big failure rate that the government has decided to provide support to new SME's in the following areas and in particular key to this paper:

- Access to training,
- Access to appropriate technology and encouragement of inter-firm linkages
- Access to educational institutions for assistance to improve the competitiveness of SMEs in several sectors.

Despite these efforts many SMEs still lack the necessary access to information channels, the ways of obtaining and absorbing information from internal and external information channels.

4. RESEARCH QUESTIONS

The research questions to be answered by this study were as under:

- What are the effects of information channels and networks on knowledge generation and technology transfer in SMEs?
- How do SMEs generate collaborative knowledge and transfer technology through information channels and networks?

5. SIGNIFICANCE OF THE RESEARCH

The main aim of this study was to determine how are SMEs generating collaborative knowledge and transferring technology through information channels and networks, and how can they sustain their existence and what they need to do to become innovative.

6. ETHICAL ISSUE

All the information collected from the companies is only used for research purposes. The identity of the companies and personal information of all the participants are confidential.

7. RESEARCH METHOD

7.1 The Case Study as a Research Method

The case study research method is defined as an empirical inquiry that investigates a contemporary phenomenon within its real-life context; when the boundaries between phenomenon and context are not clearly evident and in which multiple sources of evidence are used (Yin, 1994).

The following technique is suggested for organizing and conducting the research successfully.

- ✰ Determine and define the research questions
- ✰ Select the cases and determine data gathering and analysis techniques
- ✰ Collect data in the field
- ✰ Evaluate and analyze the data.

7.2 Sampling and Data Collection

Data collection of this study consisted of interviews and questionnaires that occurred with staff and management, as well as data from observations collected from 18 companies based in the Western Cape and Kwa-Zulu Natal in South Africa. The data were collected from the 18 companies relating to the following four categories:

1. Knowledge Generation (KG)
2. Collaboration and Technology Transfer (C&TT)
3. Information Sharing (IS)
4. Information Channels and Networks (IC&N)

The following questions (Table 8.1) were posed during the interviews:

7.3 Data Analysis

A qualitative content analysis was utilized to analyze the data. Krippendorff (1980:18) defines a content analysis as a research technique for making replicable and valid inferences from texts (or other meaningful matter) to the contexts of their use; as a research technique, content analysis provides new insights, increases a researcher's understanding of particular phenomena, or informs practical actions.

The authors calculated the overlapped frequencies of the four categories from the responses of the 18 companies to determine how these SMEs generate collaborative knowledge and transfer technology through information channels and networks, and what they need to do to become sustainable and innovative.

Table 8.1: Codes of Interview Questions

Categories	*Questions*	*Code*
Knowledge Generation	How does your company generate collaborative knowledge?	KG
Collaboration & Technology Transfer	Are you collaborating and initiating technology transfer with your customers?	C&TT[1]
	How does collaboration and technology transfer take place in your company?	C&TT[2]
Information Sharing	How does information sharing affect new products in your company?	IS
Information Channels and Networks	Where do you get ideas for new product or production process innovations?	IC&N

8. RESULTS AND DISCUSSION

The results of the analysis of the eighteen companies' responses, to the questions aimed at establishing the status of the company relative to the research questions, are discussed here. The responses were categorized as either positive or negative as indicated in Table 8.2. The spreadsheet indicates the companies' negative responses to the questions in the various categories. The status of the companies is indicated as a percentage overlap in a column on the right hand side of the spreadsheet. Overlap

Table 8.2: Responses from the 18 Companies

Questions / *Companies*	*Categories*					*Total*	
	KG	*C&TT[1]*	*C&TT[2]*	*IS*	*IC&N*	*f*	*Overlap (%)*
Company 1						5	100
Company 2						5	100
Company 3						5	100
Company 4						5	100
Company 5						3	60
Company 6						5	100
Company 7						4	80
Company 8						4	80
Company 9						5	100
Company 10						2	40
Company 11						5	100
Company 12						5	100
Company 13						5	100
Company 14						4	80
Company 15						4	80
Company 16						4	80
Company 17						2	40
Company 18						4	80

means conformance to the positives accrued from the answers to the questions put to the companies under the four categories.

Nine of the companies had 100 per cent overlap with the questionnaire, six companies attained 80 per cent overlap, one company with 60 per cent overlap and two with 40 per cent overlap (Frequency-f):

Companies 5, 7, 8, 10, 16 and 17 need to involve more people in the knowledge generation process.

This could be done by creating incentives for the workers who have ideas that reduces expenditure or decreases the frequency of the maintenance required on a product or improves the aesthetics of the product. Any one of these are worthy of an incentive. The companies should also have dedicated innovation meetings and include all the workers at the meetings. It could be people at the company and outside of the company. When doing this, they will enhance their products and the morale of their workers.

Companies 5, 10, 17 and 18 need to collaborate with their customers by involving them in the design process or by getting them to sample the products. Dedicated people need to be tasked to do this in the companies. Customers are very good at assisting companies with the quality, price and aesthetics of a product.

Companies 10 and 15 do not have a collaborative environment and therefore technology transfer happens when necessary in the company. It is for this reason that there is very little collaboration of technology transfer in the companies. Workers should be given the opportunity to share ideas and discus products. This problem can be solved by having short meetings on a weekly basis to discus the progress on products and extract ideas from the workers. Collaboration between the different sections in the company will also improve. Workers are not given the full picture and all the relating detail of the products.

Company 14 and 17 do not have an effective manner in which to share information inside the company. Most of the information are kept by the owners and usually shared when there is a problem. The company has not yet learnt that they should trust their workers with information. The owner should learn how to trust his workers and so doing will create an environment where people work together free of intimidation. This can be done by giving the workers small tasks at a time until he is happy to trust them fully. In the interim, if the company is not comfortable with sharing all the information at once they should share some of the information to all at short weekly meetings.

9. CONCLUSION

It is clear from the information gained from the interviews that the companies are aware of the effect that information channels and networks have on collaborative knowledge generation and technology transfer. This was demonstrated when fifteen of the eighteen companies responded positively to 80 per cent or more of the questions. The other three companies (5, 10 and 17) are running the risk of not remaining viable. They will definitely have to implement the recommendations if they wish to produce competitive products and remain viable.

REFERENCES

Argyris, C. and Schon., D., 1993. Knowledge for Action, San Francisco: Jossey Bass.

Brown, J. S. and Duguid, P., 1991. "Organizational learning and communities-of-practice: toward a unified view of working, learning, and innovation," Organization Science 2: no. 1 pp. 40-57.

Grant, R. M., 1991. The Resource-Based Theory of Competitive Advantage Implications for Strategy Formulation, California Management Review 33 (3): pp.114-135.

March, J. G., 1991. Exploration and Exploitation in Organizational Learning, Organization Science 2: pp.71-87.

Teece, D. J., Pisano, G. and Shuen, A., 1997. Dynamic Capabilities and Strategic Management, Strategic Management Journal 18: pp. 509-533.

Yin, R. K., 1984. Case Study Research: Design and Methods (1st ed.), Beverly Hills, CA: Sage Publications.

Chapter 9

Factors Affecting Utilization of Information Technology in Organizations

R.W. Gakure and M. Wachira
Director, CBD CAMPUS
Jomo Kenyatta University of Agriculture and Technology (JKUAT), Nairobi, Kenya
E-mail: rwgakure@yahoo.com; rwgakure@gmail.com

ABSTRACT

Information technology (IT) and its utilization in the work place are crucial for enhancing efficiency, quality, quantity, productivity and overall profitability of organizations. Although entrepreneurs know the importance of IT and its direct relationship with profitability of business, very few organizations ensure optimal utilization of IT by their employees. Despite its crucial importance, very few employees utilize IT in Kenya. The purpose of this study was to investigate the factors, which influence employees to utilize IT in the workplace.

This was an exploratory research aimed at investigating the factors that influence utilization of information technology (IT) in a Kenyan organization with a focus on Magadi Soda Company.

The study interviewed eighty-four respondents selected using stratified random sampling. The selection was based on the various functional departments of the organization.

The instrument used for data collection was an in depth questionnaire. Although most of the employees used computers fairly frequently, the level of usage was found to be related to the following user characteristics: level of education, type of IT training received, years of experience and ownership of a personal computer. Despite the usage of IT systems, the majority of the users were not conversant with the system in place. Results indicated that employees (64) who had received higher training were most likely to use IT frequently. However, internal training, rather than previous computer use, was found to be positively co-related with level of usage. Interestingly, allocation of a personal computer was found not to be positively influence the utilization of IT. The implication is that the employees lacked the technical know-how which would enable them utilize IT optimally in order to improve their productivity and help the organization to achieve its corporate goals.

The major recommendations were: employees should be trained on IT frequently or on a continuous basis as it is very dynamic; organizations should revise the IT training approaches; more funds should be allocated to IT departments; and accessibility of the powerful IT equipment and the latest IT. This is crucial if the organizations are to enhance profitability through optimal utilization of IT adoption. It was further recommended that IT should be institutionalized in education curriculum and transformed to an examinable subject.

Keywords: *Information technology, Organizations, Magadi Soda Company, Optimal utilization, Small and Medium enterprises, Technology transfer.*

1. INTRODUCTION

Efficiency and cost of operation is key to profitability in any business. However, for a business to operate efficiently and effectively the techniques and procedures of operation must be sound.

Information technology is key in this regard. Technology combined with pragmatic business policies has proved to be very effective vehicles through which several countries such the Asian Tigers (Thailand, Singapore among others), (World Competitiveness Report, 1997) have enhanced development. Technology capability and its transfer involves both intangible assets such as design, engineering plans, models and tangible assets such as capital goods, machinery on the one hand and human resource assets which include skills and technical know-how (Viramani and Rao, 1999). IT requires capital investment more often than not in very high proportions. For business to be efficient and effective they need to employ IT. IT enhances representation of data in an efficient, usable form in addition to increasing performance of organizations, which take advantage of it.

The concept of IT is relatively recent having been in commercial use for the last 50 years but comprehensively only for the last 20 years. IT combines and enhances a combination of human capital–skills, knowledge, ability, machines and tools. Initially, IT products were very expensive but the breakthrough in the sector has rendered the products relatively cheap; yet not all firms have taken advantage at the same pace.

Businesses seem not to have realized the importance of IT and the role it can play in enhancing productivity of the organizations. Since most consumers buy physical goods and not "information", organizations tend to look at physical facilities and material processing needs first and then only at IT requirements. It is therefore not surprising that there are businesses, which have not computerized their operations, yet IT has a lot to offer in enhancing their productivity. Some of the questions which need to be answered include the reason why organizations are reluctant to take full advantage of IT, what variables can impede utilization, and whether such variables are in place. Some of these variables include (*a*) infrastructure–telephone, internet, power, equipments, computers, and (*b*) training–many organizations have a propensity to disregard training and to hire part-time employees and spend very little resources, if any, for updating the skills of their employees.

There is a feeling that when organizations want to cut costs, the first cost to go is the training cost. Yet it is known that IT is very dynamic and requires constant training in order to keep abreast with the current skills requirement. What's the implication?

Since organizations are hard pressed for cash, projects are always competing. The methods used to explore an organization's requirements tend to look at material processing needs first hence IT requirements are considered last, yet IT projects are expensive both on equipment and skills and therefore tend to miss out.

1.1 Problem Statement

IT is now a powerful driving force for business with technologies such as e-commerce having a dramatic impact on the way national and international business is being conducted. The IT revolution is fundamental in changing the way we do business, yet majority of the Kenyan organizations are not embracing it at the same place. In line with Andrew Blau (May, 2001), there are four important factors that affect adaptation of IT by organizations: technology structure, training and upkeep, funds and work environment. He further argues that only the organizations best prepared for all the ramifications would succeed best.

There is very scanty literature concerning factors, which impede utilization of IT although the benefits of utilizing IT are well known. This study aimed at investigating the factors that hinder utilization of IT in organizations. Specifically, the study examined the infrastructure, training, environment and funds.

1.2 General Objective

The overall objective was to investigate factors that hinder utilization of IT in organizations.

1.3 Specific Objectives

The specific objectives of the study were to:

1. Find to what extent infrastructure affects utilization of IT.
2. Determine the effects of staff training on utilization of IT.
3. Find out if working environment affects the utilization of IT.
4. Determine if financial resources affect the utilization of IT.

1.4 Research Questions

The study was guided by the following research questions

1. What is the role of infrastructure in IT utilization?
2. Does staff training contribute to utilization of IT?

3. Is work environment a factor affecting utilization of IT?
4. Does availability of financial resources affect the utilization of IT?

1.5 Justification

IT is becoming more and more crucial for survival of organizations in the global environment. It offers great potential for the organizations that embrace it. Since organizations are driven by the human resource, the utilization of this technology by employees is crucial. Gaining knowledge about the factors that influence user utilization of IT is important so as to determine the path organizations need to follow so as to maximize the IT investment. Understanding how these factors influence usage of the technology by employees will enable stakeholders formulates appropriate strategies to counter existing shortcomings. Magadi Soda Company is the leading Soda ash producer in Africa and the biggest, single foreign exchange earner Kenyan private company. Having traded in the international market for about 100 years and yet having not adopted IT fully prompted the researchers to focus on it. This will no doubt be helpful to the Kenyan organizations and especially the small and medium enterprises (SMEs), which are struggling to survive. The findings would have implications on IT adoption and utilization in other organizations.

2. LITERATURE REVIEW

This study examined factors affecting utilization of information technology in organizations. This section highlights literature on factors hindering IT utilization with specific focus on a manufacturing company in Kenya-Magadi Soda Ltd. To understand the factors that hinder adoption of technology among organizations (large organizations and small and medium enterprises – SMEs), relevant empirical studies and secondary literature were reviewed. The section comprises of a conceptual framework that forms the basis for comparison of data analysis and model, relevant theories forming the conceptual framework and past studies.

2.1 Conceptual Framework

For the purpose of this study, both empirical and secondary data were reviewed. Technology has been defined by Commission del Acuredo de Cartegene, Decision, (1984) as: 'The combination of types of knowledge indispensable to carrying out the necessary operations for transforming the factors of production into products, the use of that knowledge or provision of services'.

Stewart (1992) posits that the term 'technology' refers to new and better ways of achieving economic ends that contribute to economic development and growth. Thus technology development should not be seen just as the development of the ability to undertake in order to absorb and build upon the knowledge that has to be utilized in production. Similarly, technology transfer has been given many meanings but essentially involves four main stages of negotiating, setting up and sustaining growth. According to Virmani and Rao (1999), technological capability and its transfer involves intangible assets such as design and engineering plans, models (for process and product) and tangible assets such as capital goods, machinery and pilot plants on the one hand and human assets which include skills and know how on the other.

Technology can be used to refer to new and better ways of enhancing growth of organizations (Stewart, 1992).

IT is now a powerful driving force and every management's task is to drive that technology to the organization's advantage and not to be driven by it. Extant literature indicates a significant amount of work on Information Systems (IS) that has looked at the adoption of technological innovations; in general: Kwon and Zmund (1987); Rodgers (1983); and Tornatzky and Klein (1982). Most of the previous studies in Kenyan organizations have focused on information technology interventions and installations (Nyambati: 2001, and Nyambura: 2000). These have emphasized ways to improve productivity and profitability for organizations by lowering costs. According to Infonetics Research (1993), today's organizations are continually re-engineering their processes to increase their competitive advantage and must have an infrastructure that is strategic and adaptive to change.

Conceptual Framework Model

The conceptual framework depicts the relationship of variables – dependent (phenomenon being investigated) and independent variables, which are hypothesized to influence utilization. The study followed the following conceptual model. The model indicates some of the independent variables, which affect utilization of information technology (dependent variable) in business.

Training

The objective of training are to develop competencies of employees and improve their performances, help them grow within an organization in order that as far as possible its future needs for human resource can be met from within while reducing the learning time for employees (Armstrong, 1999).

Training strengthens technical skills and capacity within the organization (Fisseha and McPherson, 1991). It seeks to address the uncertain and ever changing business environmental needs occasioned by the dynamic social, economic, political and technological factors. The quest for improved profitability of organizations makes it therefore imperative for organizations to keep on training their manpower to enhance their capabilities and competencies.

Technology is very dynamic as it is constantly generated, absorbed, applied, adapted, exchanged, bought and sold and upgraded. In this process, institutionalization of technological generation and innovation, its market orientation,

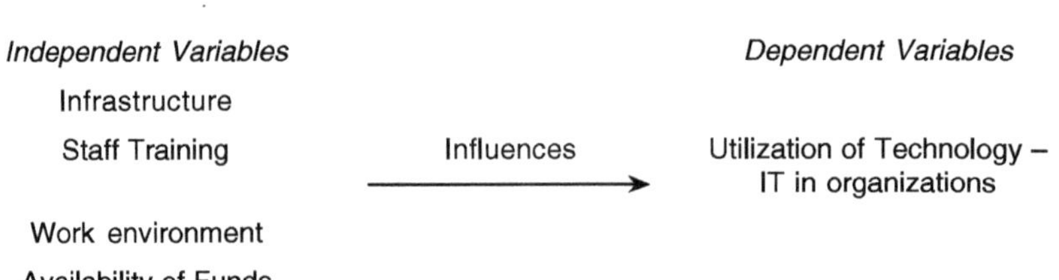

Figure 9.1: Factors Influencing Utilization of Technology

and enhancement of intellectual property rights takes place and subtly alters the methods of diffusion of technology as also its acquisition for commercial production.

Technology combined with pragmatic business policies have proved to be very effective vehicles through which several countries such as the Asian Tigers (Thailand, Singapore, and Malaysia) promote their economic growth both at small-scale enterprises (SSE) and large-scale industry level (World Competitiveness Report, 1997).

IT utilization of the respondents in this study was investigated through examination of their IT skills, awareness and rating how they use IT in their daily activities. Their enthusiasm was measured through their willingness to attend IT related training. The study showed there were several factors that influence IT utilization such as adopting of IT by the organization (Blau, 2001), and the need to understand social networks (Warshauer, 2003). Adoption of technologies involves changing the way things are done and produced in organizations. However, changes are resisted and 'Nobody likes changes except wet babies' as Micheal Schartz (2001) puts it and IT, viewed as a new way of working, therefore needs time for acceptance.

Factors Affecting Utilization of Technology

Information technology has its own constraints such as software crashes, hardware failure and viruses, which are likely to influence users negatively (Onduso, 2001).

Demographic variables such as education level, area of specialization, prior computer experience, availability of IT equipment, anticipated future use have influence on how employees use IT.

Most of the previous studies in Kenyan organizations have focused on information technology interventions and installations (Nyambati, 2001). The studies emphasize ways to improve productivity and profitability for organizations by lowering costs.

3. RESEARCH DESIGN

3.1 Introduction

The purpose of the studies determines the structuring of the entire activity (Singleton *et al.*, 1988 and Babie, 1983). Researchers in social sciences distinguish among three broad types of research designs as exploratory, descriptive and explanatory. The study used descriptive research design and involved a case study where the researchers sampled the population of interest to address issues related to the problem under study. The advantage of this design was that it allowed the researchers to be flexible in data collection with no undue influence on respondents. Also, since the deployment of the questionnaire was e-mail based, interview bias, guiding, and cues that can impact on the validity and reliability of data collection were avoided

3.2 Population

The population of this study was the employees of Magadi Soda Company (MSC). There were 504 employees out of whom 334 were permanent, working in nine functional areas with main offices located in Mombasa (MSC Organizational Records, 2004). The permanent employees formed the sampling frame.

3.3 Sample

A stratified random sampling technique was used to select a sample of eighty-four (84) respondents. This was appropriate because of the heterogeneity of the respondents. The stratification was based on the functional departments of Magadi Soda Company as illustrated Table 9.1.

Table 9.1: Sample Design

Characteristic		*Respondents n=26*	
		Frequency	*Percentage*
Age	30-39	23	50
	40-49	8	31
	Over50	5	19
Sex	Male	16	62
	Female	10	38
Hierarchy	Support	8	31
	Supervisory	12	46
	Management	6	23
Length of Service	1-2yrs	6	23
	3-5yrs	4	15
	Above 5yrs	16	62
Distribution	Administration	2	8
	Sales and Marketing	3	16
	Finance	3	12
	Manufacturing	8	33
	Human Resources	2	8
	Engineering	2	8
	Supply Chain (MSA)	1	4
	Supply Chain (HQ)	4	15
	Hospital	1	4

3.4 Instruments

The instruments used to gather data was a questionnaire. The questionnaire focused on factors affecting the utilization of IT in Magadi Soda Company (MSC). The questionnaire comprised of open-ended and closed questions aimed at gathering relevant information impending the IT utilization.

3.5 Pilot Test

Five employees were used to pre-test the instrument for the purpose of validating and testing the reliability of the instruments. The feedback was used to modify and enhance the clarity of the instrument.

3.6 Data Analysis

Data was coded and stored in a database ready for analysis. Data was thereafter analyzed using IT programs such as spreadsheets, word processors and statistical packages. The main application used for this analysis was Statistical Package for Social Sciences (SPSS).

4. FINDINGS AND DISCUSSION

4.1 Introduction

This section presents the findings and discussion of the study. The study presents the factors impeding utilization of information technology and discussion of the factors.

4.2 Characteristics of the Respondents

Most of Magadi Soda Company (MSC) IT users were in the 30-39 age groups, which was approximately 50 per cent of the respondents. A greater percentage (62 per cent) of the users was male which was consistent with employment distribution. Most of the respondents were in manufacturing; probably indicating IT usage is high in this area. This is not surprising because studies conducted on integration of science and technology in micro finance institutions revealed that women tend to shun IT (Gakure, 2003). Details of the characteristics are indicated in Table 9.2.

Table 9.2: Demographic Data of Magadi Soda Company Employees

Characteristics		*Respondents n=26*	
		Frequency	*Percentage*
Education	No formal education	2	4
	Primary level	1	4
	Secondary level	3	12
	Tertiary level	20	80
Type of training	Professional	6	23
	Management	3	12
	Technical	3	12
IT related training	Vendors	10	24
	In house	23	55
	Self sponsored	18	19
	Others	1	2
IT training adequacy	Yes	19	76
	No	6	24
Ownership of PC	Yes	12	48
	No	13	52
Use of PC for routine work	Yes	22	85
	No	4	15

On hierarchy, it was found that supervisory staff had the highest representation, 46 per cent. This could be explained by the technical/competency requirements for this category and the accessibility of personal computers. The duration of service varied from 1-5 years and above with the majority of the respondents, 62 per cent, above 5 years.

4.3 Education

A majority of the respondents (80 per cent) had reached tertiary level of education, followed by secondary level at 12 per cent. Greater percentages (62 per cent) of the users were males, which was consistent with employment distribution. A higher number of the respondents, 42 per cent owned a personal computer (PC).

These findings may indicate the reason for high exposure to IT despite the high number of employees sharing computers (24 per cent) and the frequent usage of IT. It would therefore appear that availability or accessibility of personal computers influences utilization of IT.

On formal education, the higher the level of education the higher is the utilization rate. For, example the highest users were found to be those who attained college education at 71.4 per cent followed by those who attained college education at 19 per cent. Education therefore appeared to be one of the key factors influencing utilization of technology. It can be argued that the respondents were probably exposed to and they probably had access to PCs (Table 9.2).

4.4 Training on IT

The study measured both formal and computer literacy training of the respondents. The study revealed a great significance for both on the utilization of IT.

4.5 In-house Training

One of the interesting findings was the effectiveness of in –house training at 100 per cent on the impact of utilization as opposed to vendor/consultant training at 83 per cent and self-study at 88.9 per cent (Table 9.2). The majority, 55 per cent attained training through in house programmes. This corroborates extant literature, which indicates that formal training may influence technological capability (Massaquoi, 1992). These findings can be interpreted to indicate that training is very crucial and can enhance utilization of IT.

Organizations need to realize the importance of training and its effect on utilization of IT. It is a common knowledge that the skills requisite for use of IT are very dynamic and volatile and change very frequently. Training and development therefore becomes crucial for effective utilization of IT.

4.6 Knowledge of IT Systems

These findings revealed that a majority of respondents, 55 per cent, are not conversant with IT systems in use in the organization. This is not surprising because majority of IT users are not trained on IT systems and they may probably be interested in learning the basic IT skills, for instance how to use MS Word and other less demanding IT packages. However, employees who specialize in IT would be more inclined to delve deeper in IT systems.

4.7 Infrastructure

A majority of the respondents reported availability of the platform for information technology processing needs. This is not consistent with the findings in the literature which indicate inadequate resources mainly in the areas of: company local area network–only 60 per cent of the functional work areas are fully covered telecommunications–the currents link is not reliable and has frequent outages impacting on the utilization of the technology; personal computers (PC)–only about 50 per cent of the total employees had a PC allocated for individual use. This is inconsistence and could be explained by the research findings on knowledge of organizational systems where only 42.3 per cent of the respondents reported awareness.

4.8 Environment

There was no correlation between utilization and the environment variable.

4.9 Funds

The study established the significance of availability of funds for the utilization of IT. Evidence revealed that only 21- 40 per cent of IT projects are implemented. The study established that the challenging operating environment for MSC compel every employee to utilize information analysis tools for is impacted upon by a number of factors including those investigated. This study determined that computers are used more for coordination tasks, making a networked environment core for utilization of IT. Users communicate more frequently through e-mails, exchange information through computers and share resources such as printers very often.

4.10 The Role of Technology on Small and Medium Enterprises

The development of competitive and resilient small and medium enterprises (SMEs) forms an integral component of Kenya's initiatives to be globally competitive and prosperous nation with a high quality of life by 2030. The challenges posed by increased liberalization, new entrants to the market, increased standards requirements and technological developments require SMEs to raise efficiency levels, strengthen inter-firm linkages and respond timely to market changes. At the same time, greater integration into the global economy provides opportunities for SMEs to participate in the international value chain and supply chains networks. This will enable SMEs to move up the value chain and adopt new technologies, particularly information and communication technology (ICT). Only SMEs that are capable of harnessing and adopting technology and knowledge to develop high value-added products of superior quality will be able to compete globally (GoK, 2007). This can only be feasible through technological transfer and adoption. International businesses penetrating all economies of the world are highly sophisticated in their business management which is made possible by their utilization of technology. For SMEs in the developing economies to develop and become competitive in the international world arena, it is crucial that they also adopt and use technology to enhance their efficiency, productivity and competitiveness.

Technology is a broad concept that deals with a species' usage and knowledge of tools and crafts, and how it affects a species' ability to control and adapt to its environment. Chen (1995) offers a useful framework for characterizing technology. In this framework, technology is characterized by three elements- products, processes and practices. 'Product' element of technology includes the knowledge and processes that are involved in the design and development of new models and variants/upgrades of the existing products. Similarly, 'process' element refers to the manufacturing processes- choice of processes, layout and organization of the shop floor, equipment machinery, skill level of work force and so on, required to manufacture the product.

Finally, 'practices' include managerial systems employed to manage the process. Therefore just-in-time (JIT), planning and control systems, total quality management (TQM), quality management systems constitute the 'practices' component of the technology. By interfacing between products and processes, practices innovations have the potential to enhance firm-level (SMEs) capabilities either in products or processes or both. For example, adoption of lean manufacturing practices in production enables the firm to improve its processes. Extending application of lean philosophy to design and entire organization will result in enhanced product technologies as well (Greenway, 1994). Design for manufacturability involves departure from traditional design of the product. Besides leading to development of superior products and enhancing ability to frequently introduce new products and enhancing ability to frequently introduce new products, this technology results in more manufacturability product and increases production efficiencies as well (Tirupti, 2008). This is in line with Schumpeterian model of entrepreneurs being the agents of creative destruction. This model posits that entrepreneurs need to be innovative by coming up with new methods, new processes, new markets, new products to add value to their products. To accomplish this in a highly liberalized and dynamic world, SMEs will need to adopt technology to facilitate innovation capable of adding value to their products or services.

SMEs in Kenya produce goods which require refinement for them to meet the quality standards required by consumers. Large firms which have adopted technology in Kenya buy the products from SMEs and use technology to improve the quality and end up selling the products at very high prices. SMEs reap only marginal benefits. This plight can be reversed through technology transfer whereby SMEs also use technology to improve their quality management systems, their designs, processes, and products to meet the needs of 21st century customers needs and be able to compete with others.

The products of Kenyan SMEs can be enhanced through adoption of technology. This would improve their quality, efficacy and output. Technology transfer can improve quality management systems in the SMEs thereby equipping them with crucial tools of managing time, enhancing outputs and enabling them to keep track record of performance through utilization of transferred technology.

In order to penetrate international markets, achieve and maintain competitiveness SMEs must embrace modern technologies that enable them develop efficient production (Greenway, 1994). Maintaining consistent quality of products and reducing human content are major factors affecting a firm's decision to upgrade manufacturing technology. Advanced manufacturing technology can improve quality throughout the entire manufacturing process in areas such as materials handling, inventory control, time management and production planning and scheduling. Advanced systems lead to quality improvements in the design stage because errors are discovered earlier in the process and more quickly. This allows adjustments to be made much faster and more accurately than without advanced manufacturing technology, helping to ensure quality in the manufacturing process (Ariss, Raghunathan and Kunnathar, 2000). Adoption of technology by small manufacturers gives them advantages over traditional manufacturing systems, such as lower cost, quality improvements, higher productivity, and less working capital tied up in inventory (Phillips and Ledgerwood, 1994). In his study of 20 companies equipped with advanced manufacturing technology, Zairi (1993) observed that quality was a major competitive objective of advanced manufacturing technology users. He concluded that the introduction of technological innovation is considered to be a facilitator in moving toward competitive excellence.

Technology is mainly concerned with production automation, flexible manufacturing and advanced processing equipment. Technology contributes to the competitive advantages of product quality, flexibility and low cost (Chen, 1999). Sousa and Voss (2001) found that managerial experience and prior experience in a similar or the same type of business as well as the adoption of new manufacturing technologies, the availability of resources to adopt new technology and the development of a competitive advantage were strongly related to growth. Moreno-Luzon (1993) examined the factors that were responsible for success with quality initiatives like TQM in small manufacturing firms in the Valencia region in Spain. They found that the firms experiencing greatest success with TQM placed more emphasis on innovation, in products, markets, processes and production equipment.

Small firms have been considered to be an important element of the industrial structure in Japan since the turn of the century. Under the regime of rapid industrial restructuring in Japan, small firms are identified as developing their own diverse management strategies, especially with respect to international trade and the implementation of information and high technology (Sato, 1998). SMEs can respond flexibly under difficult and changing conditions because they do not depend heavily on infrastructure and because their typically low level of technology allow product lines and inputs to be charged at relatively low cost (Steel and Webster,1992).

The globalization of the marketplace and the rapid improvement in high quality products and services has brought about high levels of market pressure across the world. It is clear that quality has emerged as a strategic competitive tool for organizational success (Yong and Wilknson, 2002). In today's business environment, organizations cannot afford to ignore the strategic implications of quality for its competitive position. In the light of this, it is vital for firms to develop or adopt an effective Quality Management System (QMS) very often associated with quality initiatives such as ISO 9000 series (Rohitrana and Boon-Itt, 2001). Quality Management Systems (QMS) are needed in all areas of activity, whether large or small businesses, manufacturing service or public sector. No matter where it is implemented QMS will improve process control, reduce wastage, lower costs, increase market share, and facilitate training. Increasing product quality results in higher profits because costs are decreased and productivity and market share are improved (Ryan, Deane, Ellington, 2001; and Gupta, 2004). Technology has a major role to play in improving the products and management systems of SMEs.

Small firms need to adapt technology if they are to survive the global competition posed by large organizations. Central to this is the application of technology as a way for adopting and surviving (UNIDO, 2004), and by extension, it is not possible for small firms to grow and become competitive without technological change and accumulation of knowledge (Buainaina, 2002).

Without access to technology, small firms lack capacity to produce efficiently, meet deadlines, upgrade products quality and evolve new products design. It is only small firms with the capacity to initiate improvements in products, processes and production that take advantage of the emerging opportunities (Moyi and Njiraini, 2005). However, as argued by UNIDO (2004), acquiring new technology and applying it to improve quality and thus get the advantage of competition and sustained productivity would require basic capacity to assimilate the technology, to manage it and to control results with it. The purpose of technology is to improve productivity of enterprises and enhance the quality of goods produced by enterprises to help them with-stand local and international competition (ILO/UNDP, 2000).

4.11 Role of Information Technology (IT) on SMEs

All businesses need information technology to survive the competitive environment in the 21st Century. The role played by IT in international business is particularly crucial as it enables entrepreneurs owning large and SMEs to transact businesses more efficiently through the use of IT to network, order materials, identify clients, monitor and evaluate the progress of business through creation of reliable record keeping systems. IT can revolutionize and transform SMEs by equipping SME entrepreneurs with important tools for efficient quality management systems. As Stewart (1992) posits that the term 'technology' refers to new and better ways of achieving economic ends that contribute to economic development and growth, IT would enable the entrepreneurs to come up with new ways of tracking, a) how resources of the business are utilized, b) financial requirement and how they are utilized c) time, and to measure outputs against identified objective or business targets d) investments and return on investments among other benefits, and e) accountability and transparency in the way the SMEs are managed. This would fit very well with Schumpetarian view of entrepreneurs being innovators in all aspects of an SME which add value to them and enhance their profitability. IT can therefore be used to refer to new and better ways of enhancing growth of organizations including SMEs (Stewart, 1992).

As Infonetics Research (1993) noted, today's organizations including SMEs are continually re-engineering their processes to increase their competitive advantage.

Re-engineering involves among other thing "creative destruction" of old ways of doing things (processes) and adaptation of new innovative ways of doing the same things more efficiently. IT is a powerful tool which can contribute to enhanced performance of SME as it improves processes and helps businesses to improve quality management systems.

5. SUMMARY, CONCLUSION AND RECOMMEDATION

This section summarizes the study and comprises three sections: summary, conclusion and recommendations.

5.1 Summary

The following factors were found to influence utilization of information technology (IT): infrastructure, training, financial resource, accessibility of a PC and funds. There was negative correlation between work environment and utilization of IT as predicted. The infrastructure, staff training and availability of funds were found to be statistically significant with the utilization of IT (dependent variable).

Infrastructure

A majority of the respondents reported availability of the platform for information technology.

Training

Education and training were found to be some of the major factors influencing utilization of IT. On formal education, the higher the level of education the higher the utility rate, for example, the highest users were found to be those with university education at 71.4 per cent followed by those who attained college education at 19 per cent. One of the interesting findings was the effective of in-house training at 100 per cent on the impact of utilization as opposed to vendor/consultant training at 83 per cent and self-study at 88.9 per cent

Environment

The study indicates non-significance of work environment to utilization of IT.

Funds

The study established a significance of availability of funds to utilization of IT with only 21-40 per cent of IT projects having been implemented. The study established that the challenging operation environment for Magadi Soda Company compel every employee to utilize information analysis tools for supporting business decisions. However, the utilization of these tools impacted on a number of factors including those investigated.

The study determined that computers are used more for coordination tasks, making a networked environment core for utilization of IT. Users communicate more frequently through e-mails, exchange information through computers and share resource such as printers very often. Therefore, infrastructure, staff training, and availability of funds were found to be the key factors for better utilization of IT in the workplace.

Based on the above findings and literature reviewed, it is obvious that technology transfer has a major role to play in enhancing efficacy, output and competitiveness of not only large firms but also SMEs.

5.2 Conclusions

Based on the findings, it can be concluded that:

1. Infrastructure, staff training, availability of funds, exposure to computers or accessibility to computers especially at formal education level particularly in higher learning institutions (universities) are some of the key factors which influence utilization of IT in the workplace.
2. The training offered by vendors/consultants did not have as much benefit towards utilization of IT as other training attended elsewhere such as in-house training; self sponsored training, training achieved through university and colleges.
3. IT is critical for enhancing productivity in organizations and should therefore be treated just like other capital projects-new production line etc.
4. In-house training for IT skills up gradation is significant.
5. Adequate funds, skills and other competencies in IT are valuable and it is only after careful consideration of these factors that IT utilization can be widespread.
6. Information liquidity is crucial for organizations to compete effectively in the global economy.

5.3 Recommendations

1. The government should therefore include study of IT in the education policy and make it a compulsory subject that is examinable. This would expose students to information technology at an early stage and help them overcome traumas associated with learning new ways of working at a mature stage
2. Organization, which relies on IT vendors for conducting their training, need reassess the benefits.
3. Organizations should put more emphasis on in-house approach of training their employees on IT.
4. Since IT is very dynamic, organizations should train their employees frequently to enhance their competencies and IT skills to enable them take advantage of the benefits of IT for attainment of corporate goals.

Suggestions for Further Research

The referent group respondents identified the factors which affects individual's decision to adopt IT applications. It would therefore be of interest to study to what extent this affects the utilization of IT by organizations including SMEs in Kenya.

It would also be interesting to determine to what extent government policies determine the use of IT and technology transfer in general.

REFERENCES

Andrew, Blau, (2001). More than bit players, Surdna Foundation Inc, http://www.surdna.org/documents/morefinal.pdf

Ariss, S.S., Raghunathan,T.S. and Kunnather, A. (2000). "Factors affecting the adoption of advanced manufacturing technology in small firms. *S.A.M. Advanced Management Journal.* Vol 65, 2:pg.14.

Armstrong, M. (2001). Strategic Human Resource Management, Great Britain, (2nd Ed.) Stylus Pub Llc Ltd Comision del Acuredo de Cartegene, Decision 84.

Chen, W. (1999). "The manufacturing strategy and competitive priority of SMEs in Taiwan: A case survey." Asia Pacific Journal of Management. Vol.13. pg.331.

Gakure, R.W. (2003). Analysis of policies for integration of science and technology into small scale Enterprises. A Study conducted for UNESCO, Nairobi, Kenya

Greenway, D. (1994). The diffusion of new technology. *The Economic Journal,* Vol 104, pp. 916-17.

Gupta, P. (2004). *Six sigma Business Scorecard.* NY: McGraw-Hill

Morenzo- Luzon, M.D. (1993). Can total quality management make small firms competitive? *Total Quality Management.* 4 (2).pp. 165- 181.

Moyi, E. and Njiraini, P. (2005). *Towards Technology Models for MSEs in Kenya: Common Principles and best practices.* KIPPRA discussion paper No.51.

Nyambati, R. (2001). Information technology planning practices in Kenya banks, University of Nairobi (UON)

Ondus, T. S. (2001). A survey of ethical issues in the use of information technology among commercial banks in Kenya, UON

Phillips, T.E. and Leedgerwood, J.R. (1994). Running with the pack: JIT and Automation for small manufacturers. *National Public Account.* 39, 26- 28.

Republic of Kenya. (2007). Ministry of Trade and Industry. Final Draft National Industrial Policy. Nairobi

Rohitratana, K. and Boon-Itt, S. (2001). "The implementation of ISO 9000 in Thai seafood processing industry: an empirical study." *Integrated Management: Proceedings of the 6th International Conference on ISO 9000 and TQM.* Pp.477-482.

Ryan, C., Deane, R. and Ellington, N.P. (2001). Quality management training in small to midsized manufacturing firms. *Quality Management Journal,* Vol.8 No. 2

Sato,Y.(1998). Small Business Economics. Amsterdam: Springer Netherlands.

Sousa, R. and Voss, C.A. (2001). Quality management: Universal or context dependent? *Production and Operations Management,* Vol. 10, Iss. 4, pp.383-404.

Steel, W.F. and Wester, L.M. (1992). How small enterprises in Ghana have responded to Adjustment. *The World Bank Economic Review,* Vol 6, No. 3 September 1992 pp 423-438.

Stewart, F., Lall, S. and Wangwe, S. (eds) (1992). Alternative development strategies in Sub-Saharan Africa, London: Macmillan

Tirupati, D. (2008). "Role of Technological Innovations for Competiveness and Entrepreneurship." *Journal of Entrepreneurship.* 17; 103.

Virmani, B. R. and Rao, K., (1999). Economic restructuring, technology transfer and human resource development. Response Books, New Delhi 110048

Warschauer, Mark. (2003). Technology and social inclusion; Rethinking the Digital Divide, Cambridge.

UNIDO.(2004). Integrated Institution *Capacity Building Services and Programme for Technology Centres and Parks Venue.* United Nations Industrial Development Organization.

Yong, J. and Wilkinson, A. (2002). "The long and winding road: the evolution of quality management". *Total Quality Management.* Vol. 13. No. 1.pp.101-21.

Zairi, M. (1993). Competitive manufacturing: Combining total quality management with advanced technology. *Long Range Planning.* 26(3), 123-132.

UNIDO.(2004). Integrated Institution *Capacity Building Services and Programme for Technology Centres and Parks Venue.* United Nations Industrial Development Organization.

– Section II –

Technology Transfer and SMEs in the Developing Countries

Chapter 10

The Strategy of Egypt in Enhancing the Small and Medium Enterprises

Janet Ibrahim Yousef
President, Invention and Innovation Agency, Academy of Scientific Research and Technology (ASRT), 101 Kasr Al Aini Street, Cairo, Egypt
E-mail: janet_ibrahim2008@yahoo.com

ABSTRACT

This paper gives a brief description of some of the policy initiatives and plans and programmes in Egypt for the development of micro, small and medium-sized enterprises

Keywords: *SMEs, MSMEs, Egypt, Enterprise, Strategy, Governmental role, Policy, Technology, Incubator.*

1. INTRODUCTION

Nowadays, there is no objective discussion about the importance of small and medium-sized enterprises (SMEs) for the development of economies of the developing countries and Middle East region and even less about their labour absorption skill.

In Latin America, the Caribbean, India and China SMEs play a strategic role. Their ability to adapt quickly to changes and to identify market niches, apart from their innovative potential, has placed them at a key position within the labour market. They have proliferated as a consequence of the increase in unemployment and poverty that has forced men and women to enter this sector as a unique means of survival, rather than as a new labour opportunity.

In 1998, there was a General Conference held in Egypt by the Egyptian government in which it was recognized that Small and Medium-Sized Enterprises were a vibrant sector, employment creator and particularly suitable to fight against poverty.

The recommendations in the conference include crucial references to several issues such as: protection and promotion of female and male workers; advisory activities for organizations of employers and workers, gender equality and non-discriminatory measures; social protection; measures and enforcement mechanisms to safeguard the interests of workers in SMEs, and the need to promote respect for ILO's (International Labour Organization) Conventions related to freedom of association, discrimination and child labour.

The year 2005 has been called the International Year of Micro-credit with the intention that the international community becomes more aware of the importance of micro-credit and micro-finance in order to eradicate poverty and to improve the already existing programmes that support inclusive and sustainable financial sectors around the world. Working together for the assessment and promotion of microfinance is a challenge for the UN (United Nations) State members, the UN agencies and the main financing partners to change the lives of people that live under poverty conditions and to contribute to the achievement of the Millennium Development Goals.

The role and potential of small enterprises of the Middle East region as well as the importance and particular significance of all such factors are closely related to their recent history. It is therefore essential to establish a conceptual framework that includes a reasonable perspective, characteristics, role and challenges of the enterprises in this sector.

This paper gives a brief description of some of the efforts, plans and programmes implemented in Egypt in order to strengthen micro, small and medium-sized enterprises, the workers and entrepreneurial sector; and to generate and support new enterprise projects.

2. POLICY EFFORTS

2.1 The Invention and Innovation Development Agency (IIDA), ASRT

The Invention and Innovation Development Agency (IIDA) under the Academy of Scientific Research and Technology (ASRT), Egypt, emphasizes on a short-term policy development and longer-run capacity building to support the process. Given the critical challenges currently facing SMEs, early action by ASRT in Egypt initiated improvement in work plan. However, the success of the Academy's policy development efforts will depend on its ability to strengthen its in-house capacities, to improve access to outside sources of research and information, and to develop effective linkages with governmental and non-governmental stakeholders.

Policy and development planning are thus the central theme of the Academy Project, around which other activities revolve. Training, research and networking activities are all designed to support achieving the technology development and scientific service sector goals. Formal adoption policies, regulations or legislation, of course, depends on a range of factors beyond the topic of this paper. The focus is therefore limited to the identification, analysis and presentation of options for possible action by the Academy.

In dealing with the issue of SMEs policies, the Academy starts from the premise that for SME specific policies and programs to bear fruit, they must be placed within the larger context of the country's economic policies. Otherwise, policy or program interventions will either eventually die out or become a burden on the normal functioning of the economy. Moreover, for policies to bear fruit, they do not necessarily have to entail sizeable investments or administrative burdens that are difficult to undertake. In several instances, streamlining regulations and minimizing paperwork, or the simple dissemination of information, has proven to be much more beneficial than complex and expensive assistance schemes. Likewise, experience with MSME development efforts worldwide suggests that the most successful efforts were those that sought to stimulate intermediaries to provide assistance to MSMEs. Thus, the approach for policy formulation that the Academy adopted has relied on tools that guaranteed wide participation of stakeholders in the formulation process, reliance experimentation and best practices, creation of mechanisms of implementation, and evaluation of impact. The following diagram describes the policy formulation process that the Academy have developed and adopted.

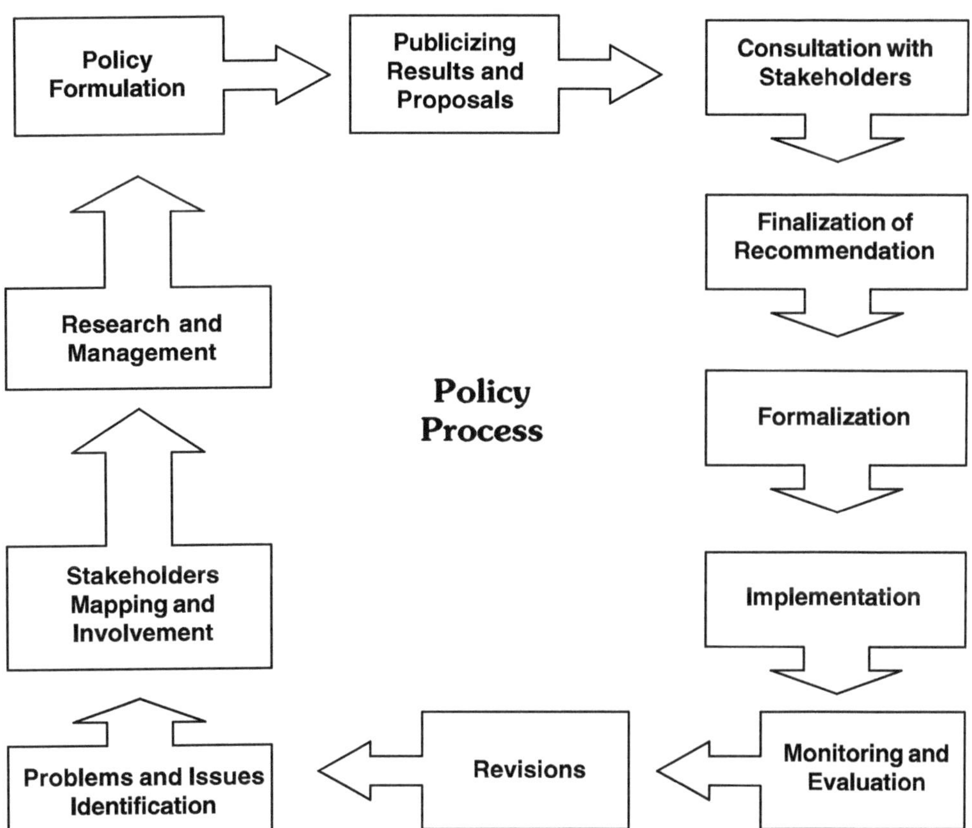

2.2 Activities on Technology Development and Scientific Services for SMEs

The areas targeted by technology development and scientific service sector for SMEs are:

- Developing/updating a general policy framework for MSME development in Egypt.
- Developing operational definition/s for SMEs in Egypt.
- Enhancing the access of SMEs to financial services.
- Regulatory reform to reduce the cost of compliance with regulations.
- Promotion of SMEs

A statement outlining the objective of the proposed intervention is put forward, followed by comments of some aspects of the policy area that are specifically related to SME development. Achievements are then briefly presented in the conclusion.

2.2.1 Updating the General Policy Framework for MSME Sector

Objective

The adoption by the Egyptian Government of a consistent national policy framework for SMEs development which will enable better planning, coordinating, and managing SMEs development efforts.

Discussion

In June 1998, the Egyptian Ministry of Economy and Foreign Trade (then the Ministry of Economy) issued a "Draft National Policy for Small and Medium Enterprise Development in Egypt." The document was probably the first attempt at proposing a national policy framework in response to SME constraints.

While far from being complete, it is believed that the Draft National Policy constituted a good start for an efficient process of consultation, refinement, and updating that culminated in a document that was endorsed by all stakeholders the former Prime Minister Dr. Atef Ebied, and current Prime Minister Dr. Ahmad Nazif. "Enhancing Competitiveness for SMEs in Egypt" is a General policy framework and Action plan was the product of an extensive consultative process that took over two years. A National Policy Framework, it should be stressed, is not the property of a single entity. It constitutes a frame of reference for policy-makers and service providers alike. It outlines the parameters and the basic direction along which the desired consistency can be achieved.

Achievements

The document "Enhancing Competitiveness for SMEs in Egypt" released by the Ministry of Economy in June 1998 was initiated as part of the Policy component as an update for the "Draft National Policy on Small and Medium Enterprise Development in Egypt".

The document was a product of an extensive consultative process that involved every stakeholder in Egypt who works in the SMEs development field, Over 180 organizations contributed in the development of the document.

The purpose of this document is to develop a general policy framework that would constitute the broad parameters within which SME development policies and initiatives are to be implemented.

Based on the above, the proposed policy framework embodied an inclusive, well-informed, planned and executed participatory process that established and nurtured effective communication between various stakeholders active in the field.

The final action plan forms the basis of discussion now in the Policy Committees, which embodies decision makers from various Governmental bodies, in order to activate and implement the activities laid in the action plan.

2.2.2 Definition of MSME

Objective

The development of a unified operational definition of the MSME sector, thus enabling better planning and targeting of MSME-related policies and assistance programs.

Discussion

Throughout the world, providing a definition for MSMEs has proven problematic. Size (micro, small, medium, and large) is a relative thing that varies across countries, as well as across business sectors. Countries, especially those with policies and programs targeting MSMEs, have operational definitions for the sector.

Several criteria are used to define small businesses. The most widely used include labour, capital, and annual receipts. Definitions vary from using a single criterion or more than one. Whereas a government can provide certain incentives or services to its MSMEs, the absence of clear definitions and criteria can lead to inefficiency in service delivery.

In developing a definition for Egypt, several considerations should be noted. First, there is no such thing as a correct definition. A definition's adequacy however is a result of several factors, including the availability and reliability of data, the size structure of enterprises in the country (or within business sectors), and the government's purpose (access to certain services or privileges), among others. Second, while more complex definitions that use several variables may seem to be more accurate, some experts warn against using certain variables. Because assets and receipts are measured in currency, and the value of almost all currencies changes from time to time, employment is the preferred measurement dimension. In fact, data, relating to annual receipts and capital investments are not easily accessible poses an additional problem when it comes to utilizing these indicators.

Achievements

Research adopted by IIDA has been undertaken as follows:

1. "Operational Definition for M/SMEs in Egypt – International Experiences"
2. "Report on Operationally Defining the Micro, Small and Medium Enterprises Sector in Egypt"
3. "The Definition of Micro, Small, and Medium Enterprises"

Through a consultative process and through a 3 phase study (International Experience Study, Local experience study that was divided into 2 phase study) a definition was formulated that depended on three criteria: Number of Workers, Fixed Assets, and Annual Turnover.

2.2.3 Financial Policies

Objective

To enhance and sustain the access of SMEs to a variety of financial instruments along market lines.

Discussion

Worldwide, access to finance continues to be one of the most serious impediments to SME development. Most countries have either sought or are establishing measures by various means to fill what has come to be known as the finance gap. The problem is naturally more acute in developing countries where the process of financial deepening is weak and financial instruments (short- and long-term credit, leasing, venture capital etc.) are not sufficiently developed or widely used. The expansion of credit services on a significant scale requires several measures, including the following:

- ✰ Developing Institutional Infrastructure for SMEs Lending
- ✰ Developing Supporting Institutions for entrepreneurs.
- ✰ Creating an Enabling Legal and Regulatory Environment
- ✰ Better Regulation and Supervision by law

Achievements

There are ongoing studies that look into the experience of Egypt in sufficient Credit and Leasing, as tools to increase accessibility to finance for small businesses. The outcome of these researches will lead to action plans that suggest definite actions to be implemented through the concerned stakeholders.

In the export finance study for SMEs the recommendations were presented to the Export Development Bank in order to be implemented.

2.2.4 Procurement

Objective

To enhance the involvement of SMEs in government and public procurement in an economically sound and mutually beneficial manner, through providing necessary SMEs related inputs to the Ministry of Finance.

Discussion

Potentially the most important constraint facing SMEs is marketing. Governments can constitute an important market for SMEs. In most EU countries, for example, public procurement constitutes 10-15 per cent of the GDP. Twenty percent of Latin America's GDP is spent on procurement. Nevertheless, almost worldwide, the general rule is that these procurements remain largely the monopoly of larger firms, simply because the transaction costs for both the government (buyer) and the seller are much higher.

Achievements

Research has been undertaken and completed and results distributed to stakeholders as follows:

- "Public Procurement Systems for SMEs"
- The recommendations from the study were met by the SME Law that came to allocate a 10 per cent of government procurement to SMEs (also in the SME law).
- A Committee comprised from advisors from the concerned Ministries and the Ministry of Finance was formed to discuss the methods of implementation.

2.2.5 Regulatory Reform

Objective

To reduce the cost of compliance with regulations covering enterprise establishment, operations, and growth.

Discussion

While regulatory constraints are common for all private businesses in general, they weigh more heavily against MSMEs. This is primarily due to the latter's lack of adequate education, sufficient institutional mass, as well as its limited ability to tap into the legal and other pertinent consulting services that large firms can easily access. Although legal and regulatory constraints vary by sub-sector, it is a well-known fact that MSMEs face significant and numerous problems with regards to establishment and licensing, operation, taxes, registration, export and import permits, and compliance with the different directives of various governmental entities.

Achievements

Research has been undertaken and completed and results distributed to stakeholders as follows:

- "Streamlining the Regulatory Procedures for Small and Medium Enterprises.
- "Legal and Regulatory Framework Governing SMEs Establishment, Operation and Growth"
- An ongoing Research Study on streamlining the Egyptian laws, regulations and procedures governing SMEs establishment, growth.

2.2.6 Promotion

SMEs are generally under performing when it comes to promotion. The vast majority of SMEs did not even attempt to export. There was a consensus on what constituted the main obstacles hindering promotion, marketing and exporting. Those were:

- ✰ Inaccessibility of export channels.
- ✰ Lack of awareness.
- ✰ High input prices.
- ✰ Scarcity and difficulty of finance.

While Egypt currently focuses on labour-intensive sectors for local marketing and export promotion, efforts should also be undertaken to promote the "winning sectors" that have been enjoying increased dynamism (*e.g.*, medium and high technology manufacture and information and communication technologies). In addition, emphasis should be directed towards:

- ✰ Move within the existing labour intensive sectors to link in the value chains where value is concentrated (*e.g.*, through developing its textile design capabilities, rather than mainly focusing on textiles production, within the textiles value chain).
- ✰ Address market failures and distortions affecting current actors in the value chain.

Achievements

Research completed by IIDA (Invention and Innovation Development Agency) and results distributed as are follows:

- ✰ "International Experience in SMEs Export Development"
- ✰ "Promoting Trading Houses in Egypt"
- ✰ "SMEs Export Strategy". The export promotion study is about to be finalized. The strategy will indicate the sectors and the possible markets for SMEs to export.

Table 10.1: Number of Patents that Succeeded in being Marketed by Small and Medium Enterprises

Sl.No.	*Patents*	*Patent Number*	*Inventor's name*
1.	A car Seat Belt Against Sinking	21965	Awad Mohamed Abd –El Ghafaar
2.	A moving Chair (Manual Automatic)	21279	Soliman Mohamed Soliman
3.	Petrol Electronic Injectors cleaner Machine	21683	Salah Hamed Ramadan
4.	Plastic Antena	21818	Ashraf Mohamed Hanafi
5.	Machine That Protects Breaks	22649	Adib Mitry Sedrak
6.	The Ellipatical Solar Energy Heater	22989	Dr. Abd El Fattah Montasser
7.	Antiflame and Fire For Waste	23104	Gamal Hassan Nada
8.	The Egyptian KH-Y Lithoclast	21150	Khalid Abd Allah Yassin
9.	Solar Cooker with a truncated cone Shape	22767	Dr. Kassim Kadry Matrawy

3. INSTITUTIONAL SUPPORT

3.1 Social Fund for Development (SFD)

Objectives

Since the establishment in 1991, SFD has played different roles at the development arenas of Egypt, Initially; it was designed as a social safety net associated with the government of Egypt's agreement to undertake its extensive Economic Reform and Structural Adjustment Program (ERASP). Therefore, the Fund was considered essential to the actual success of the reform program. Serving as a vanguard of economic empowerment, quality human resources, and an enabling environment for human development, SFD has successfully developed into a full-fledged organization with a new development perspective: it consolidates mechanisms for better understanding of the impact of globalization and mobilizes efforts to minimize risks of social exclusion, help alleviate poverty and combat unemployment. In this capacity, it creates employment opportunities for start-up entrepreneurs and provides them with credit, technical assistance, skills, and technological know-how both Egyptian and international.

Major Themes of SFD Activities

SFD operates according to several major themes based on:

- ✰ Developing and financing small enterprises serving as feeding industries to medium and large enterprises in all governorates with a view to creating permanent and stable job opportunities,
- ✰ Encouraging micro enterprises to improve income and ameliorate living standards in Egyptian villages and hamlets
- ✰ Creating the climate appropriate for supporting human/community development/public works enterprises
- ✰ Promoting self employment and entrepreneurial culture, and providing an integrated package of financial/non-financial services for small/micro enterprises.

3.2 Small Enterprise Development Organization (SEDO)

SEDO aims at providing every possible assistance to small and medium enterprises (SMEs) and consequently, SFD allocates about 50 per cent of its resources for that purpose. The Organization develops various funding mechanisms and upgraded technical support systems.

3.3 Small Enterprise Development Group

Objectives of the group are:

- ✰ Designing and implementing small enterprise development sector programs and providing technical support.
- ✰ Proposing legal, organizational, economic, and financial measures required for developing these enterprises.

- ☆ Establishing and managing an information network for small enterprises in conjunction with the Marketing and Management Group.
- ☆ Providing various credit packages, developing funding mechanisms, and expanding the specialized windows for lending.
- ☆ Employing developed tools to bear credit risks, and smooth loan guarantees, monitoring and developing credit performance, and proposing policies for settling default cases.
- ☆ Assessing loan portfolio and tripartite contracts.
- ☆ Setting up and encouraging business/technology incubators that sponsor non-traditional enterprises and creative ideas.
- ☆ Designing replicable packages of successful enterprises to expand their base
- ☆ Establishing business development centres in all Egypt's Governorates

3.4 Marketing and Management Group

Objectives of the group are:

- ☆ Developing marketing and planning systems in accordance with the market needs to verify that products are financed second to conducting the necessary marketing studies
- ☆ Enhancing the competitive edge of small enterprises, improving their quality and opening new markets for their products,
- ☆ Financing and updating marketing studies and research,
- ☆ Developing technology support and transfer systems,
- ☆ Training the youth in technical vocational training centres to promote their entrepreneurial skills,
- ☆ Cooperating with different ministries and authorities to establish joint ventures with a view to employing the youth and generating new job opportunities,
- ☆ Raising awareness among entrepreneurs to apply quality systems and cope with the requirements of the World Trade Organization (WTO),
- ☆ Creating opportunities for business match making and devising tools for developing exports,
- ☆ Setting the appropriate climate for cooperation with large enterprises,
- ☆ Implementing an integrated strategy that observes the economic, social, and environmental dimensions.

3.5 Business and Technology Incubators

Business and Technology Incubators are the Development Institutions providing support to foster the growth of start-up and early stage enterprises, and generally improve their chances of success in an increasingly competitive global environment.

3.5.1 Mission and Objectives of the Incubator

- To provide assistance to the projects inside and outside the Incubator premises.
- To promote the pioneer ship spirit and to assist the small pioneer institutions in facing their difficulties at their start-ups in facing their difficulties
- Develop Business skills and Managerial efficiency
- To nurture early-stage enterprises from start-up to survival and succeed
- Assist the entrepreneurs in obtaining all the necessary permission and licenses before start-up
- Help achieving high developing rates through provided assistants–Incubate new business projects in their start-ups and growth to sustainability
- Create new job opportunities (permanent/non permanent and direct/indirect).
- To link small entrepreneurial companies to medium and large organizations/companies in order to develop their activities (feeding industries)
- Adapt the projects based on technology which are related to Universities and Scientific research centres to Convert their researches to Marketable products
- Offer a full package of assistance such as Quality Control Units, Special Specifications, Technical and commercial Data base.

3.5.2 Assistance Provided by Incubators

Business incubator is being seen as a significant micro economic tool to foster the growth of start-up and early stage enterprises, and generally improve their chances of success in an increasingly competitive global environment. In this context, Incubators offer a full package of assistants to help overcome the problems which usually faces the Projects in their start-p phase, these assistant could be offered directly through the managerial crew of the incubator or through external Specialist.

Administrative Assistance

Business plan, day to day managerial skills training, secretarial works, fax, computer, internet etc.)

Technical Assistance

Specialized local and global consultants, product quality control labs, technical training etc.)

Financial Assistance

SFD loans, feasibility studies, balance sheets, etc.

Marketing Assistance

Exhibitions and advertisement

3.6 SFD Incubators Program

3.6.1 Within the over all concern of SFD to increase employment opportunities in the small and micro enterprises sector, it encourages the creation of new businesses, expanding existing production activities and providing budding entrepreneurs with an integrated productive packages (finance, training, technical assistance marketing and know how). The Incubators Department has been established to create a national network of incubators though-out, choosing the most appropriate sites for incubators according to a scientific site criteria which should by provided with the main public works utilities such as electricity, water etc. In addition, for promising a business community (new industrial zones) near universities and scientific research centres, SFD finances all the construction and rehabilitation needed and also provides finances required for the Management of the incubators (covers the deficit in the Budget) till the incubator reaches sustainability.

3.6.2 Types of Incubators

The Idea of incubators since its start in the early fifties have been through many stages of development which resulted in a wide variation in the types and specialty of incubators, which led to specify the types of incubators according to the goals and the kinds of special assistance provided by each

General Incubators

Promote formation of new enterprises based on a local environment needs.

Technology Incubators

Promote the development of new enterprises taking advantage of the influence such as universities, institutions, science parks etc.

Specialized Incubators

They focus on a specific field or sub sector *i.e.* biotechnology, information technology.

3.6.3 Methodology of Incubators

In Simple terms the Business Incubator is a micro-facility with a small management staff that provides the physical workspace, shared facilities and access to technical and business support services in one integrated and affordable package. Such caring and sharing have been adapted to facilitate business start-ups by reducing initial costs, and diminish the chance for failure of fledgling enterprise. That means the incubator will provide all kind of assistance to youth man/woman to start their own business when they only got a good idea of a project and strong potential. Incubators tenants are to pay a gradually increasing rent fees/subscription rate which would be lower than the surrounding market in the first period.

The incubators provide assistance to its tenants through out in the following phases.

- ☆ Preliminary discussion and Feasibility study.
- ☆ Preparing the business plan
- ☆ Joining the incubators premises and start-up

- ☆ Growing up to stability.
- ☆ Graduation from the Incubator
- ☆ After graduation

3.6.4 Working Incubators

Following incubators are presently functioning in the country

- ☆ Tala Business Incubator – Monofia Governorate
- ☆ Tabbin Technology Incubator – Cairo Governorate
- ☆ New thebes Business Incubator – Luxor City.
- ☆ Mansoura Business Incubator – Dakahlea Governorate.
- ☆ Doweka Virtual Incubator – Cairo Governorate
- ☆ Assuit Technology Incubator – Assuit Governorate.
- ☆ Port Said Virtual Incubator–Port Said Governorate.

Furhter, following incubator projects are under implementation

- ☆ 10 of Ramadan Virtual Incubator. – Sharkia Governorate.
- ☆ El- Daowea Virtual Incubator – Bani Soweif Governorate
- ☆ Abou Rawash Business and Technology incubator–Giza Governorate

Feasibility study of three more projects is being carried out:

- ☆ El-Menia University Business Technology Incubator – El-Menia Governorate
- ☆ Bani Ebeid Agricultural Technology Incubator–Dakahlea Governorate.
- ☆ Technology Incubator in cooperation with the academy of scientific Research and Technology

REFERENCES

Paper presented in a conference about strategy of supporting small and medium enterprises, Jan. 2008 (www.mof.gov.eg)

Paper published by social fund for development, 2008, (www.sfdegypt.org)

Egyptian Patent office (www.egypo.gov.eg).

Chapter 11

Policies and Programmes for Technological Capacity Building of Small and Medium Enterprises in India

M. Bandyopadhyay

Senior Expert, Centre for Science and Technology of the Non-aligned and Other Developing Countries (NAM S&T Centre), India Habitat Centre, Lodi Road, New Delhi –110 003, India

(Formerly, Adviser, International Cooperation, Department of Science and Technology, Government of India, New Delhi)

E-mail: bandodst@google.com

ABSTRACT

The Small and Medium Enterprise (SME) sector is recognized for its contribution to employment, innovation and economic dynamism, and is considered as an engine of growth and an essential part of a healthy economy. By virtue of their inherently greater flexibility and responsiveness than large firms, SMEs are in a better position in developing or adopting new technologies, in creating specialised products and processes and also, in moving into niche markets.

The Government of India has taken several initiatives for the promotion, protection and development of SMEs according to the provisions of the economic and industrial policy guidelines. Over the years, the small-scale industry sector has attained an impressive growth and is helping generate new jobs, supplying a wide range of products, contributing to exports, and facilitating more equitable distribution of national income in the country. A number of initiatives have also been taken to promote technology-based industries and encourage and support SMEs to introduce new technologies, and develop new products and processes to improve their technological competitiveness.

In view of the commonalities in strength and weaknesses of the SME sector, it would be of mutual benefit if other developing countries collaborate with India on technological capacity building of SMEs through transfer of technologies, consultancy services, institution-industry linkage and sharing of information and expertise.

Keywords: *Enterprise, Institution-industry linkage, MSE, MSME, Modernisation, NTBF, R&D, SME, S&T, Technological upgradation, Technology transfer.*

1. INTRODUCTION

The strategic importance of Small and Medium Enterprises (SMEs) in the economy in terms of their contribution to Gross Domestic Product and employment, and their impact on income distribution is well recognized. SMEs provide the industrial leaders of the future, improve the competitive edge of the economy by maximising the range of choice available through market provision, and challenge the dominance of existing large industrial units thereby forcing them to innovate. It would not be an exaggeration to say that the overall health of the economy depends to a large extent on the health of the SME sector in a country. Various studies conducted on the importance of small firms demonstrate that the bulk of new jobs are emanated from small enterprises.

Recognizing the distinctly positive impact of small enterprises on the economy, the governments of many industrially advanced countries have taken several policy initiatives not only for the growth and expansion of this sector but also for improving its capability. Among the developing countries, India is an extremely successful example for an unparalleled growth of small-scale industry. The government of India has been encouraging and supporting micro, small and medium enterprises (MSMEs) through policies for infrastructural support, technological upgradation, preferential access to credit, reservation of products for exclusive manufacture in the small scale sector, preferential purchase policy etc. Several schemes and programmes have been instituted to help MSMEs by providing facilities for transfer, absorption and adaptation of technologies, training of personnel, product development, testing and calibration, and fiscal incentives and venture capital funding for introduction of new technologies. As a result of these proactive policies, the MSME sector plays an important part in the Indian economy in terms of its contribution to the industrial production, exports, employment and creation of an entrepreneurial base in the country.

This paper presents a comprehensive overview of various policy initiatives, schemes and programmes of different arms of the Government aimed at facilitating adoption, development and transfer of technologies for technological capability building, and improving technological competitiveness of small and medium enterprises in India. In the concluding part, as an initiative on south-south cooperation, a few possible areas of collaboration between India and other developing countries have been recommended.

2. SMALL AND MEDIUM ENTERPRISE (SME)

2.1 Concept and Definitions

The concept and scope of a Small and Medium Enterprise (SME) are different in different countries and the distinction between 'large' and 'small' is often arbitrary. The abbreviation, SME is commonly used in the European Union and other international organisations, while in the USA it is designated as small business encompassing manufacturing, services and also trading in some limited areas. In India, the emphasis was on small scale industry (SSI) earlier, which included manufacturing, and repair and maintenance activities, but recently, the scope has been expanded to Micro, Small and Medium Enterprises (MSMEs), or sometimes simply Micro and Small Enterprises (MSEs), that also include the services.

The definition of SME, small firm, small business or small industry thus varies from country to country. The smallness of a unit is recognized on the basis of either maximum number of employees or maximum investment or maximum investment as well as employment or maximum employment as well as annual turn-over/balance sheet total. The basis on which SMEs are defined depends on the stage of economic development of the country and the broad policy purposes for which the definition is used.

2.2 Definition of SME in India

In India, Micro, Small and Medium Enterprises (MSMEs) are classified in two categories *viz.* (*a*) Manufacturing enterprises, defined in terms of investment in plant and machinery and (*b*) Service enterprises, defined in terms of investment in equipment. According to the revised definition adopted under the Micro, Small and Medium Enterprises Development (MSMED) Act of October 2006, the limit for investment in plant and machinery/equipment for manufacturing and service enterprises are given in Table 11.1.

Table 11.1: Definition of MSME in India

Classification	*Manufacturing Enterprises*	*Service Enterprises*
	Investment in Plant and Machinery	*Investment in Equipment*
Micro- Enterprises	Does not exceed INRs. 2.5 million	Does not exceed INRs. 1.0 million
Small Enterprises	More than INRs. 2.5 million but does not exceed INRs. 50 million	More than INRs. 1.0 million but does not exceed INRs. 20 million
Medium Enterprises	More than INRs. 50 million but does not exceed INRs. 100 million	More than INRs. 20 million but does not exceed INRs. 50 million

Source: MMSME 2009

~50 INRs. (Indian Rupees) = 1 US$

3. DEVELOPMENT OF SMEs IN INDIA

3.1 Policy Initiatives

The progress of industrialisation over the last few decades, particularly in the post-liberalisation period has been a striking feature of India's economic development. In pursuance of a conscious and deliberate policy of industrialisation, a large investment has been made to build up capability over a wide spectrum of industry. Industrial production has made rapid strides in terms of variety, quality and quantity. Impressive indigenous capabilities have now been established to the point of self-sufficiency in sectors like mining, irrigation, power, chemicals, transport and communication etc. With the help of a proactive policy and liberal fiscal incentives, knowledge-based industries in biotechnology, drugs and pharmaceuticals, food processing, information and communication technology including information technology enabled services have brought a revolution in the economy of the country. India today operates the largest programme for the development of small industry in any developing country. The government of India has taken a number of steps for the promotion, protection and development of this sector. A separate Policy Statement for the promotion and strengthening of small, tiny and rural industries was made in 1991 (*GOI, 1991*). Subsequently, the Micro, Small and Medium Enterprises Development (MSMED) Act, 2006 was adopted in October, 2006 (GOI, 2006).

The government of India's encouragement to SME is reflected in the fact that a separate Ministry of Small Scale Industries, and Agro and Rural Industries (SSI and ARI), was established in October 1999 as the nodal Ministry for policy formulation and implementation and coordination of schemes and programmes on promotion and development of these industries. This Ministry was bifurcated into two separate identities *viz.* Ministry of SSI and Ministry of A&RI in September 2002. Later, all the activities were again brought under a single Ministry of Micro, Small and Medium Enterprises (MSME) in May 2007. The Office of the Development Commissioner, MSME (earlier DCSSI) functioning within the Ministry of MSME is entrusted with the responsibilities of providing techno-economic and managerial consultancy, common facilities and extension services and facilities for technology upgradation, modernisation, quality improvement and infrastructural support to MSMEs.

3.2 Current Status of SMEs in India

The micro, small and medium enterprises, particularly, the small scale industries have played a vital role in the economic progress of India. The MSME sector has grown rapidly over the years. However, since no reliable data for the medium scale industries is available, and because the total number and the impact of the medium enterprises are insignificant as compared with the Micro and Small Enterprises (MSEs), the statistical data has been compiled for the MSEs only, *i.e.* after excluding the medium sector from MSMEs.

The number of MSEs has increased from an estimated 0.87 million units in the year 1980-81 to over 3 million in the year 2000, and presently to over 13 million. It has been estimated that a million INRs of investment in fixed assets in the small-scale sector produces INRs 4.62 million worth of goods or services with an approximate value addition of 10 per cent. Some important features of the MSE sector in India are highlighted below (*MMSME, 2009; DCMSME*).

1. The MSEs have grown rapidly over the years with overall growth in terms of the Index of Industrial Production (IIP) (base: 100 in 2001-02) at 12.32 per cent in 2005-06 and 12.60 per cent in 2006-07.
2. The annual growth in the number of units has been ~4 per cent.
3. The MSE units produce an amazing variety and types of products. More than 6000 products are manufactured in more than 20 major industry groups, particularly in consumer and household products. The important product groups are: Food products (22 per cent), chemicals and chemical products (12 per cent), basic metal industry (10 per cent), metal products (8 per cent), electrical machinery products (6 per cent), rubber and plastic products (6 per cent) and others (36 per cent).
4. The total annual production in the sector accounts for almost 40 per cent of the gross industrial value added and contributes ~6 per cent of GDP.
5. The cumulative employment provided in the sector is about 32 million, which is second only to the Agriculture sector and is about 35 per cent of total employment in manufacturing and services in the country. The average increase in employment in the sector is 4.37 per cent per annum.
6. MSEs play a major role in India's export performance. Direct exports from the sector account for nearly 35 per cent of total exports. Further, MSEs contribute around 15 per cent to exports indirectly. Non-traditional products account for more than 95 per cent of the MSE exports.
7. A majority of the units are engaged in the production of basic, conventional, low technology items. 96 per cent of the units are in the tiny or micro sector and 21 per cent of the units are engaged in job work and there are very little technological inputs in both of these categories of units.

A summary of performance of the MSE sector in the last five years is presented in Table 11.2.

Table 11.2: Performance of Micro and Small Enterprises in India

Year	*Number of Enterprises (Million Nos.)*			*Empl. (Million Persons)*	*Growth of Production Rate (Per cent)*	*Share in GDP (Per cent)*
	Registered	*Unregistered*	*Total*			
2002–2003	1.591	9.358	10.949	26.349	8.68	5.92
2003–2004	1.697	9.698	11.395	27.530	9.64	5.79
2004–2005	1.753	10.106	11.859	28.755	10.88	5.84
2005–2006	1.871	10.471	12.342	29.985	12.32	5.83
2006–2007	2.098	10.746	12.844	31.252	12.65	5.94
2007–2008 (Projected)	2.468	10.899	13.367	32.228	13.00	NA

Source: MMSME, 2009; DCMSME.

According to informal sources, there are 10,000 to 15,000 medium units, and their production accounts for 5 per cent of total industrial output, 2 per cent of GDP and contributes 10 per cent to total exports of the country.

4. TECHNOLOGICAL CAPACITY BUILDING OF SMEs–POLICIES AND PROGRAMMES IN INDIA

4.1 The Need for Technological Capacity Building of SMEs

Technological developments in the recent years in the new and emerging areas such as information technology, computers and telecommunications have had a revolutionary impact on the competitiveness of nations. Building a sound technological infrastructure, development and transfer of technologies and promoting academic research–industry links have thus become the key policy issues for improving the future competitiveness of a country.

SMEs in the developing countries suffer from several inadequacies such as technological backwardness, low production efficiency and poor quality of products. SMEs suffer from a number of disadvantages compared with larger firms in responding to and adapting new technological advances. They are often not aware about the recent scientific advancements in absence of a proper information dissemination system. Further, there are no motivational inputs and incentives for them for innovation and technological upgradation. They lack suitable qualified specialists and are largely unable to provide in-house R&D support required for development of new products and processes and adoption of new technologies.

Therefore, it is imperative for the developing countries to take appropriate policy measures for building capabilities of SMEs to develop, acquire and adapt new technologies; enhance efficiency of production and reduce production costs; and undertake total quality control, quality assurance and good manufacturing practices. Undertaking R&D and being the first to present a new product or service to the market give a competitive advantage to a company. This has particularly become more important for India because of the policy of economic liberalisation and increasing foreign investments, as a result of which the indigenous SMEs are facing stiff competition from the multi-national corporations due to their superior quality of products, high production efficiency and better product design.

Firms that grow up on the edges of a university or research institution are often called New Technology-Based Small Firms (NTBFs). NTBFs help to encourage more positive attitudes among scientific researchers and among the academic community generally towards industry and towards becoming entrepreneurs. The areas usually picked out as the industries based on new technologies include genetic engineering and biotechnology, electronics, information and communication technology, computers, scientific instruments, aerospace, chemicals, office machinery, automobiles, etc. The OECD emphasises that small enterprises, especially NTBFs play a distinctive and increasing role within innovation systems, and beyond their direct contribution to the creation and diffusion of new goods and services, NTBFs help instil a culture of innovation, encourage investments in skills and improve economy-wide dynamic allocative efficiency (*OECD, 1999*).

In this context, development of institution-industry links is widely recognized to be of increasing significance in the overall technological and industrial competitiveness of SMEs. Besides their obvious importance of being the repository of technical manpower, academic and research institutions also have their indirect impacts on the economy through technology transfer, creation of high technology spin-off firms, employment generation, continuing education and providing research and technical advice to existing industrial units. It is therefore necessary to reorient the S&T and industrial policies of a developing country to encourage strategic links between industry and institutions. According to the studies on successful mechanisms in the industrially advanced countries, the types of linkages that may be adopted in the developing countries for assisting the SMEs and creating new small enterprises can be classified as, i) institution-based linkages *e.g.* industrial liaison office, technology transfer office, research consortia, industry cooperative, technology brokerage company, teaching company scheme; ii) linkages through the creation of additional organizational infrastructure *e.g.* cooperative research centre, centre of excellence, incubation centres; and iii) large property based initiatives in/around institutions *e.g.* science park, innovation centre, technology park, research park, industrial park, technopolis etc. (*Bandyopadhyay, 2008*).

In order to tackle the problems faced by the SME sector and improve its technological competitiveness, several policy measures and programmes have been initiated by the Government of India for helping the small scale industrial units in introducing new technologies in their products and processes. While the major schemes and programmes are being implemented by the nodal Ministry *viz.*, the Ministry of Micro, Small and Medium Enterprises (MMSME) and its agencies, various Science and Technology Ministries and departments and other line Ministries have also taken complementary measures on the development of infrastructural facilities, technological services, human resource development and providing fiscal incentives for technology development and transfer.

4.2 Programmes of the Ministry of MSME for Technological Capacity Building of SMEs

As one of the policy initiatives, the Government of India has set up the National Manufacturing Competitive Council (NMCC), under which a National Manufacturing Competitive Programme (NMCP) has been announced particularly to support the MSMEs in their endeavour to become competitive. A ten component scheme has been launched under NMCP that includes, (1) marketing support/assistance (barcode); (2) support for entrepreneurial and managerial development through incubators; (3) improving competitiveness through Quality Management Standards and Quality Tech. Tools (QMS/QTT); (4) building awareness on IPRs; (5) lean manufacturing competitive schemes; (6) mini tool rooms; (7) design clinic; (8) marketing assistance and technology upgradation; (9) technology and quality upgradation support; and (10) promotion of ICT in manufacturing sector.

The Ministry of MSME, Development Commissioner, MSME and other organizations of the Ministry (*MMSME, 2009; DCMSME*) operate following schemes and programmes for providing techno-economic and managerial consultancy, common facilities and extension services, and facilities for technology upgradation, modernisation, quality improvement and infrastructural support for MSMEs.

4.2.1 Micro and Small Enterprises Cluster Development Programme (MSE-CDP)

This is the flagship scheme of the Ministry for the development of clusters for capacity building, technology upgradation, quality standardization and testing, and also skill development, marketing and export support, improved credit delivery, setting up of common facility centres, etc. Till October 2007, 400 micro, village and small enterprise clusters have been developed by the Ministry of MSME, and other Ministries have also promoted 800 more clusters.

4.2.2 Credit Linked Capital Subsidy Scheme (CLCSS)

The scheme aims at facilitating technology upgradation by providing upfront capital subsidy to MSMEs, on institutional finance availed of by them to promote new and appropriate technology, assessing present level of technology and their forecasting, setting up technology information centres/data banks and IT portal for information dissemination, carrying out detailed technology audit, modernisation of production equipment and techniques. So far, 2620 units have been assisted with a total amount of INRs. 743 million.

4.2.3 Support for ISO-9000/14001 certification fee reimbursement

To encourage technology upgradation, quality improvement and better environment management, the fees paid by SMEs for obtaining ISO certification are reimbursed. So far, 13,433 units have been supported with an amount of INRs. 709 million.

4.2.4 Promotion of Rural Industries

Measures taken on technology upgradation, development and transfer for the promotion of rural industries include, (*i*) Product Development, Design Intervention and Packaging (PRODIP) scheme–for improving the quality of khadi (hand spun/ hand woven) products and also to diversify into new products. (*ii*) Scheme of Fund for Regeneration of Traditional Industries (SFURTI) for comprehensive development of traditional industry clusters of khadi, village industry and coir products. (*iii*) Scheme for rejuvenation, modernisation and technology upgradation of the coir industry. (*iv*) Khadi and Village Industries Commission (KVIC) undertakes R&D projects with the help of national level institutes such as IITs and National Institutes of Technology; and has set up Mahatma Gandhi Institute for Rural Industrialisation (MGIRI) and in-house testing laboratory for the promotion and technology support for rural industries.

4.2.5 Technological Services

The Ministry of MSME has several programmes to provide technological services and technological support facilities to the SME sector as given in Table 11.3.

Table 11.3: Technological Services and Support Facilities for SMEs

Programme	*Objectives*	*No.*
Technology Resource Centres (TRCs) and Small Ind. Service Institutes (SISIs)	To help SMEs to upgrade and modernize technology, and provide information on latest technologies, machinery raw material suppliers, quality standards, etc	30 TRCs/SISIs. 28 branch SISIs
Product-cum-Process Development Centres (PPDCs)	For R&D, product design and innovation, product and process improvement and development of improved packaging techniques, common facility centre, HRD	6 PPDCs
Regional Testing Centres (RTCs) Field Testing Stations (FTSs)	To provide testing facilities to enable SMEs to produce stores conforming to specifications. Also, capital grants to industry associations to set up international level testing centres.	4RTCs 7 FTSs
Mini Tool Rooms	Assistance for setting up new mini tool rooms, or upgradation of existing tool rooms.	–

Source: MSME, 2009.

4.2.6 Programmes of National Small Industries Corporation (NSIC)

NSIC, a public-sector undertaking of the Ministry has set up five technology service centres and three technology service extension centres for technology sourcing and dissemination of information, institutional linkages with R&D institutions; one Technology Business Incubator; two Software Technology Parks etc.

4.3 S&T Policies and Programmes for Technological Capacity Building of SMEs

The objectives of the successive Science and Technology Policies in India include, encouraging research and innovation in areas of relevance for the economy and society, particularly by promoting close and productive interaction between private and public institutions in science and technology and strengthening enabling mechanisms that relate to technology development, evaluation, absorption and upgradation from concept to utilization (*GOI, 2003*).

The Steering Committee on S&T for the Eleventh Five Year Plan (2007-12) had recommended following measures on science and technology for SMEs (*GOI, 2006a*):

1. Setting up 170 technology business incubators and 50 technology innovation centres.
2. Entrepreneurship development training, training of trainers on entrepreneurship and creating a climate of entrepreneurship.
3. Using polytechnics and Industrial Training Institutes for producing multi-skilled human resources with the involvement of industry.
4. A Mission mode approach for technology intervention in clusters to help SMEs to increase productivity and exports.

5. Preparing sectoral technology profiles for SMEs to critically examine and address their technological needs in line with their business requirements.
6. Identifying ICT needs of SMEs and providing resources for implementation.
7. Development and import of Special Purpose Machines (SPM) to improve productivity of SMEs.
8. Promoting academia-SME interaction for innovation.

The Government has also recently proposed to enact a Public Funded R&D projects (Protection, Utilization and Regulation of Intellectual Property) Bill, which is often referred to as India's Bayh-Dole Act (BDA). The bill aims to organize, promote and regulate the public availability of the intellectual property (IP) originating from government-aided R&D.

Various departments and agencies under the Ministry of Science and Technology have taken the following initiatives.

4.3.1 Promotion of New Technology-Based Small Firms (NTBFs)

The National Science and Technology Entrepreneurship Development Board (NSTEDB) was set up by the Ministry of Science and Technology in 1982 as an institutional mechanism to help establish knowledge driven and technology intensive enterprises and facilitate S&T entrepreneurship development. Through the activities of NSTEDB, a total of 10,000 micro-enterprises have been promoted at 40 locations in the country, 35,000 persons benefited through self-employment skill development training, and awareness on entrepreneurship created among about 100,000 S&T students. The objectives and performance of individual programmes of NSTEDB are given in Table 11.4.

4.3.2 Technology Development Board (TDB)

The Government of India set up a Technology Development Board in 1996 as a statutory body to promote development and commercialization of indigenous technologies and adaptation of imported technologies for wider application. The Board provides financial assistance in the form of equity soft loans to industrial concerns at a simple interest rate of 5 per cent per annum, or grants to R&D institutions. Up to 2005-06, the Board has supported 156 projects with financial assistance of INRs. 5.60 billion (*TDB*).

4.3.3 Technology Information, Forecasting and Assessment Council (TIFAC)

The TIFAC operates the Home Grown Technology (HGT) programme to support SME-led technology efforts in early stages in the innovation chain for encouraging them to carry out innovations at pilot production level. So far, 75 projects, most of which are SMEs have been supported, 15 technologies commercialised and 10 Patents filed (*TIFAC*).

Table 11.4: Programmes of NSTEDB

Activity	*Objectives*	*Performance*
Science and Technology Entrepreneurs Parks (STEPs)	To forge academic-research-industry linkages and promote new technology based small enterprises and spin-off firms	15 STEPs. 788 SMEs set up with annual turnover: Rs. 1.30 bill. creating 5000 jobs
Technology Business Incubators (TBIs)	To provide infrastructural and support facilities to start-up enterprises	36 TBIs
Entrepreneurship Development Cells (EDCs)	To create an entrepreneurial culture in S&T academic institutions and to foster techno-entrepreneurship	22 EDCs
S&T Entrepreneurship Development (STED) scheme	To create micro-enterprises in rural areas by utilizing local resources and rural technologies	STED in 48 districts. 2600 micro-enterprises, 7000 employed
Entrepreneurship Dev. Programmes (EDPs)/ Technology based EDPs (TEDPs)	To train S&T qualified persons in setting up of industry	530 organized.
Skill Development programmes.	Training rural people in S&T skills for setting up self-employment ventures	35,000 trained. 100 new products and technologies developed

Source: NSTEDB.

4.3.4 Department of Scientific and Industrial Research (DSIR)

The primary aim of DSIR is to promote R&D by industries, support a larger cross section of SMEs to develop state-of-the art globally competitive technologies, catalyze faster commercialization of lab-scale R&D, and strengthen technology management capabilities. DSIR is operating several schemes for technology development, demonstration, innovation and management and provide several fiscal benefits to encourage private R&D as shown in Table 11.5

Table 11.5: Programmes of DSIR

Programme	*Objectives and Achievements*
Technology Development and Demonstration Programme (TDDP)	To catalyze and support activities on absorption, adaptation and demonstration of technology. 166 projects supported, 30 technologies commercialized/prototypes developed.
Technopreneur Promotion Programme (TePP)	To support/encourage innovators to convert original ideas into working models and prototypes. 122 projects supported.
Technology Management Programme (TMP)	To facilitate industry and consultancy organisations, and S&T institutions in developing and learning tools and techniques.
Fiscal incentives	Duty free imports, weighted tax deductions, tax holiday, income tax exemption etc. provided to companies for undertaking industrial R&D

Source: DSIR, 2008.

4.3.5 Council of Scientific and Industrial Research (CSIR)

CSIR has New Millennium Indian Technology Leadership Initiative (NMITLI) scheme, the largest Public-Private Partnership effort to catalyse innovation centred technological development in selected niche areas by synergising the competencies of publicly funded R&D institutions, academia and industry. The programme has evolved 42 projects involving 287 partners with an estimated outlay of INRs 3 billion (*CSIR*). The projects are setting new global paradigms in the areas such as nano material catalysis, biotechnology, bio-informatics, industrial chemicals, gene-based new targets for advanced drug delivery systems, low cost office computers, improved liquid crystal devices and others.

4.3.6 National Research Development Corporation (NRDC)

NRDC provides support for commercialisation of laboratory know-how, licensing indigenous technologies, technology development loans for setting up pilot plants to prove/scale-up laboratory processes, development and promotion of rural technology, etc. It has licensed over 2000 technologies for commercial exploitation, of which nearly 1000 are in production with a current annual turnover of about Rs 12 billion. NRDC has also set up 52 rural technology demonstration cum training centres in various parts of the country to give hands-on training and demonstrations to artisans and technicians living in rural areas (*NRDC, 2008*).

4.4 Technology Programmes for Specific Industry Sectors

4.4.1 Biotechnology Industry

Department of Biotechnology (DBT) has launched Small Business Innovation Research Initiative (SBIRI) to bring biotechnology industry at the forefront of the technological revolution and provide financial support to SMEs to help early phase of product development. Further, DBT has established Micro-propagation Technology Parks (MTP), biotech parks and biotech incubation centres for providing interface between research institutions and industry and promote biotech, agro and food industries. Biotech Consortium India Ltd. (BCIL) under the department is engaged in the promotion and development of biotech industry. DBT has also very recently launched the Biotechnology Industry Partnership Programme (BIPP), to give a fillip to innovation and R&D in Biotechnology by promoting government partnership with industry. BIPP would strictly promote high risk, transformational technology/process development and is directed at path breaking research in frontier futuristic technology areas having major economic potential and making Indian industry globally competitive and focussed on IP creation with ownership by Indian industry and collaborating scientists, as relevant. The emphasis of the programme would be to fulfil major unmet national technology needs in health, agriculture, energy and environment friendly/green manufacturing areas (*DBT, 2009*).

4.4.2 Information and Communication Technology (ICT) Industry

As a part of the reforms agenda, the Indian Government has taken major steps to promote ICT, with a focus on software development for export; telecommunications policy reform, privatisation of the national long-distance and mobile phone markets;

and development of a more comprehensive approach to ICT. The Department of Information Technology (DIT) has set up a Technology Development Council (TDC) to facilitate R&D in IT, promote free and open source software and to promote applications of IT for indigenous, efficient and cost effective solutions for product and process development in the industrial sectors including Bioinformatics. Under the scheme on innovation promotion in IT the department provides support to specified institutions to promote start up companies. The Software Technology Parks (STPs) of India (STPI) of the department has set up 35 STPs including 19 international gateways to boost software exports using high speed data communication links. DIT also has programmes on Standards, Testing and Quality Control (STQC), DOEACC etc. to provide infrastructure and support facilities for quality control and testing and HRD for ICT industry (*DIT 2009*).

4.4.3 Textiles

The Government adopted the National Textile Policy in 2000, which had the major thrust on technological upgradation, enhancement of productivity, quality consciousness, strengthening of the raw material base, product diversification, etc. The textiles industry gets support under several programmes *e.g.* Technology Upgradation Funds Scheme (TUFS), Technology Mission for Cotton (TMC), Jute Technology Mission (JTM), Powerloom Schemes, Jute Manufacturers Development Council (JMDC), Integrated Textiles Parks etc.

4.4.4 Renewable Energy Industry

The Integrated Energy Policy of 2008 recognizes the role of renewable energy sources to meet the increasing energy demand in the country. The Ministry of New and Renewable Energy has been implementing several schemes and programmes to support commercial ventures on solar photovoltaic, solar thermal, hydro-power, wind energy, geothermal energy, ocean energy, bio-energy etc., R&D support is provided for renewable energy industry by the Solar Energy Centre (SEC), Centre for Wind Energy Technology (C-WET), Indian Renewable Energy Development Agency (IREDA) and other R&D organisations. A package of incentives which includes fiscal concessions such as accelerated depreciation, customs duty concessions, excise duty exemption, and income tax exemption have been provided for projects on renewable energy. Renewable energy technologies, in the last two decades have matured enough to not only supplement conventional power, it also provides viable solutions in the off-grid mode to meet the electricity needs of remote villages in the country. The grid interactive and off-grid renewable power installed capacity as on 30.09.2008 was about 13,451 MW and 279 MW respectively.

4.4.5 Food Processing Industry

Food processing and agro industries have been accorded high priority with a number of important relieves and incentives on regulation and control, fiscal policy and taxation; export promotion etc. As a result of several policy initiatives undertaken since liberalisation in 1991, the industry has witnessed fast growth in most of the segments. As per a recent study on the food processing sector, the turnover of the total food market is approximately INRs.2500 billion (US $ 50 billion) out of which value-added food products comprise Rs. 800 billion (US $ 16 billion). Ministry of Food Processing industries has launched a mega Food Parks Scheme in order to provide infrastructure facilities for the food processing industries along with value chain from the farms to the market. Several other promotional measures are being taken by the Ministry for technological upgradation, quality improvement, standards and safety etc.

Besides the above, other Ministries such as, Ministry of Environment and Forests, Ministry of Agriculture, Department of Animal Husbandry etc. have their sector specific programmes on technology development and transfer relevant to SMEs. Development financing institutions such as the Small Industry Development Bank of India (SIDBI) and commercial banks have set up several funds to provide support to SMEs for modernisation and technology/quality upgradation. These initiatives include (*i*) Technology Development and Modernisation Fund (TDMF), (*ii*) National Venture Fund for Software and Information Technology (NFSIT), (*iii*) Credit Grant Fund Trust for Micro and Small Enterprises (CGTMSE), (*iv*) SME Growth Fund and (*v*) Microfinance Programme. SIDBI has also set up a Technology Bureau for Small Enterprises (TBSE) in collaboration with UN-APCTT to provide a platform for small enterprises to tap opportunities at the global level for acquisition of technology or establish business collaboration. Further, several Venture Capital Funds (VCF) have been set up to encourage knowledge based industries such as ICT, biotechnology, food technology, entertainment, education, new materials etc.

5. SUMMARY AND OBSERVATIONS

The Government of India has taken several policy measures and initiated a large number of schemes and programmes for improving the technological competitiveness of SMEs, details of which have been given in sections 4.2 to 4.5. The impact of some of the more important policies and activities is discussed below.

5.1 Evolution of Industrial Policy–Protectionism to Liberalization

India has achieved impressive success in industrial development during the last six decades after its independence as a result of changes in industrial policy from time to time that have been necessitated due to changing developmental needs. Several complementary initiatives were also taken for the creation of S&T infrastructure and human resources, promotion of R&D, and the development, transfer and adaptation of technologies for industry. By and large, Indian industrial policy (and also the S&T policy) has evolved over this period in the following four phases:

1. The Formative Phase (up to 1960)

Huge investments were made by the State in heavy industries in order to create a reasonable industrial base in the country. A Scientific Policy Resolution (SPR) was passed in 1958 and a large infrastructure on S&T education and research was created.

2. The Protectionist Phase (1960 to 1980)

Industrialisation was heavily state controlled and under stiff licensing and regulatory requirements which resulted in technological obsolescence and non-competitiveness in various sectors. The development and application of appropriate technology and utilisation of indigenous technology for industries were emphasized.

3. Transition to Liberalization (1980 to 1991)

Transfer of technology was given due importance and a number of measures were proposed for technological modernization, and for the simplification of procedures for industrial approvals, de-licensing, promoting foreign collaboration, simplifying technology imports etc. Also, the Technology Policy Statement (TPS) was adopted in 1983. In order to give greater thrust on new and emerging technologies, separate administrative structures were created in the Government in various sectors of S&T. A large number of schemes and programmes were also started for the promotion, development and transfer of industrial technologies during the Sixth (1980-85) and Seventh (1985-90) Five Year Plans.

4. The Post Liberalization Phase (1990 Onwards)

A new Industrial Policy Statement (IPS) was announced in 1991, under which the distortions of the pre-liberalization period were corrected by dispensing with industrial licensing and facilitating foreign investment and technology imports, and at the same time taking steps to strengthen indigenous technological capability. India ratified the agreement establishing the WTO, which came into force from 1st January 1995, and as per the commitments under the TRIPS Agreement, the Patent Act was relaxed to allow product patents, which had positive impact on sectors like pharmaceuticals industry. Further, a Science and Technology Policy was adopted in 2003 with the objectives of encouraging research and innovation (*GOI, 2003*).

Since the very beginning, a common feature of the successive industrial policy decisions of the government has been to encourage, promote and support small scale industry. The IPS 1991 was accompanied by a separate policy statement for the promotion and strengthening of small, tiny and rural industries.

The Industrial Policy of India in the post-1990 period gave a rapid thrust to various sectors of the economy, and Indian industry registered an impressive growth during last two decades when the.number of industries including SMEs increased manifold (ref. section – 3.2). With the liberalization and globalization of the Indian economy, the SMEs in India have unprecedented opportunities on the one hand while facing serious challenges on the other. With the entry of multinationals, immense opportunities have been created for the SMEs for outsourcing, sub-contracting and ancillarization of products manufactured by large companies, particularly in non-core sectors like automobiles, engineering, consumer electronics, IT enabled services (ITeS), business process outsourcing (BPO) etc.. But at the same time, SMEs face many challenges due to their low scale of operation, technological backwardness, lack of access to institutional credit and market competition.

5.2 Programmes of the Ministry of MSME

The Micro and Small Enterprises – Cluster Development Programme (MSE-CDP) (ref. section-4.2.1), which was upgraded in 2006 from the erstwhile Integrated Technology Upgradation and Management Programme (UPTECH) of 1998, and then the Small Industry Cluster Development Programme (SICDP) of 2003 with revised guidelines and increased financial provisions, is the flagship programme of the Government of India for holistic development of MSEs in clusters through cost-effective deployment of resources, introduction of technology, collective bargaining, common marketing, common brand building, common procurement etc. The cluster approach has been a key strategy for enhancing productivity, competitiveness and capacity building of MSEs in the country. This is a broad-based programme for the promotion of marketing, exports, skill development and technological upgradation. Since its inception, the programme has been assisting MSE clusters through diagnostic studies, counselling, training, facilitating credit, market research and development, and information dissemination. Further, under the revised guidelines of 2006, provision was made to assist MSEs for the improvement of technology, quality control, standardization, and testing with additional financial support for creating common facilities, for which the MSEs are not able to arrange the required large investment on their own. Encouraging results have been achieved through cluster development in terms of employment generation, increased turnover, improvement in product quality, minimizing wastages, exploring new markets, awareness about new technologies, better marketing techniques, etc. The facilities extended under the programme have proved to be a boon to the MSE sector and thus MSE-CDP has provided the only 'safety net' for the MSE sector in the country (*Ittyerah A. C., 2009*).

However, despite the large number of clusters in India, the majority of them have not been able to exploit their full potential. It has been observed that while vast scientific infrastructure, manpower and technologies are available in the country that are suitable to various sectors of MSEs, the industry clusters often suffer from technological inadequacies for want of relevant scientific information, guidance and support.

Another noteworthy initiative is the ten component programme of the National Manufacturing Competitive Council (NMCC) (ref. section-4.2) to support the MSMEs in their endeavour to become competitive. This aims to benefit over 10,000 units in more than 500 clusters with infusion of technology. However, appropriate mechanism and guidelines for implementation of the programme have not been put in place, and physical targets not clearly spelt out, as a result of which, monitoring and evaluation of the programme have been difficult.

5.3 Science and Technology Programmes

Among various schemes and programmes of the S&T Ministries and departments (ref. section-4.3.1), the activities of the National Science and Technology Entrepreneurship Development Board (NSTEDB) have shown good results in creating an entrepreneurial culture in and around academic institutions and establishing knowledge driven and technology intensive small enterprises. The important activities being implemented under NSTEDB include STEPs, TBIs, EDCs in educational institutions and STED projects at the district level. The STEPs offer a no. of facilities and services that are crucial for the promotion and sustenance of new and small enterprises. The STEPs are successfully working with significant results in terms of no. of start-up ventures created, new technologies developed and employment generated. The initial results of STEPs and TBIs are encouraging though the gestation period is quite high. The leadership role in enterprise building in these initiatives is very prominent that ensure the success of an enterprise in the commercialization of technologies involving risks (*Nagayya 2006*). An evaluation study conducted by National Council of Applied Economic Research, New Delhi (*NCAER 1998*) shows that NSTEDB programmes have resulted in a social yield of Internal Rate of Return (IRR) of around 30 per cent, and in view of the high yield on investments, the study highlighted that the programmes need to be expanded significantly.

The performance of the Technology Development Board (TDB) (ref. section-4.3.2) has demonstrated concerted efforts and proper management techniques, and the success is well acknowledged by all the stakeholders, *i.e.* the government, the industry and the R&D institutions. The TDB receives fund from the reserve collected through 5 per cent cess levied on all payments made for technology imports according to the R&D Cess Act, 1986, which ensures a continuous flow of funds for its operation. Further, being located in DST, the TDB always has access to the technical expertise required for project evaluation, carry out market study and monitor progress. Notable successful ventures supported by TDB are vaccines (M/S Shantha Biotech, Hyderabad), healthcare products (M/S Bharath Biotech, Hyderabad), battery operated car (Reva, Bangalore) and Indica car (M/S Tata Motors, Mumbai).

Another important initiative is the NMITLI scheme of CSIR (ref. section-4.3.5), which has emerged as a unique step of the government carving out a niche in the innovation space. The scheme enjoys a good brand image and is looked upon as a role model as India's largest public-private partnership (PPP) R&D programme. The industrial partners have highly benefited not only by achieving their project objectives but also through high quality technological inputs from their R&D partners.

Due to the first changing global business scenario after the economic liberalisation, the Indian IT industry has become the fastest growing sector in the country. As a result of several proactive policy measures of the.Department of Information Technology (DIT), including fiscal incentives, enhanced credit flow, providing infrastructure and incubation facilities, SMEs have occupied a major space in the IT industry primarily through tremendous opportunities in ITeS and BPO, in the technology sector, products and components, and in e-Governance, education, transportation, etc. However, like other sectors, the IT-SMEs will also have to keep updated and enhance their technological capabilities in order to increase and sustain their business opportunities and develop global competitiveness.

The Indian Government has taken several initiatives in the field of IPR, including amendments in patent laws, modernisation of patent offices, and awareness programmes on IPR and related issues. But the SMEs in India have not come forward to adopt IPR as a business strategy and a means for enhancing competitiveness in the global market, primarily due to the lack of awareness about the role of IPR. The Indian SME sector needs more information, orientation and facilities for protecting their intellectual powers.

In spite of the favourable policy measures and availability of vast institutional set up, and S&T network and resources, the SMEs in India are facing several problems.

Apart from financial and managerial constraints, the SMEs suffer from the shortcomings of technological backwardness. There is a lack of proper coordination between the S&T related Ministries and agencies on the one hand, and the user Ministries and the industry on the other, There is no mechanism for providing forward and backward linkages through which the experts in the respective fields may address specific technological problems of SMEs and suggest solutions to the same Various S&T institutions and agencies seem to be working in isolation, thereby leading to multiplicity and overlapping of efforts, and dissipation, and poor and inefficient utilization of funds.

There is not much awareness among SME entrepreneurs about the benefits of modernisation and innovation, and even if there is any, there is a lack of dissemination of information about the sources of technology and guidance for transfer of technology. The academic and research institutions have not come out in a big way to solve the problems of the SMEs. In the absence of the active involvement of the R&D community, the modernisation and technological upgradation initiatives of the governments have suffered. This has largely resulted in poor utilization of S&T expertise and technologies that are available in the country for commercial purposes.

6. CONCLUSIONS

Innovation for new product development and the improvement of productivity and quality of existing products are essential to the future of small and medium enterprises in their need to keep pace with the changing market economies and preferences. Technology transfer from academic and research institutions and efficient product engineering are necessary to achieve this goal. But a small scale entrepreneur has no means to equip his unit with adequate facilities on R&D, product development, testing and calibration etc. which are required to solve his problems. As such SMEs should be helped by providing these facilities, and by instituting schemes and programmes for transfer, absorption and adaptation of technologies and training of personnel. SMEs should be encouraged by providing fiscal incentives and awards, and venture capital funding for introducing new technologies. SMEs need to be supported for establishing technological cooperation with larger enterprises, R&D and academic institutions and other centres of technology development.

In order to enable SMEs tide over the problems of technological backwardness and enhance their access to new technologies, it is imperative to offer them a conducive environment, in which, knowledge plays a predominant role. There is a need to understand and assess the real needs of the SMEs and accordingly devise mechanisms to ensure their sustainable growth. The need today is also to exploit modern technologies and utilize S&T expertise and human resources through increased communication, cooperation and linkages, both within an enterprise, and across enterprises, and also with technology producing organizations.

Nevertheless, by its less capital intensive and high labour absorption nature, the SME sector in India has made significant contribution to employment generation and rural industrialisation. The policies and programmes of the Government in the last decade or so have enabled more and more small enterprises to improve on their products and processes through modernization and adoption of technologies and diversify into new and emerging technology areas. This sector is ideally suited to build on the strengths of traditional skills and knowledge, by infusion of technologies, capital and innovative marketing practices.

Considering the initiatives taken for promoting SMEs towards socio-economic development and sustainability, India has achieved considerable success as compared to other developing countries. For example, policy measures to support the SME sector in their endeavour to become competitive; stable credit policy to ensure capital flow to the SME sector by the commercial banks, development financial institutions and venture capital funds; cluster development programme; setting up technology information centres/data banks; IT portals for information dissemination; fiscal incentives for promoting industrial R&D and technology transfer; STEPs and TBIs etc., could be taken as lessons of learning for other countries in the South.

In view of the commonalities in the opportunities and challenges before the SME sector, it would be of mutual benefit for other developing countries to collaborate with India on improving technological competitiveness of SMEs through various programmes. The areas of such South-South cooperation may include the following:

1. Consultancy on technology, and policy and institutional framework;
2. Cluster development studies, sickness studies, surveys;
3. Product and process development and prototyping services, setting up of testing, calibration, quality control facilities, CAD/CAM design, machine tools etc.;
4. Human resource development–technology-based entrepreneurship development, training in capacity building of entrepreneurs and technical manpower, skill upgradation training in identified areas, trainers' training programmes;
5. Institution-industry linkage – feasibility studies and preparing project proposals for setting up of technology business incubators, innovation centres, and S&T Parks.;
6. Advice and guidance on development, transfer and adaptation of technology for the promotion of NTBFs/spin-off firms;
7. Sharing of information and expertise – exchange visits of experts, organization of seminars and workshops, dissemination of information on development and transfer of technologies suitable to SMEs.

ACKNOWLEDGEMENT

The author extends his grateful thanks to Professor Arun P. Kulshreshtha, Director, NAM S&T Centre for his encouragement and guidance for preparing this article.

REFERENCES

Bandyopadhyay, M., 2008, 'Collaborative links between academic and research institutions and industry for stimulating technological innovation and economic development: need for science and technology policy initiatives and networking amongst NAM and other developing countries': in "Science and Technology policy and indicators for development–Perspectives from developing countries"; Ed. Seetha I. Wickremasinghe and V. K. Gupta; Daya Publishing House, Delhi, 2008 pp 254-72.

CSIR, Council of Scientific and Industrial Research (CSIR), Government of India, New Delhi (http://www.csir.res.in) – Accessed on 12 December 2009.

DBT, 2009, 'Annual Report 2008-09', Department of Biotechnology, Ministry of Science and Technology, Government of India, New Delhi (http://www.dbtindia.nicin)

DCMSME, 'Micro, Small and Medium Enterprises in India–An Overview', Development Commissioner, Micro, Small and Medium Enterprises, Government of India (http://www.dcmsme.gov.in–Accessed on 12 December 2009.

DIT, 2009, 'Annual Report (2008-09)', Department of Information Technology, Government of India, (http://www.mit.gov.in)

DSIR, 2008, 'Annual Report 2007-08', Department of Scientific and Industrial Research, Government of India (http://www.dsir.nic.in).

GOI, 1991, 'Policy Measures for Promoting and Strengthening Small, Tiny and Village Industries', Department of Small, Agro and Rural Industries, Ministry of Industry, Government of India, 1991.

GOI, 2003, 'Science and Technology Policy', Department of Science and Technology, Government of India, 2003.

GOI, 2006, 'The Micro, Small and Medium Enterprises Development (MSMED) Act, 2006', Ministry of Micro, Small and Medium Enterprises, Government of India, 2006.

GOI, 2006a, 'Report of the Steering Committee on Science and Technology for the Eleventh Five Year Plan (2007-12)', Office of the Principal Scientific Adviser to the Government of India, New Delhi, December 2006.

Ittyerah A. C., 2009, 'Evaluation study of micro and small enterprises cluster development Programme', Indian Institute of Public Administration, New Delhi, prepared for the Ministry of Micro, Small and Medium Enterprises (MSME).

MMSME, 2009, 'Annual Report, 2008-09', Ministry of Micro, Small and Medium Enterprises, Government of India.(http://www.msme.gov.in)

Nagayya, D., 2006, 'Perspectives of entrepreneurship development, and innovation and business incubators' in "Banking Sector and Human Resources (Changing Scenario)" Ed. T. Sreenivas, Discovery Publishing House, New Delhi.

NCAER, 1998, 'Evaluation of NSTEDB Programmes and its impact on Entrepreneurship Development, Employment Generation and Technology Transfer', Study sponsored by Department of Science and Technology, Government of India, New Delhi

NRDC, 2008, Annual Report 2007-08', National Research Development Corporation, New Delhi (http://www.nrdcindia.com)

NSTEDB, National S&T Entrepreneurship Development Board, Government of India, (http://www.nstedb.com) Accessed on 12 December 2009.

OECD, 1999, 'Managing Innovation Systems'; DSTI/STP/TIP (99)1

TDB, Technology Development Board, Government of India, New Delhi (*www.tdb.gov.in*) Accessed on 12 December 2009.

TIFAC, Technology Information Forecasting and Assessment Council, Government of India, New Delhi (http://www.tifac.org.in) Accessed on 12 December 2009.

ABBREVIATIONS

BCIL:	Biotech Consortium India Ltd.
BIPP:	Biotechnology Industry Partnership Programme
BPO:	Business Process Outsourcing
CGTMSE:	Credit Grant Fund Trust for Micro and Small Enterprises
CLCSS:	Credit Linked Capital Subsidy Scheme
CSIR:	Council of Scientific and Industrial Research
C-WET:	Centre for Wind Energy Technology
DBT:	Department of Biotechnology
DIT:	Department of Information Technology
DSIR:	Department of Scientific and Industrial Research
DST:	Department of Science and Technology
EDC:	Entrepreneurship Development Cell
EDP:	Entrepreneurship Development Programme
FTS:	Field Testing Station
GOI:	Government of India
HGT:	Home Grown Technology
IIP:	Index of Industrial Production
IIT:	Indian Institute of Technology
IPR:	Intellectual Property Right
IPS:	Industrial Policy Statement
IREDA:	Indian Renewable Energy Development Agency
ITeS:	Information Technology enabled Services
KVIC:	Khadi and Village Industries Commission
MSE-CDP:	Micro and Small Enterprise Cluster Development Programme
NFSIT:	National Venture Fund for Software and Information Technology
NMCC/NMCP:	National Manufacturing Competitive Council/ Programme
NMITLI:	New Millennium Indian Technology Leadership Initiative
NSIC:	National Small Industries Corporation
NSTEDB:	National S&T Entrepreneurship Development Board

NTBF: New Technology Based small Firms
PPDC: Product and Process Development Centre
PRODIP: Product Development, Design Intervention and Packaging
RTC: Regional Testing Centre
SBIRI: Small Business Innovation Research Initiative
SEC: Solar Energy Centre
SISI: Small Industry Service Institute
MSE: Micro and Small Enterprise
MSME: Micro, Small and Medium Enterprise
SICDP: Small Industry Cluster Development Programme
SFURTI: Scheme of Fund for Regeneration of Traditional Industries
SIDBI: Small Industry Development Bank of India
SME: Small and Medium Enterprise
SPR: Scientific Policy Resolution
SSI: Small Scale Industry
STED: S&T Entrepreneurship Development
STEP: S&T Entrepreneurs Park
STP: Software Technology Park
STQC: Standards, Testing and Quality Control
TBI: Technology Business Incubator
TBSE: Technology Bureau for Small Enterprises
TDB: Technology Development Board
TDDP: Technology Development and Demonstration Programme
TDMF: Technology Development and Modernisation Fund
TePP: Technopreneur Promotion Programme
TIFAC: Technology Information, Forecasting and Assessment Council
TMC: Technology Mission for Cotton
TMP: Technology Management Programme
TPS: Technology Policy Statement
TRC: Technology Resource Centre
TUFS: Technology Upgradation Fund Scheme

Chapter 12

Technology Innovation in Indian Small and Medium Enterprises (SMEs): The Role of Clusters and R&D Alliances

Rama Swami Bansal[1] and V.P. Kharbanda[2]

[1] *Scientist, Council of Scientific and Industrial Research (CSIR), Rafi Marg, New Delhi – 110 001, India*
E-mail: rsb@csir.res.in; ramabansal@gmail.com
[2] *Publications Adviser, NAM S&T Centre, New Delhi, India*
Formerly, Senior Scientist, National Institute of Science Technology and Development Studies, CSIR, New Delhi – 110 012, India
E-mail: kharbandavpk@yahoo.com

ABSTRACT

The present day knowledge economy demands knowledge intensive enterprises which only can survive in the ongoing process of globalization and increased international competition. Constant up-gradation and transfer of technologies with high knowledge contents has become crucial for the small enterprises. Knowledge as a factor for competitive advantage has replaced traditional factors such as labour and capital. As knowledge resides only in the human mind, it can be harnessed only by focusing on increasing human capabilities through the process of increased Communication, Cooperation and Competition, both within the enterprise as well as across enterprises and knowledge producing organizations. This paper dwells upon a few case studies, how enterprises in India are facing this challenge, and particularly, how small scale enterprises (SMEs) are moving towards clusters and association with R&D institutions for sustaining international competition. To learn lessons from the other countries, a few case studies from advanced countries like Germany, UK and Italy have been cited.

Keywords: *Small and medium enterprises, Clusters, Alliances, Innovation, Technological capacity, Knowledge acquisition, Human cantered systems, Technology transfer.*

1. INTRODUCTION

In most developing countries Small and Medium Enterprises (SMEs) constitute the bulk of the industrial base and contribute very significantly to their exports as well as to their GDP or GNP. For instance, India has more than three million SMEs (Table 12.1 and Figures 12.1–12.3) which account for almost 50 per cent of industrial output and 34 per cent of India's total exports. It is the most important employment generating sector and is an effective tool for promotion of balanced regional development. These account for 50 per cent of private sector employment and 30-40 per cent of value-addition in manufacturing (Thyagarajan, 1998). It produces a diverse range of products (about 8000) including consumer items, capital and intermediate goods.

Table 12.1: Status of SMEs in India

Year	*Number (million)*	*Employment (million)*	*Exports (Rs. billion)*	*Per cent of Total Exports*
1992	2.1	13.0	139	31.5
1993	2.2	13.4	178	33
1994	2.4	14.0	253	36
1995	2.6	14.7	241	35
1996	2.7	15.2	365	34
1997	2.8	16.0	312	33
1998	2.9	16.7	444	35
1999	3.1	17.2	490	35
2000	3.2	17.2	542	34
2001	3.3	18.6	698	34
2002	3.4	19.2	712	34
2003	3.6	20.0	860	34

Source: Bhagaban Das *et al.*: Strengthening SMEs to make Export Competitive. MPRA paper No. 7800, March 2008. http://mpra.ub.uni muenchen.de/7800) Notes Data has been rounded to the nearest decimal.

However, the SMEs in India, which constitute more than 80 per cent of the total number of industrial enterprises and form the backbone of industrial development, are as yet, in technological backwaters vis-a-vis advances in science and technology. These suffer from problems of sub-optimal scales of operations and technological obsolescence. While most of the large companies, even in developing countries, have financial as well as technical capacity to identify technological sources and evaluate alternate technologies that would suit their requirements, unfortunately, this capacity is conspicuously missing in most SMEs. It is these features of SMEs that make them an ideal target for technological upgradation through technological cooperation with foreign and local enterprises, with R&D institutions and centres of technology development.

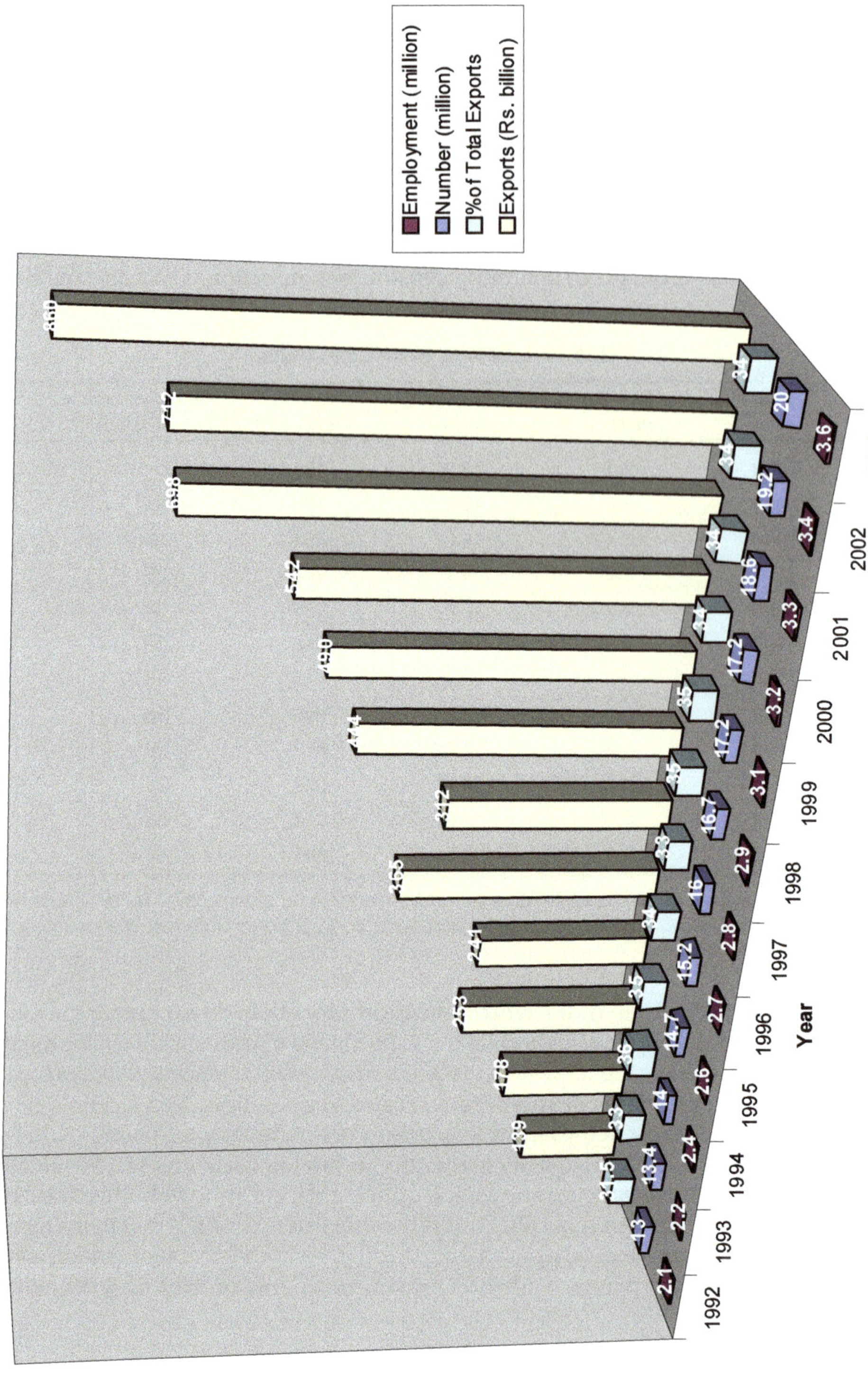

Figure 12.1: Graphical Representation of the Status of SMEs in India

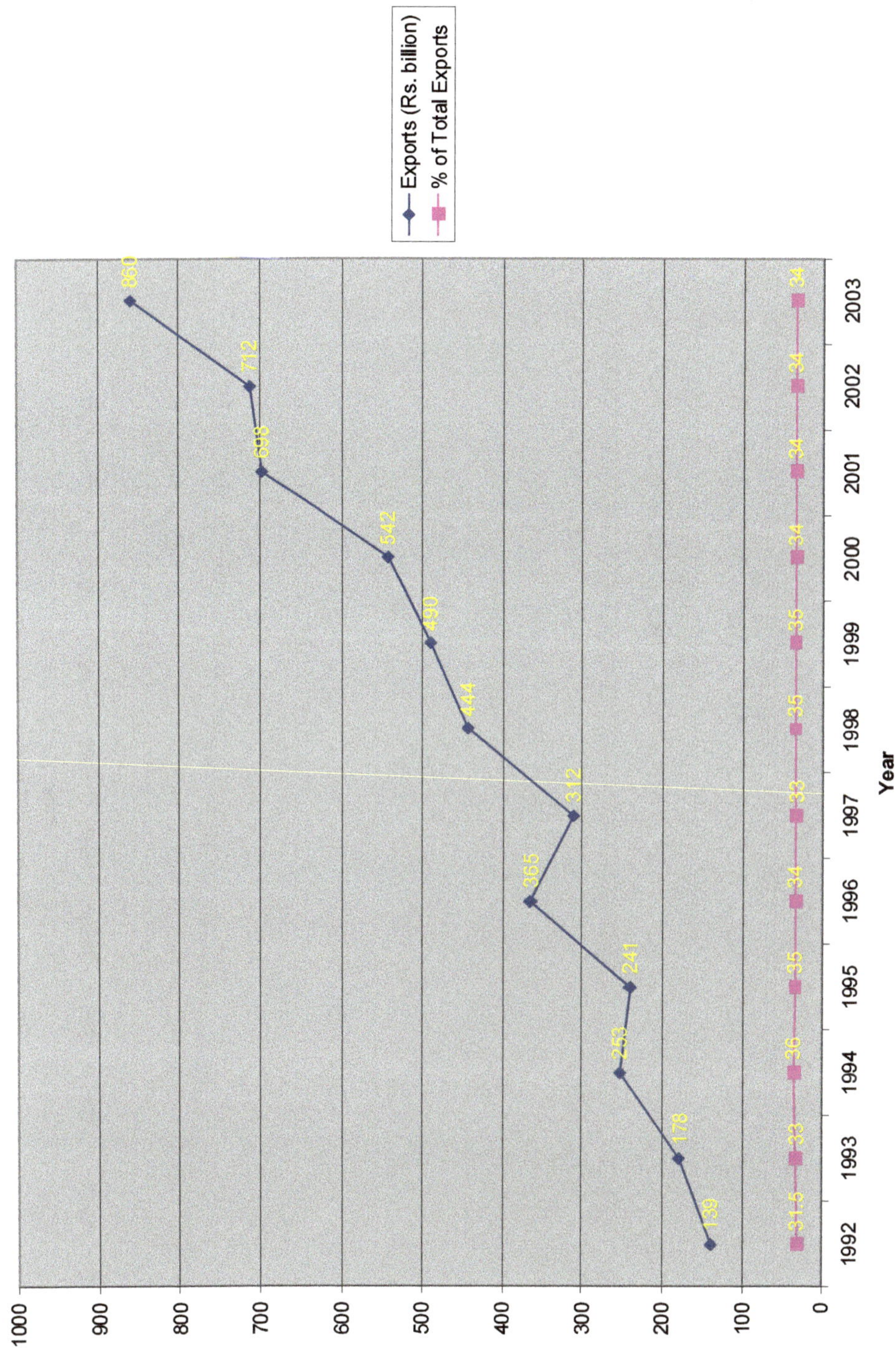

Figure 12.2: Exports of Indian SMEs and Percentage of Total Exports

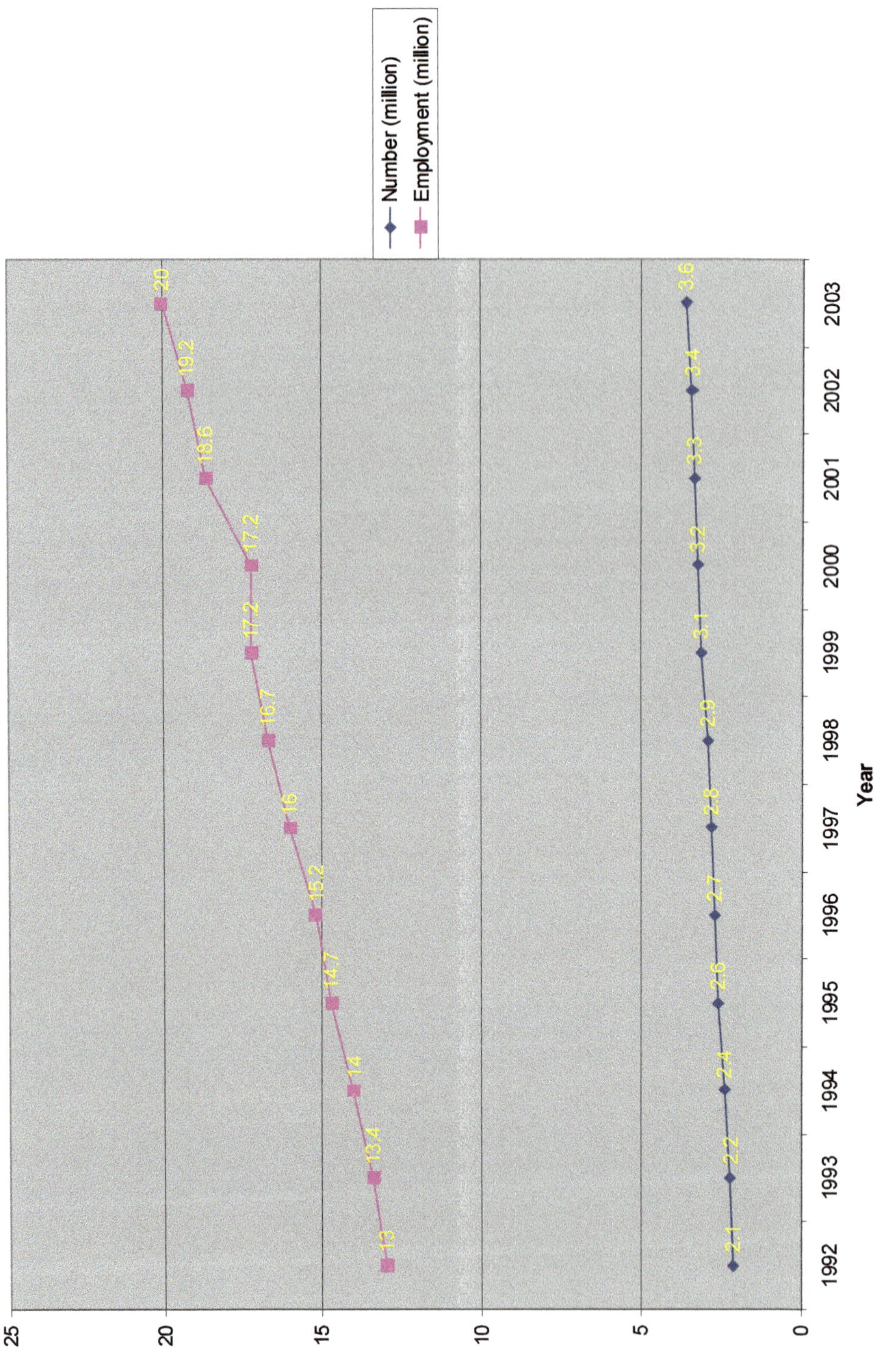

Figure 12.3: Number Of Indian SMEs and Employment therein

As the countries integrate into the global village, these SMEs will have to respond accordingly and thus deserve special attention. To enable SMEs to mitigate problems of technological backwardness and enhance their access to new technologies, it is imperative to give them a conducive environment, which in the present context of globalization calls for a human cantered approach with tacit knowledge playing a predominant role. As knowledge resides only in the human mind, it can be harnessed only by focusing on increasing human capabilities through the process of increased communication, cooperation and linkages, both within the enterprise as well as across enterprises and knowledge producing organizations. How are SMEs in India facing this challenge? How are these moving towards industrial clusters for sustaining international competition? What policy measures are required? These are some of the questions, which are dealt with in this paper taking a few examples from the Indian scenario.

2. NEED OF SMEs FOR ACCESS TO NEW TECHNOLOGY

Small enterprises in India, with their dynamism, flexibility and innovative drive are increasingly focusing on improved production methods, penetrative marketing strategies and modern scientific management capabilities to sustain and strengthen their operations. They are poised for global partnership and have the potential to absorb latest technologies in diverse industrial fields.

Small in India is more than beautiful–it is efficient, adaptable and adds value to economic and social spheres. As the country integrates into the global village, the small and medium sector will have to respond accordingly. It deserves special attention as it plays a pivotal role in the country's socio-economic development.

The problems faced by the SMEs particularly in accessing technology and maintaining competitiveness have been formidable. The main reasons for the inability of SMEs to identify their technology needs are:

- Poor financial situations and low levels of R&D;
- Poor adaptability to changing trade trends;
- Desire to avoid risk;
- Non-availability of technically trained human resources;
- Emphasis on production and not on production costs;
- Lack of management skills;
- Lack of access to technological information and consultancy services;
- Isolation from technology hubs.

A recent study carried out by Rama Bansal on 'R&D Agency – Industry Partnership for Technology Development and Transfer in Indian Context' confirms this and highlights major weaknesses in the effort to upgrade technological capabilities of Indian SMEs, although there have been some successes too, which can not be negated. But these successes are not sufficient for them to compete at the global level and much needs to be done. SMEs do approach the R&D institutions directly or through intermediary agencies for technological solutions through consultancies and R&D partnerships. Some of the main reasons for failure of such partnerships include:

- ✰ Lack of experience of the R&D agency in identifying and developing technologies into products for dynamic commercial markets;
- ✰ Lack of foresightedness on availability of raw material in sufficient quantity and market;
- ✰ Slow speed of administrative actions in the R&D agency/industry;
- ✰ Transfer of unripe/unproven technology;
- ✰ Lack of mutual understanding;
- ✰ Improper written agreement;
- ✰ Lack of information and training
- ✰ Distrust of an enterprise on the technology transfer programmes of the intermediary agencies and unwillingness to share the benefit to be accrued from transfer of technology with them.

All this shows that much more needs to be done apart from simple alliances between SMEs and R&D institutions. This in fact needs a symbiotic approach where there is constant effort for exchange and upgradation of knowledge not only for technology but also for human and material resources and market conditions *i.e.* for commercialization.

To enable SMEs to mitigate the above problems and enhance their access to new technologies with a view to increasing their competitiveness in the international market, it is imperative to give them a conducive environment which includes:

1. Formulation of appropriate national policies and programmes,
2. Building up indigenous technological capacity and R&D alliances,
3. Knowledge flows and technology databases,
4. R&D and inter-firm linkages.

2.1 Formulation of Policies and Programmes

Conducive policy environment is a pre-requisite. Major policy reforms aimed at substantially deregulating industrial sector and liberalizing foreign investment as well as technology imports, have been the most significant development since 1991. The post-liberalization era in the Indian economy has enhanced opportunities and challenges for small industry sector. The ultimate objective of promotional policies is to enhance SMEs capacity to grow so as to become viable units. The national policies and programmes may be oriented to:

- ✰ Setting up Industrial District/Technology Parks/Clusters to promote sourcing of new technology, innovation and effective transfer;
- ✰ Organizing local level information services, data banks and seminars in collaboration with professional bodies;
- ✰ Sending experts to SMEs to assist them with the introduction of new technologies;

- To establish training centres for human resource development for SMEs;
- Establishing business centres;
- Promote strategic alliances with R&D institutions, universities and other enterprises at national, regional and international level.

This calls for promoting endogenous industrialization with emphasis on self-reliance through use and creation of local resources and at the same time adoption, absorption and diffusion of the imported technologies. This strategy involves an expanded role of SMEs for global competition in exports which at the same time can alleviate high levels of unemployment and income inequality.

The National Small Industries Corporation (NSIC) has pioneered several schemes for the growth and development of small scale sector. In the initial stages of development of small scale industries after independence, it was the innovative and novel schemes of the NSIC such as Government Purchase, Hire Purchase, Development of Prototypes, Technical Training etc., which led to the establishment of new enterprises, development of appropriate manufacturing technologies and creation of a strong first generation entrepreneurial base. These schemes of the Corporation acted as a catalyst for this sector. It has helped the small units in identification, transfer, adoption and absorption of technology, entrepreneurial and technical training, establishing common production facilities and marketing and extended financial support to these units. With the successful implementation of NSIC schemes, a number of financial institutions and banks have come forward to assist this sector thus creating an important constituent of the Indian economy. Further, the Council of Scientific and Industrial Research (CSIR) is also an important technology generator in the country and has a lot of S&T inputs to offer to small scale industrial sector. To further invigorate this sector, the NSIC, CSIR and Asia and Pacific Centre for Transfer of Technology (APCTT) have pooled their resources and have formed a strategic alliance to help upgrade technology in the small scale industrial sector. APCTT has a comprehensive Data Bank of information on appropriate technologies in diverse industrial sectors with access in the ESCAP region and elsewhere. It has been assisting technology buyers and technology sellers from different countries to come together. The alliance of these three organizations assists the SSI units with emphasis on the following aspects–improve and secure access of small scale industries to advanced technologies, strengthen linkages between R&D institutions and create and strengthen the local innovative system. To begin with eight technology acquisition and modernization workshops (two days each) in selected industrial centres (New Delhi, Bangaluru, Chennai, Mumbai, Ahmedabad, Ludhiana, Indore, Kolkata) were organized as far back in 1994 to deliberate on specific technologies in identified sectors. These workshops provided an opportunity to identify specific technological needs and capabilities of small scale units and the nature of technical assistance sought from promotional and financial agencies. Main initiatives of the Indian Government for promoting technological partnerships are listed in the Table 12.2.

Table 12.2: Major Government Initiatives Promoting R&D Agency – Industry Partnership

Sl.No.	*Name of Agency/ Programme*	*Name of Programme/ Purpose*	*Launch Year*	*Accomplishments*
1.	NRDC	Transfer of Technologies	1953	Forged strong links with Indian/foreign R&D agencies; has large repository of wide range of technologies in almost all industrial R&D sector; has successfully exported technologies to Brazil, Burma, Bangladesh, Germany, Indonesia, Kenya, Madagascar, Malaysia, Nepal, Philippine, Senegal, Sri Lanka, Vietnam and USA
2.	NSTDEB of DST	S&T Entrepreneurship Programme (STEP)	1984-85	17 STEP's established near educational and research infrastructure to facilitate continued closer ties between R&D agencies and industry
3.	NSTEDB of DST	Entrepreneurship Development Cell (EDC)	1986	EDC set up in 55 academic institutions to provide information to budding technopreneurs; creating entrepreneurial culture, fostering better parent institute – industries–R&D agencies linkages.
4.	DOS	Technology Transfer and Industrial Consultancy Group	~1990	~ 268 technologies of ISRO have been transferred to industries for commercialisation and more than 265 consultancy assignments have been undertaken by ISRO for small, medium and large-scale industries
5.	DBT	Biotech Consortium India Ltd. (BCIL)	1990	Provides linkages amongst res. institutions, industry, government and funding institutions to facilitate accelerated commercialisation of biotechnology. ~9 technologies transferred to industry and ~ 300 trainees trained for industry during last 5 years.
6.	TIFAC	Home Grown Technology (HGT)	1993	>50 projects supported resulting in development of 30 technologies (10 commercialised)
7.	DSIR	Programme Aimed at Technological Self-Reliance (PATSER) (now called Tech. Development and Demonstration Prog.)	1992	165 projects supported. 65 projects completed resulting in commercialisation of 30 technologies/prototypes; strengthening linkages of industries with over 30 research institutes.

Contd...

Table 12.2~Contd...

Sl.No.	Name of Agency/ Programme	Name of Programme/ Purpose	Launch Year	Accomplishments
8.	DST	Drugs and Pharma Research Programme (DPRD)	1995	70 projects supported resulting in development of 6 products and filing of 13 process patents
9.	TDB	Converting the fruits of indigenous research into commercial products or services	1996	131 projects of 107 industries resulting in development of many industries, rise of new industry
10.	DBT	Micropropagation Technology Parks (MTP)	1997	2 MTPs at TERI and NCL; State-of-the-art tissue culture production facility with an annual production capacity of 2 million plants at TERI
11.	CSIR	New Millennium Indian Technology Leadership Initiative (NMITLI)	2001	33 projects supported resulting in 3 major technologies developed, 3 commercialised, many minor under development and networking of 167 R&D institutes with 55 industries
12.	NSTEDB of DST	Technology Business Incubator (TBI) Scheme	2001	16 TBIs established to provide hand holding, mentoring, specialised support services and networking during start-up phase of an enterprise
13.	DST	Pharmaceuticals R&D Support Fund (PRDSF)	2005	5 projects being supported

Source: BANSAL, Rama Swami (2005): 'R&D Agency – Industry Partnership for Technology Development and Transfer in Indian Context', Ph.D. Thesis, Birla Institute of Technology and Science, Pilani, Rajasthan.

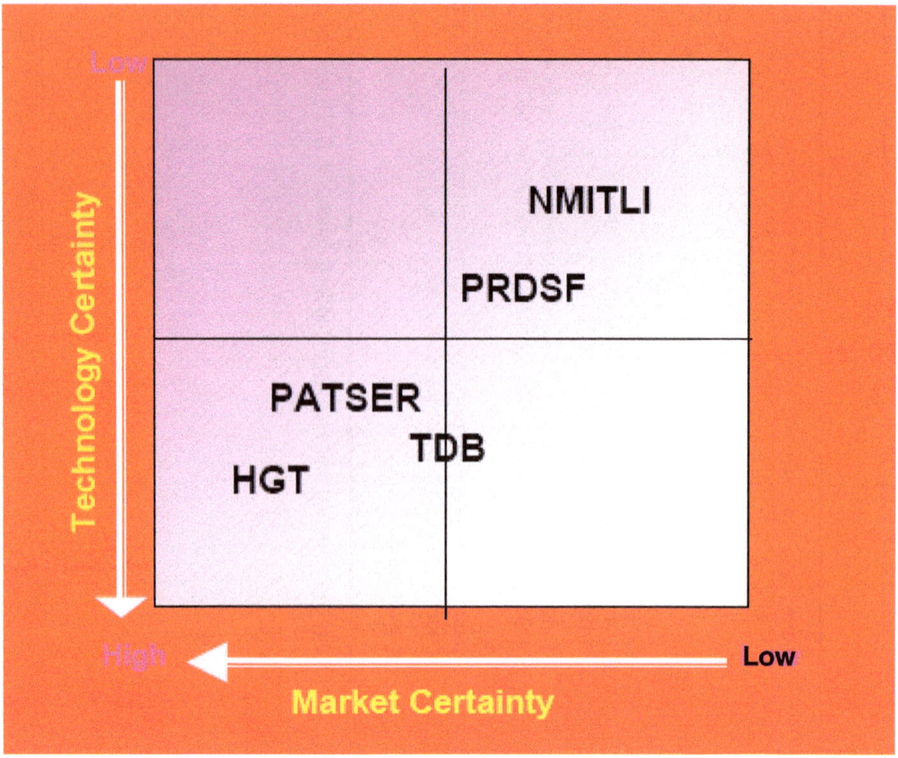

Figure 12.4: Level of Market Certainty and Technology Certainty Supported Under GOI Programmes Supporting Partnerships

***Source*: BANSAL, Rama Swami (2005): 'R&D Agency – Industry Partnership for Technology Development and Transfer in Indian Context', Ph.D. Thesis, Birla Institute of Technology and Science, Pilani, Rajasthan.**

2.2 Building up Indigenous Technological Capacity and R&D Alliances

Access to advanced technologies and assistance in their adoption is crucial to build up Indigenous Technological Capacity (ITC) to face international competition. In the present scenario of globalization, knowledge of and access to latest advances holds the key to international competition. In this venture the industrial sectors which are able to identify their technology needs and adopt the required technologies in time, will benefit the most. In India, the traditional sector which has shown this remarkable ability to adapt to technological change is Leather. While a large number of CSIR laboratories all over the country help meet the technological needs of the SMEs, Central Leather Research Institute (CLRI), Chennai, India, has successfully introduced microprocessor control in tannery wet operations at industry level and promoted cleaner processing, quality consistency and international equivalence. Anticipating the impact potential of computer-aided technique in footwear and garment design, CLRI has trained the SMEs in a way that CAD is today widely employed by export manufacturers of footwear and garments. As at present CLRI is associated to the SMEs so strongly that it is both an advisor to the policy making bodies and a technology partner to the user industry. For example, through synergy, the artisans engaged in the fabrication of Kholapuri chapels have been technologically empowered by CLRI.

In another example, under NISTADS[1] Bankura Project, a software package called as 'MADHU' (Modernization of Artistic Design for Handloom Unit) was developed in collaboration with Indian Institute of Technology, Kharagpur. This CAD software package facilitates computerized digitization of the picture/design produced by the artist/designer. This package has been successfully demonstrated to designers and weavers of Bankura. As a result a large number of craftsmen have adopted blending of this new technology to increase efficiency, quality, flexibility and cost effectiveness.

In another example, Mechanical Engineering Research and Development Organization (MERADO) Ludhiana, was established in 1965 to nurture and assist the growing industrial clusters in Ludhiana in the field of knitwear, agro industrial machine tools and bicycle industry. Since then it has contributed significantly towards design, development and standardization of industrial machinery, agriculture machinery, machine tools, special purpose machines, consumer durables etc. Some of the notable examples are Design and Developments of 10 TPD Modern Oil Expeller, Sugarcane Harvester, Electric cloth cutting machine, High speed over-lock machine, Leather shaving machine, Flat bed and Post bed Leather sewing machines, Button hole attachment for domestic sewing machine, wool knitting machine, Friction welding machine etc. The technologies developed by MERADO, Ludhiana have benefited not only the SMEs in the state but also in other states *e.g.* in West Bengal, Haryana and U.P. etc. It also caters to major testing facilities for the SMEs in Ludhiana (MERADO, 1999). However, in the present scenario of globalization and liberalization, stronger linkages still need to be promoted not only between technology support institutions and SMEs but also between SMEs for horizontal transfer of technologies. In this direction human resource development in the technology support institutions is one area of crucial importance, which needs attention.

Apart from CSIR, a number of other R&D agencies in the country are actively involved in strategic partnership with industry (including both small and large industry). Table 12.3 shows some of the successes achieved by these partnerships in 10 years till 2005.

In yet another example, Central Glass and Ceramic Research Institute (CGCRI) Naroda Centre, Ahmedabad, one of the constituent laboratories of CSIR, is actively serving the ceramic clusters in the region. This is a small institute with 18 scientists primarily engaged in the improvement of traditional ceramics and rural pottery.

---

1 National Institute of Science, Technology and Development Studies (NISTADS) has set up a S&T field station at Bankura, West Bengal, India to upgrade the technologies for artisans and craftmen by blending the traditional technologies with new technologies and promote commercialization.

Table 12.3: Number of Technologies Developed by Industries through Partnership with R&D Agencies during 1995–2005

Type/Discipline	*Name of Industry*	*R&D Investment (Per cent of Annual Turnover)*	*New Tech Developed through Partnership in Last 10 yrs*	*Tech Inducted/ Commercialised from R&D Agencies in last 10 yrs.*
Pharma	Nicholas Piramal India Ltd.	>100 Cr.	—	—
Biotech	Shantha Biotechnics Ltd., Hyderabad	10-15 per cent	2	0/1
Drugs	J. Mitra and Co. Ltd., Delhi	Rs. 1.55 Cr. during 2003-04	3	3/3
Biotech	Dalmia Centre for R&D, Coimbatore	Rs. 100 lakhs	15	15/10
Microbiological and Chemical Sc.	Arbro Pharmaceuticals Ltd., New Delhi	17.59 lakhs	—	NA
Biopharma/Biotech	Millipore India Pvt. Ltd., Bangalore	NA	NA	NA
Agriculture	Sungro Seeds Ltd., Delhi	—	100 new products	100/100
Petrochemical Production	Indian Petro-chemicals Corp. Ltd., Vadodra	Rs. 10 Cr.	8	8/8
Chem (Agro)	Rallis Research Centre, Bangalore	Rs.10 Cr.	Nil	Nil
Chemical	Godavari Sugar Mills Ltd., Mumbai	Rs.80 lakhs	2	
Chemicals, drugs	Rubamin Limited, Vadodra	—	—	—
Chemicals and Paints	Mathur Corr-Tech (P) Ltd, Coimbatore	Rs 3 lakhs	6	—
Steel	Tata Steel Mines Division, Jharkhand	~ Rs 150 Cr.	1	1/0
Engineering	Tata Refractories Ltd., Orissa	~Rs.15 Cr. (T/O Rs. 400 cr.)	Nil	Nil/Nil
Electro-chemical	High Energy Batteries India Ltd, Pudukottai	~Rs 434.01 lakhs as on 31.3.2005	3	1/1
Consumer Electronics	Ahuja Radios, New Delhi	Rs.44 lakhs	Many	Nil/Nil
Electronics	Omtek Electronics (P) Ltd., Bhubaneswar	Rs. 50 lakhs (T/O Rs.1.50 cr.)	Nil (36 Self developed Tech.)	36/13

Contd...

Table 12.3–Contd...

Type/Discipline	*Name of Industry*	*R&D Investment (Per cent of Annual Turnover)*	*New Tech Developed through Partnership in Last 10 yrs*	*Tech Inducted/ Commercialised from R&D Agencies in last 10 yrs.*
Engineering	Basic Technology (P) Ltd., Kolkata	Rs. 20 lakhs	10	6/4
Mines	Satna Cement Works, Satna	~Rs.1.5 lakhs	Nil	Nil/Nil
Engineering	St. Josephs Tiles, Ernakulam	—	1	1/1
Engineering and Infrastructure	Shriram Energy Systems Ltd. Hyderabad	—	1	Nil/Nil
Engineering	Abaqus Engg. India, Chennai	Nil	Nil	Nil/Nil
Engineering	Ador Fontech. Ltd., Bangalore	Nil	Nil	Nil/Nil
Atomic Energy	Indira Gandhi Centre for Atomic Research, Kalpakkam	—	—	—

Source: BANSAL, Rama Swami (2005): 'R&D Agency – Industry Partnership for Technology Development and Transfer in Indian Context', Ph.D. Thesis, Birla Institute of Technology and Science, Pilani, Rajasthan.

There are 950 ceramic units in Gujarat. The Institute was established in 1977 to serve these cluster units and to cater to the needs of the developing ceramic industry in the small-scale sector of Gujarat and adjoining areas. Most of these clusters are poor in knowledge but have high entrepreneurial spirit. To improve the skills of these units, training and manpower development programme were initiated in 1978.

Since then CGCRI has been conducting training-cum-demonstration programmes and capsule courses on various topics of current interests to the ceramic industry. The Institute till 2003 had conducted 32 technology development programmes and has trained about 600 persons from the industry. It has also conducted three entrepreneurship development programmes for encouraging new enterprises. The clusters at Morbi, Himmat Nagar and Mehsana etc. have benefited a lot in technology development from CGCRI in improving the quality of the raw materials for the manufacture of ceramic tiles and in improving process control parameters. One of the important achievements of the Institute has been that it has been able to use the Fly ash (a highly polluting waste product of thermal plants) in the manufacture of Ceramic tiles. Ceramic tiles now contain 30 per cent of the raw materials as fly ash. It has also been able to produce blue ceramic tiles, which are free from lead. Ceramic filter candles have been developed which also contain 20-30 per cent of fly ash. Cotton wick of the kerosene oil lamp has been replaced by the permanent ceramic wick which need not be replaced at all. It has developed the technology for production of Bone China utilizing the China clays of Gujarat. This technology has been transferred to 21 units in the small scale sector including M/s. Anil Ceramics, Mori; M/s. Hitari Ceramics, Himmat Nagar; and M/s Ideal Ceramics, Delhi. The training of the workers is carried on the site which has received very encouraging response. However, there is a need to build up linkages with the university sector, to benefit from the generation of new knowledge.

One of the case examples of academia – enterprise linkages is the 'Sonya Ceramics' in Ahmedabad. The Managing Director of this enterprise Rupesh G Shah-a chemical engineer from USA, took over the business from his father, who set up this business in 1959, manufacturing sewage pipes. Over the years, with the cooperation of the CGCRI and his own R&D efforts, he has been able to improve the process technology and consequently improve the products. He diversified his business and today the enterprise is manufacturing about 560 products used in a wide area of social activity–from thermal insulators to sewage pipes. It has taken 9002 ISO standard and works in close cooperation with the customers. The products are modified with the demand from the customers as required by them. The enterprise has no in-house R&D of its own but has strong linkages with the R&D institutions. Today the products of this enterprise are being exported to 15 countries in USA and Europe. Some of the R&D –Industry partnerships leading to technology developments are listed in the Table 12.3.

In India, most of the SMEs are building up ITCs through the process of Learning by Doing. The process of Learning by Doing is central to incremental innovation and technological change. It has been pointed out that engineers on the shop floor play a vital role in supporting the operator's effort to acquire new skills and come up with new ideas (Okuda, 1983). This is a must for a learning organization. In a learning organization, everyone in a group is an expert and can give his/her knowledge to the process of complex problem solving (Honecker, *et al.*, 1999). The fact that proper linkages between the managerial staff, including engineers and workers at the shop floor within an enterprise, are an important factor in the information flow and the innovation process is exemplified by the studies on firms in the electronics sector in India (Bowonder and Miyake, 1988; Kharbanda and Jain, 1997). The studies show that learning by doing and entrepreneurial capabilities have been instrumental in strengthening humanware and technoware at the enterprise level.

2.3 Knowledge Flows and Technology Databases

Dynamism of SMEs is crucial for their long term competitiveness. For this they need to focus on systems of knowledge accumulation rather than just production system (Bell, and Albu, 1999). The SMEs should have capabilities for generating, accessing and diffusing knowledge and also have access to external sources of knowledge including new technology data bases. For this efficient national information services and networks are essential.

In developing countries, most of the SMEs do not have access to well researched technology databases that provide information pro-actively on a regular basis. The information usually remains in the data-banks untapped, while the purveyors of the information wait for the targeted beneficiaries to request it.

In India, an important technology information data bank is the National Research and Development Corporation (NRDC), which serves as an important link between research and SMEs. NRDC acquires, evaluates, develops and transfers all worthwhile technologies generated at various national institutions. The technologies available with NRDC for commercial exploitation cover a wide range of products, namely, drugs and pharmaceuticals, pesticides and herbicides, plasticizers, resins, electro-chemical products, metals, paints and varnishes, leather chemicals and auxilliaries, electrical and electronic goods, building materials, etc. It provides a very comprehensive international patent search and also has a few renowned international data bases on line for carrying out search.

An important database on local innovations is the "Honey Bee Database" supported by SRISTI[2], which is one of the biggest stores of knowledge on local innovations by farmers and artisans covering about 72 countries. It is a knowledge network, which pools the technological solutions developed by people around the world. Networking of these databases can be of immense use to mitigate the numerous technological problems of the SMEs.

2.4 R&D and Inter firm Linkages

Presently, in India there are 3960 R&D institutions in total, of which 2020 are in the private sector and 115 in the public sector. These employ over 45,000 scientific and technical personnel and incurring expenditure of the order of 12 billion rupees

2 Society for Research and Initiatives for Sustainable Technologies and Institutions (SRISTI) is a voluntary organization in India, which has received national and international recognition as a Centre of excellent for research on people's innovations.

per annum (DST, 2006). However, on the contrary, the small scale sector is largely devoid of R&D facilities and is mainly supported by public R&D for acquisition of new technologies. In general, SMEs are encouraged to invest in localized technological development. As most SMEs lack financial resources, the best way is to enter into linkages and partnerships with other enterprises, academia and/or R&D institutions.

Over the past decade, emphasis in the concept of technology transfer has gradually shifted towards a new set of strategies which can be described as "creative partnerships" or "strategic alliances" to form learning organizations (Niosi, 1991). This underlines the idea that forms of technological cooperation are no longer "one-way" but involve a longer-term mutual benefit beyond short-term financial success. According to Cooley, (1991) learning by doing is not enough. Those who know what they are doing-crafts people with skill and experience- are the ones who best understand what needs to be done. This depicts an inherent feature of entrepreneurial capabilities, which help pushing the organization towards a learning organization. "Learning by interaction' is an important feature of technological upgradation and can only happen when firms cooperate and interact through strategic alliances and other forms of linkages. For this to happen most viable systems are the clusters, which offer good opportunities for interaction and industrial upgrading (Knorringa and Meyer-Stamer, 1998). The common feature of these new forms lies in the sharing of knowledge in technological capabilities and in cooperative intentions, which may or may not include government incentives. The diverse forms of alliances include:

- ☆ Long-term multi-project partnerships that pool the production, research and marketing divisions in a single country or across several countries to accomplish strategic goals;
- ☆ Flexible networking mechanisms that promote research consortia at the pre-competitive and development phases.

Such partnerships have technical education, training and retraining, innovation, and sustainability as inherent features.

3. NETWORKING/INDUSTRIAL DISTRICTS/CLUSTERS

The approach encompassed by the concept of industrial districts/technology parks/clusters, offers new insights into the potential role of SMEs in enhancing their access to new technology.

A simple definition of a cluster suggests that it is a network made for information and resource sharing between people. Clusters are defined as concentration of activities belonging to the same sub sector. An old Indian proverb says that two hands together are better than one. The same is true for small entrepreneurs also. Cluster or network of companies have inherent advantages of both the large and small companies.(Briley, Sue and Muzyka, 2000). Network has no centralized control. Network retains the agility, inherent in a small company yet it has competitiveness of a large company in financial and infrastructural depth.

In India, in mid nineties itself, there were about 138 industrial clusters which are engaged in specialized industries such as: locks at Aligarh, leather footwear at Agra and Kanpur; cotton hosiery at Calcutta and Delhi; blankets in Panipat. Sector-specific and geographically bounded clusters seem to be a common phenomenon for small-scale manufacturing in developing countries (Nadvi and Schmitz, 1994). Most of them have been very successful which are primarily based on tacit and accumulated knowledge through learning by doing and learning by interaction. For example in India there have been a large number of "spatial clusters" of small firms engaged in specialized industries such as: locks at Aligarh, leather footwear at Agra and Kanpur; cotton hosiery at Calcutta and Delhi; blankets in Panipat; power looms at Bhiwadi; diamond polishing in Surat. Space bound "dense clusters" related to a specialized industry are even more pronounced in the State of Punjab with woollen garments, bicycle and bicycle parts, sewing machine parts and machine tools in Ludhiana; printing and printing goods, water pipes and bathroom fixtures in Jallandhar; foundries in Batala, etc. Ludhiana clusters make 95 per cent of country's woolen knitwear; 85 per cent of sewing machines and 60 per cent of bicycle and bicycle parts. Agra cluster makes 0.15 million pairs of shoes per day with a production value of 1.3 m US$ and exporting shoes worth US$ 57.14 million per year (Juneja, 1998). Knitwear cluster in Tiruppur, Tamil Nadu is responsible for 85 per cent of Indian Market and its export earnings have expanded from US$ 25 million in 1986 to US$ 636 million in 1997. What is interesting about Tiruppur cluster is that it is organized in a web of small work places through which the entire town works like a living industrial organization (Chari, 2000)

What follows are some salient features of the case studies from India, Germany, Italy and UK, highlighting the feasibility of cluster approach in making technologically capable and competitive in the global market..

3.1 Case 1: Diesel Engine Cluster-Rajkot, Gujarat, India

Rajkot Diesel Engine Industry is the leader in Indian Diesel Engine market with more than 60 per cent of India's total diesel engine production. It accounts for around 0.3 million no. of diesel engines per year valued around Rs. 2.5 billion with sizes from 3.5 HP to 20 HP. Majority production is in the range from 3.5 HP to 8 HP. State wise production is given in Table 12.4. The industry is made up of small-scale manufacturers and has about 400 foundry units in the city. Their annual production is more than one hundred thousand tones of casting. It employs more than 40,000 workers. The development of the Indian diesel engine industry after Independence has passed through three distinct phases in the last three decades.

Table 12.4: State wise Concentration of Diesel Engine Manufacturing Units

State and Place of Concentration	*Percentage of Concentration of Unit*
Gujarat – Rajkot	60 per cent of production consisting of high speed and slow speed diesel engines
Maharashtra – Pune, Kolhapur	15 per cent
U.P. – Agra	15 per cent
M.P. – Indore	5 per cent
Other states (Tamilnadu – Coimbatore, Punjab – Few pockets)	5 per cent

The first stage, during the Seventies, saw the demand phase wherein the supply was short. Efforts were concentrated on indigenizing and increasing production. Investments were made in production technology developed during the sixties. This period also witnessed fuel crisis, which led to the sharp rise in prices of petroleum products as a result of which there was a need to improve the fuel efficiency of engines. The industry was mainly supplying goods to the commercial vehicle and the agriculture sectors.

The second stage, in the Eighties, saw the manufacturers' attention focused on improving fuel efficiency following the fuel crisis of the seventies. The central government also offered some lucrative incentive schemes for the large-scale manufacturers as well as for small-scale sectors.

The third stage began in early Nineties, when the rest of the developed world was changing towards the rotary pump and the electronically controlled engine management systems, India continued to rely on the indigenized diesel engines, and was also able to export to developing countries due to high demand of the diesel engines made by Indian companies.

Interviews conducted with the selected entrepreneurs (fifteen) associated with this cluster highlighted a number of problems faced by the industry. These are as follows:

The Rajkot Diesel Engine industry is facing severe problems of competition and new product innovation. The problem is that the entrepreneurs are working without any support from the outside sources. They try to find solutions to their problems from inside. They have created an association called Rajkot Engineering Association (REA), which takes care of raw material supplies and bulk buying. The association also helps to keep them updated with the latest information. However, in spite of this the growth rate of this industry decreased. After enjoying a growth of more than 22 per cent for more than a decade in the international markets, the industry is now growing at a pace of 8 per cent only.

According to Pillai (2000), the problems of clusters worldwide seem to be eroding the fundamentals of the clusters. To what extent has the cluster an assured availability of key inputs like raw materials, components and parts from within the cluster itself? Adequate availability of inputs on site substantially reduces the transactions cost. In case of the components and machineries, the Rajkot cluster used to enjoy near self-sufficiency. By 1990s the situation changed and the long-standing relationship between traders and producers in the cluster which was earlier characterized by mutual trust, efficient timely delivery and good quality of the inputs, slowly started disappearing mainly because of payment delays, lack of enough orders, inventory accumulation, etc. As a result, the industry is losing its market share in the international markets and even in the domestic market. In the international market the industry is facing competition from cheap Chinese goods. In the domestic market the competition is somewhat different. It is not from the competitive goods but arises from a different basic source and that is the availability of electricity to rural farmers who now no longer need diesel engines to pump water for their farms. They now use electric motors rather than diesel engines. There are thus two different grounds on which this industry needs to survive.

3.2 Case 2: The German Model of Entrepreneurial Innovations

In Germany the Aachen Region has been described as the Learning Region, as it has a unique network of academia and the enterprise for continuous innovations and transfer of technologies from academia to the enterprise. The Aachen region has a total population of 0.35 million. In the region there are at present 8000 companies, four universities, 20 R&D institutes having 50-100 researchers each. Main research areas are lasers, ceramics, car engines, plastics, software and communications. The two main universities, out of the four, are University of Technology and University of Applied Sciences.

3.2.1 Aachen University of Technology (RWTH)

In 2001, about 29,000 students were enrolled at the Aachen University of Technology in over 65 degree courses. The largest single group of students (approximately 40 per cent) is studying for an engineering degree; about 21 per cent students are studying the natural sciences; and about 10 per cent each are enrolled in the Faculties of Arts, Economics and Medicine. In all the multifarious academic activities conducted by the 260 institutes, chairs, readerships and central institutions of the University, there is a clear across-the-board emphasis on practical relevance, interdisciplinary and internationality – all aspects with a strong element of student participation. As an application-oriented university, the RWTH has in the last few decades come to be a significant economic factor in the city and the region. With a total staff of over 10,000–approximately 400 professors, 1,900 non-professional academic staff 5,300 non-academic staff, almost 1,000 trainees and staff on temporary placement – the University is the largest employer and training facility in the Aachen region. Its budget now exceeds Euro 650 million, of which approximately 120 million are funded by non-governmental sources. It is in fact one of the world's largest universities with a technological focus. It is not, however, just budding graduates who are completing their application-oriented studies at the RWTH, there are also a total of 1,000 trainees and temporary staff on placement being given their practically oriented vocational training in the trade-technical and scientific and commercial sectors as well as in paramedical and non-medical practice. The total range of 43 training occupations offered are in the academic, central and medical institutions of the University extending from doctor' assistant through photographic laboratory assistant to dental technician.

3.2.1.1 Linkages with Industry

An outstanding feature of RWTH is its linkages with the enterprises. Industrial internships are for instance a condition of admission to the final examinations for engineering students, whilst all the University's engineering curricula include project work which usually comprises the investigation of a practical problem using scientific methods. Professors in the engineering faculties typically have strong personal links with industry, often going back to work experience of their own in industry prior to their University appointments, and these are often invaluable in setting up applied research projects of mutual interest to both parties. On the basis of contractual co-operation agreements with the Aachen Chamber of Industry and Commerce as well as with the Aachen Chamber of Handicrafts, a multifarious network of University-specific business contacts has evolved in the course of time. Aachen University of Technology is also anxious to intensify collaboration with firms in the region. To this end, the Regional Industry Club for Information Technology in Aachen (REC INA) was founded. A considerable proportion of new businesses in the Aachen region are spin-offs resulting from research staff moving from the University to the private sector. Resourceful young scientists with new ideas – backed up and encouraged by start-up capital from European, national, and local government funds – are taking the plunge into self-employment. More than 80 per cent of the new enterprises launched in the Aachen Technology Centre (TZA) since 1984 have originated from the immediate environment of Aachen University of Technology.

The RWTH Aachen attaches much importance to advanced academic training. It is the very first company focusing on the marketing of advanced academic training programmes at a German university. The Aachen Global Academy supports the transfer of new scientific results and methods as well as new problem-oriented strategies from university to business and society within the scope of job-oriented advanced academic training.

3.2.1.2 Spin-off Enterprises

The spin off enterprises from the university has been the major feature in the Aachen region. The Enterprises in Aachen region are basically divided into three groups:

- ☆ High Technology enterprises,
- ☆ Art and Craft enterprises, and
- ☆ Low technology enterprises.

High technology enterprises are highly automated with lasers and computer aided automatic machines manufacturing for example, computer parts like modems and computer peripherals. The Art and craft enterprises are the enterprises with mix of modern and traditional technologies producing furniture and other household goods. The low technology enterprises are basically in the area of recycling meant to increase employment opportunities. Most of the people employed in the third group are from Asian countries. To mitigate the problems of unemployment which has increased over the years from 5 per cent in 1974 to about 12-13 per cent on average in Germany, the present emphasis is on small enterprises with close connections with the university. A large number of small companies are being created to fight unemployment. One of these companies is in the area of recycling of electronic products, particularly computers, washing machines and television sets etc.

3.2.1.3 The Schell Gruintechnik, Korneli Munster, Aachen

The Schell Gruintechnik, Korneli Munster, Aachen is the classic example of an entrepreneur emerging from scratch and promoting innovations in the manufacture of Heavy – duty lawn mowers systems through university – enterprise cooperation. Franz Josef Schell who is the owner of the company, after obtaining an Apprenticeship

Diploma in Mechanical Engineering, gained experience in another big company for maintenance of lawn mowers and thus started his career in 1971 with maintenance and repairs of the lawn mowers. Slowly he started making his own spare parts for repair of lawn mowers. With the experience gained, he started making his own lawn mowers in 1981 and in 1991 it grew to 10-21 employees. Four of the employees are of its own family members. The sales rose to 1 million US dollars and in the year 2000 it rose to 5 million US $. The rate of growth of the company has been 20-25 per cent. In 1998 it had linkages with the university to for improvements in aerodynamic and hydraulic systems. The main problem to be solved was the aerodynamics of high-speed grass cutting and removing in the maelstrom inside the housing close to the ground underneath the lawn mower. This problem was solved by the company in close co-operation with the Department of Aerodynamics and Aircraft Design of the University of Technology, Aachen. The solution comprises the cutting-up of the grass into sufficiently small bits which can be left on the lawn to rot and fertilize the ground. Today about all leading lawn mower producers world-wide buy the Schell mowing systems to attach them to their own brands of lawn mowers. The next step of technological development will be the remote-control and the un-manned automated lawn mowers.

In this direction, presently, it has contacts with the car engine department of the Technical University of Aachen and an Austrain company for the development of robotic lawn mowers which can be operated from the remote control. Thus the company has grown from low technology to high Technology with the help of University of Technology Aachen. The main features of this region are:

- It has a cluster of 12 firms
- The companies in the cluster cooperate with each other to find solutions to the problems.
- Has strong linkages with the University of Aachen for finding out high tech solutions.
- Learning by doing has been the main characteristic feature in the development of the company.
- It also gives apprentice training to the students of the university.

3.3 Case 3: The Italian Model of Innovation

Emilia-Romagna in Italy is the characteristic model of innovative SMEs in Italy, and is also known as the Third Italy. The Population of Emilia-Romagna was 3,924,352 in 2001, approximately 7 per cent of the national population. The region accounted for 8.3 per cent of national employment and nearly 9 per cent of the Gross National Product, surpassed only by Lombardia and Lazio. The region had the third lowest rate of unemployment (4.5 per cent compared to the national average of over 11 per cent) among Italy's 20 regions, surpassed only by Valle d'Aosta and Trentino Alto Adige in the far North. Emilia-Romagna boasts one of the largest business classes in Italy. The productive system is characterized by Small and SMEs (employing an average of 5.48 people per enterprise), with a huge number of artisan enterprises (126,639 out of 304,947) and co-operatives (7,923, including 2,336 farming cooperatives and 1,187 labour and manufacturing co-operatives). The region comprises of nine provinces *i.e.* Bologna, Ferrara, Forli-Cesena, Modena, Parma, Piacenza, Ravenna, Reggio Emilia and Rimini. The major companies of the region were Barilla and Parmalat in the food sector; Ima and Tetrapak in the industrial machinery sector; Lamborghini and Ferrari in the motor vehicle manufacturing sector; and Ducati Motor in the motorbike sector. In order to optimize the advantages and reduce the limits associated with the small firms, production in Emilia-Romagna region had taken on a particular structure over the years, giving rise to local productive systems known as 'industrial districts'.

3.3.1 The Industrial Districts

The regional economy is characterized by a very low level of vertical integration. It is, in fact, rare for a single firm to undertake a complete production process. Even major firms, which supply final goods, usually carry out only certain productive processes and leave the rest up to other firms. A complex network of suppliers has thus developed each undertaking a single production phase often on behalf of different firms. As a result, every supplying firm, however small, may maximize its level of specialization, knowing that it can rely on a sufficient number of orders to pay off the cost of technological investments. This particular form of organization of production has moreover created room for the development of complementary businesses, which distribute raw materials and semi-finished goods among suppliers and even larger firms which manufacture machine tools. There is constant exchange of information, so that the product may easily be made to individual specifications. There is thus high degree of flexible specialization.

3.3.2 Exports

The region is also characterized by high level of exports. In the region, in 1999, the sales from exports exceeded 29 trillion lire (Aprox.14 billion $), equal to 11.2 per cent of the national total. Emilia-Romagna region is, thus, the fourth-largest exporter after Lombardy, Piedmont and Veneto. The products include minerals and non-metallic products (32 per cent, a sector including ceramics), food products (18.7 per cent), farm machinery (18 per cent) and metal and mechanical goods (12.4 per cent). Even within the scenario of the general slow-down in Italian exports in 1998, especially in the second semester, Emilia-Romagna recorded a growth rate of 5.3 per cent equal to almost double the national average and higher than all the principal exporting regions of the North. It is particularly interesting to note that among all the regions in the so-called "Third Italy", Emilia-Romagna, along with Friuli-Venezia Giulia, are the regions which have, least of all, been hit by the effects of foreign competition compared with what has happened in regions more specialized in traditional production such as Veneto, Marche and Tuscany.

One of the main features of this success has also been that, the SMEs of Emilia-Romagna region have set up associations to promote exports and form business alliances. For example the Emilia-Romagna Regional Federation of the National Confederation of the Craft and the Small and Medium enterprises (the Emilia-Romagna CAN) represents and defends all interests of craft firms, of small and medium enterprises and atypical workers in relations with public administration, and with political, social and economic organizations. It seeks to promote their growth in an open competitive environment, permitting the best to exploit their full managerial and enterprise potential and skills. It works at local, regional, national and regional level (Europe-wide) and is rooted in both Italian and European cultures. The Emilia-Romagna CAN has set up an extraordinary range of services and benefits which include:

- ✰ Assistance in the starting up of new businesses
- ✰ Assistance and consultancy to enterprises run by young entrepreneurs
- ✰ Consultancy and financing
- ✰ Business, managerial, vocational training
- ✰ Information and consultancy in export, promotion, internationalization and marketing
- ✰ Prevention, safety and the environment
- ✰ Company consultancy
- ✰ Assistance in the area of quality control certification
- ✰ Information and consultancy on European legislation
- ✰ IT and telematic services with analysis of system requirements and design
- ✰ Tax services
- ✰ Assistance and management of tax litigation
- ✰ Pay-roll service, personnel administration and consultancy on Employment problems
- ✰ Assistance in contracts
- ✰ Assistance and consultancy in the fields of health and pensions.

The Emilia-Romagna CAN is made up of 10 provincial associations: Bologna, Modena, Ferrara, Reggio Emilia, Piacenza, Parma, Ravenna, Forli-Cesena, Rimini, Imola. To provide information and to assist enterprises in every aspect of their work, there are 226 CAN local offices spread throughout the region. The work force based in the offices totals more than 2,500 skilled and professional staff. It has also set up a Women Business Committee (Comitato Impresa Donna) to assist women to start up and establish themselves in business, while it has also set up the Young Entrepreneurs Committee (Comitato Giovani Impreditori) designed to assist and encourage young people in Business.

3.3.3 "CENTURIA" Science and Technological Park (CSTP), Cesena

Cesena is at the heart of the Emilia Romagna region, and forms a link between Northern and the Central Italy and between Italy and the Central Europe. With a population of 1,80,000, Cesena's economic prosperity looks to the future through vigorous competition with other most developed areas in Europe. Its geographic location, acting as a pivot between central Italy and north eastern Europe, offers a wide range of opportunities for trade and exchanges. Cesena's economic prospects are based on technological innovations and on the quality of both products and services. For this purpose linkages between the university, research centres, international trade fairs and science and technology parks are at the forefront. The "CENTURIA" Science and Technological Park" located in this region, in particular, has helped to bring together the leading companies in the food sector to foster research and innovation, through synergies with the Bologna University and other scientific institutions. It has brought about the most singular and effective concentration of high technology companies, specialized in the different sectors of the agro-industry. The main characteristic feature of the region is that, to enable immediate direct contact between the various components, all the production, transport, technological information, technology, research, promotion, marketing, export management and professional capacity support structures required for the growth and consolidation of investments in the agro-industrial sector, are located in the region itself. There exist leading experiences in the fields of preservation, processing and transportation of fruits and vegetables and poultry products, of sugar, of seed and animal feed production right up to robotics applied to automatic sorting and packaging, to bio-technologies and advanced information technology. A large number of research laboratories work in conjunction with the university. These are all specialized in the agro-industrial sector and represent vital instruments for the innovation of enterprises, for research within the field of bio-technologies, for agricultural experimentation, for food and processing technologies, information technology and transport. The Centuria Science and Technology Park, which benefits from the participation of major local industrial groups, is active putting companies into direct contact with scientific and technical expertise of research institutes and the university. With this, Cesena has become a basin for agro-industrial expertise and experience with no equals in Europe in terms of concentration and achievements. For example, Poultry farming has today become an integrated system whereby the industrial groups of Cesena control the entire process from feed farming, right up to meat processing. The cultivation of sugar beet is linked to the large sugar industry (the second in order of importance in Italy). The road transport sector has consolidated into a park of over three thousand specially equipped vehicles connected by satellite and sophisticated system of telecommunications. A fundamental part of this success can be put down to the extraordinary local know-how on the subject of refrigerated transport.

Another significant Cesena experience is in the field of biotechnology. A bio-factory has been established for the mass rearing of beneficial insects and for development of biological and integrated pest control techniques applied to fruits and vegetable growing – the first of its kind in Italy. Established in 1980's as a research and experimentation centre, the bio-factory is today able to produce and market a vast range of beneficial insects throughout Europe and the Mediterranean basin. Research in Cesena's agro-industrial sector not only involves specialized centres, but most of the companies participating in the "CENTURIA" Science and Technology Park also have advanced laboratory.

The main objective of this Science and Technology Park (CSTP) is to enhance the competitiveness of SMEs in the agro-industrial district in Romagna. It has been specifically created for agro-industrial technological innovations in this region of Cesena. This district has developed from a traditional agriculture region to an advanced high tech agro-industrial region with robotics, food processing machines, fruit sorting and packaging machines, logistics and transportation and application of biotechnology. The three main areas of activity are Poultry farming and animal feed: Sugar production; Fruits and Vegetables, with about 1000 SMEs and 15000 employees in total and a total turnover of 4 billion US$. About 40 per cent of the profit goes to the Consortium and 60 per cent is distributed among the farmers. The co-operative helps the farmers in all operations from plucking to packaging. This CSTP has the following special features:

1. It is a new concept of STP having no need to build up new hard wares, but to create value from the existing expertise available in the R&D institute, universities, enterprises, and trade fairs facilities already available in the region by linking them in a network.
2. It has linked up 24 leading SME companies specialized in agro-industry with the local faculties of the University of Bologna, research centres and the local government municipality, Chamber of Commerce, Local health and environment authorities. The companies own 65 per cent of the capital share.
3. It serves as a data bank for information on patents, new technologies, and partnerships in R&D projects, technology transfer and joint ventures enhancing cluster visibility towards new investors. It thus helps in the establishment and management of a system of relationship between SMEs, universities and R&D centres, banks and associations and major industrial groups. These relationships have built up on the basis of contact research between companies and universities, and through joint projects in production development and common market between different companies.
4. It helps the SMEs in project management, data mining, market analysis, technology transfer new business start up in new geographical locations and arranges for venture capital.
5. The local actors of competitiveness (companies, institutions, banks and universities) are ready to co-operate and compete
6. Through the technique of biotechnology it is able to reproduce from a germ plasm, in the course of one year, up to one million plants having perfect identity with the mother plant.
7. It has a bio-factory for mass rearing of beneficial insects for biological and pest control- the first in Italy. Established in 1980, this bio-factory markets beneficial insects to the whole Europe.
8. Information technology is playing a crucial role of, not only to create a network for better communications but also to upgrade the local companies technologically – like automation of production processes with the use of CAD – CAM techniques, and also to robotise the movements of articulated systems of fruit processing machinery. This STP has global networking, having linkages with International Association of Science Parks (IASP); Innovation Relay Centre North-East, Italy; Eurolink Partners; Associazione Parchi Scientificie Technologici Italiani; and Agro-Consortium for Human Resources Educationin Agro-bussiness.

This model of networking of players of innovations has been reported to be most successful which is practically based on the concept of continuous learning, co-operation and competition. This model of enhancing the competitiveness through the management of a system of relationship in the territory can work anywhere if:

- ☆ There is territory with a good concentration of economic activities in similar fields.
- ☆ The local actors of the competitiveness are ready to cooperate. (companies, institutions, banks, university/research)
- ☆ There is a 'bottom up' approach to the problem solving.
- ☆ The structure is market driven.

3.4 Case 4: The British Model of Innovation

The Sussex Innovation Centre, Science Park Square, Brighton, UK

The Sussex Innovation Centre, opened in May 1996, provides support for the creation and growth of technology and knowledge based companies in Sussex. The Centre is now thriving business environment for nearly 30 high growth companies. Initially, this Innovation Centre was conceived as the flagship development of the 'Sussex Academic Corridor' (SAC). Formed in 1991 SAC is a unique collaboration between the public, academic and business sectors committed to harnessing the economic potential of the extensive education and research resources available in the Brighton and Lewes area. This Innovation Centre builds on the successful experience of university based incubators in the United States and at Oxford and Cambridge Universities. Professor Walter Heriott from the University of Cambridge was a consultant for the Sussex project. The principle is that new technology based companies (especially those linked to academic research) have the potential for high growth but also have specific needs in terms of management support and profiling. Experience and research has shown that incubators increase dramatically the proportion of companies that survive the traumatic first three years (over 85 per cent survival and often higher) and provide high growth. The experience at the Sussex Innovation Centre certainly reinforces this experience.

The basic features of the business model for the Sussex Innovation Centre are:

- ☆ Through an initial public investment of over 2 million the Sussex Innovation Centre was developed as a purpose built business incubation facility based on the campus of the University of Sussex.
- ☆ Rent levels are set which, while reflecting the prime location and excellent facilities, offer the tenant companies an ideal, flexible and relatively inexpensive environment to grow their business.
- ☆ The Innovation Centre Management Company (ManCo) collects rents on behalf of the Development Company (DevCo). Any surplus is reinvested into the Incubation Support via a "Management Fee".
- ☆ The Innovation Centre supplements its income through undertaking projects, consultancy, sponsorship and running events which are

consistent with its core objectives of incubation, commercialization and to act as a focus for new technology companies in Sussex.

- The universities gain through an increased focus on commercialization, a higher profile within the local business community and a clear route for academic spinouts.
- The local authorities receive their "Investment Return" through the impact on the local economy of the growth of the new technology companies.

The Centre has also extended its services by establishing a new department: The Commercialization Support Group (CSG). CSG's services are designed to help local companies, academics and individual inventors to develop and commercialize new products and services by making the commercialization process easier and quicker. Charges for the CSG services are intentionally kept low, to ensure accessibility. In some circumstances the Innovation Centre may even make 'seed' investments to enable the company to use the CSG resources. Services offered by CSG are:

- General business advice
- Assistance for raising finance
- Advice and sources of information for protecting Intellectual Property
- Carry out Market research
- Marketing assistance
- Assistance to secure Licensing deal
- Advice and sources of information for accessing Manufacturers
- Access to basic and applied research
- Access to general Prototyping assistance and prototyping centres

This Centre is also engaged in the development of a range of technical and professional networks. These include:

- *Know How Exchange* – network of Sussex companies who have agreed to share non-commercially sensitive information and experience within an informal environment. The Centre manages the network, maintain the company databases and facilitate the exchanges.
- *Academic Expertise and Facilities* – the Centre maintains a database of academic resources and will assist companies locating the appropriate expertise, skills and facilities. Details of the research groups can be searched via the online Skills Directory.
- *Sussex Companies Expertise and Facilities* – The Centre maintains a database of Sussex Companies which includes extensive information on the available expertise and facilities. A large part of the database is available for online searching through the Skills Directory.
- *Professional Advisors* – The Centre maintains close relationships with a wide range of commercial, legal, financial and Intellectual Property professionals who provide advice to the Innovation Centre and our client companies.

The Sussex Innovation Centre gives specialist services. For example, it is directly involved with a number of initiatives aimed directly at promoting new ideas and helping innovative companies to grow. These include the running of Inventor Workshops (with Business Link Sussex), the creation of commercialization support Group, establishing a small Seed Fund, development of the Innovation Capital business angel network (with capital Match), creation of a new regional bioscience initiative – Southern Bio-Ventures (with other South East Incubators and universities).

4. ROLE OF FINANCIAL INSTITUTIONS

The availability of financial resources on affordable terms and in a non bureaucratic manner is yet another problem the SMEs confront hindering access to new technologies. It is imperative, therefore, to influence existing financial networks and, if need be, to set up a separate mechanism for funding smaller enterprises especially those categorized as micro-enterprises.

One of the important ways to promote access to new technologies among SMEs in developing countries is to provide venture capital, which in a way helps in indigenous development of technologies. Financial Institutions such as Industrial Development Bank of India (IDBI), Industrial Credit and Investment Corporation of India (ICICI), Industrial Finance Corporation of India (IFCI), and Banks are providing assistance for commercialization of indigenously developed technologies and adoption of imported technologies for wider domestic applications through venture capital companies. A large number of small innovations are also being supported by SRISTI Venture Capital Fund (Gupta, 1996). One of the main factors of the success of knitwear cluster at Tirupper is the State intervention in the supply of short term credit to facilitate networked production (Chari, 2000)

In order to create a business environment, in which adoption and assimilation of new technologies that are resource efficient and have little waste output, the supportive infrastructure also needs to be strengthened in many developing countries. There is a need for such institutions to deal exclusively with the myriad problems that SMEs face in their day-to-day work. In India, Department of Scientific and Industrial Research (DSIR) is one such institution. It is engaged in technology promotion, development, utilization and transfer activities. The Department has launched several initiatives to encourage increased utilization of locally available R&D options through its major schemes *viz.* Programme Aimed at Technological Self Reliance (PATSER), (now called Technology Development and Demonstration Programme) and Scheme to Enhance the Efficacy of Transfer of Technology (SEETOT). Technology Bureau for Small Enterprises–a result of collaboration between APCTT and Small Industries Development Bank of India (SIDBI), represents under one roof, the synergy of technology and finance which facilitates capability building and identifying market channels.

5. CONCLUSIONS

In conclusion, any effective technological capacity building exercise to enhance competitiveness of SMEs will have to address issues that can be broadly classified under the generic heads of a conductive policy framework, knowledge acquisition and dissemination, new technology demonstration and networking, and financing and institutional support systems. In each of the above categories, the barriers must be clearly identified and enabling strategies must be developed so that the technological capacity building process in SMEs becomes meaningful, in terms not only of international trade and business or of the environment, but also of making it responsive to the needs of the people. Constant learning by doing and interacting through networking and alliances is the hub of the activities for innovation with tacit knowledge playing a crucial role as exemplified by some of the Indian SMEs and clusters and more so by the examples from Italy, Germany and UK. Here human centered systems play a crucial role for flexible manufacturing and customer orientation. There have been many successful examples of industrial clusters in India, China, South Korea, Japan and European countries. Further study of these innovation models can be highly useful for greater cross-culture exchange of innovative experiences between different countries.

Acknowledgements: We are extremely grateful to DG, CSIR India and Director, NAM S&T Centre, New Delhi for facilitating writing of this paper. Thanks are also due to Mr. Gaurav and Mr. Pankaj Buttan for the preparation of various charts and slides of the write up.

REFERENCES

Amudeswari A.(1999): Technology Management: The role of CSIR in leather sector. Paper presented at 1999 R&D Management Conference-R&D as a Business. New Delhi; RADMA,UK/CSIR, New Delhi, 13-17.

Bansal, Rama Swami (2005): 'R&D Agency – Industry Partnership for Technology Development and Transfer in Indian Context', PhD Thesis, Birla Institute of Technology and Science, Pilani, Rajasthan. (360 pages +appendix)

Bhagaban Das *et al.* (2008): Strengthening SMEs to make Export Competitive. MPRA paper No. 7800, March 2008. http://mpra.ub.uni_muenchen.de/7800)

Bell M. and M. Albu (1999). Knowledge System and Technological Dynamism in Industrial Clusters in Developing Countries. *World Development*, 27, 1715-1734.

Best, M.H. (1990). The New Competition, Institutions of Industrial Restructuring. Harvard University Press, Cambridge.

Bischoff J (1998): Harnessing Technology. In *Small and Medium Enterprise-Challenges and Opportunities edited by J.S.Juneja*. New Delhi; All India Management Association, p85-101.

Bowonder B, Miyake T, (1988). Measuring Innovation of an Industry: An analysis of the electronics industry in India, Japan and Korea. *Science and Public Policy*, 15, 279-303.

Brandt D and Cernetic J(1999). Developing, Implementing and Assessing Human-centered Control and Information Technology. Paper presented at *14th World Congress of International Federation of Automatic Control*. Beijing, China, July 5-9.

Chari, Sharad (2000). The Agrarian Origins of the Knitwear Industrial Cluster in Tirupper, India. World Development, 28(3) p579-599.

Cooley M. (1991). *Architect or Bee ? The Human Price of Technology*. The Hogarth Press, London.

Directory of R&D Institutions, 2006. New Delhi, NSTMIS, DST, 2006, 346 p.

Gupta A.K. (1996). *Roots of Creativity and Innovation in Indian Society-A Honey Bee perspective.* Society for Promotion of Wastelands Development, New Delhi.

Honecker N., Gund J. and Sell R. (1999). Learning Organization-A Lasting Concept. *Human Factors and Ergonomics in Manufacturing*.. 9, 303-311.

Juneja J.S. (1998). *Small and Medium Enterprises-Challenges and Opportunities*. All India Management Association, New Delhi.

Kharbanda V.P., and A. Jain (1997). Indigenization and Technological Change at the Firm Level – A Case Study of Black and White Picture Tubes in India. *Technovation*, 17, 439-456.

Kharbanda V.P: Academia–Industry Symbiosis: Need for Technological Competitiveness in India and China. *China Report*, 2000, 36(1) 43-72.

Kharbanda V.P.: Industrial Clusters-The Hubs of Innovative Activities At Enterprise Level. *Tech Monitor (APCTT)*, Sep-Oct 2000, p36-40.

Kharbanda V.P.: Facilitating Innovation in Indian Small and Medium Enterprises – The Role of Clusters. *Current Science*, 2001, 80(3) p343-348.

Kharbanda V.P.: Learning Organizations: The Process of Innovation and Technological Change. *AI and Society*, 2002, vol 16. p89-99.

Kharbanda V.P.: Knowledge sharing in industrial clusters. Asia Pacific Tech Monitor, 2004 (Jan –Feb), pp1-17. (www.techmonitor.net), Online.

Kharbanda V.P in association with Mehta, Dhawal,.and Shukla, Paurav (2003): Networking Industrial Clusters: A conceptual Approach – The Gujarat Experience. In *Navigating Innovations: Indo-European Cross Cultural Experiences* edited by Dietrich Brandt. New Delhi; India Research Press, Vol 1. 313 p. (Two Vols.) p5-33.

Knorringa P. and J. Meyer-Stamer (1998). New Dimensions in Local Enterprise Cooperation and Development-From Clusters to Industrial districts. In:*New Approaches to Science and Technology and Capacity Building*. Unctad; New York.

MERADO, Ludhiana. (Brochure). September. 1999.

Niosi J. (1999). The Internationalization of Industrial R&D-From technology transfer to the learning organization. *Research Policy*, 28, 107-117.

Okuda K. (1983).The Role of Engineers in Japanese Industry and Education. *Journal of Japanese Trade and Industry*, 1, 23-26

Thyagarajan, G. (1998): Brainstorming on Identification of Technology Needs of Small and Medium Enterprises in Developing Countries (with focus on India), June 1,2,1998, COSTED, Chennai.

Chapter 13

The Role of Business Technology Center for the Development of Small and Medium Enterprises in Indonesia

Anteng Setia Ningsih
Ministry of Research and Technology, Republic of Indonesia,
Jl. MH. Thamrin No. 8 Center Jakarta 10340, Indonesia
E-mail: anteng@ristek.go.id

ABSTRACT

Good innovation ensures the success of product commercialization. Collaboration between researchers and industries/businesses generally leads to good innovation. A Business Technology Center (BTC) is an intermediary institution located in regions, established to help surrounding industries/SMEs in obtaining access to technology. Since 2003, the Ministry of Research and Technology have built nine BTCs in Indonesia. Innovation consultancy is the main service provided for SMEs since many of them are not aware of technological advancements. A BTC provides its services to SMEs, research institutions, local governments and industrial clusters. The main objective of BTC is to enhance the growth of local economy through technology. All BTCs are monitored and coordinated by Business Technology Center Network (BTC-Network) at the national level, which provides services such as information dissemination, website management, promotion and know-how support to all BTCs for training, and creates intermediary guidelines. These services are provided to local BTCs to enable them to provide consultancy to assist SMEs for upgrading their competencies related to product quality, human resources, information access, accessing funds, etc.

Keywords: *Business, Technology, Intermediary, Innovation, Consultancy, SMEs, Local economy.*

1. INTRODUCTION

One of the agenda of national development is to increase social welfare through a provision of conducive business and investment environment which is believed to be able to accelerate economic development, particularly the development of Small and Medium Enterprises (SME).

SME development aims to reduce economic gaps, to provide new job opportunities, to expand economic base, to improve local economy as well as to sustain the national economy. SME development, therefore, has been taken into account as one of the priorities, and considered as an essential target.

In the framework of SME development, there are several issues and obstacles to be addressed, for instance: market situation, marketing problems, lack of funding, unskilled human resources, lack of access to and knowledge on technology, and poor product quality.

In the context of access to technology and technology mastering, it is essential to discuss capacity building for organizations which provide technology/technical assistance for SMEs. One of the efforts that have been initiated by the Ministry of Research and Technology is the establishment of Business Technology Centres (BTC) at the local level. The core of BTC's activity is to involve participation of all central and regional components, including R&D agencies, universities, SMEs and other stakeholders.

2. BUSINESS TECHNOLOGY CENTRE (BTC)

2.1 Definition

A Business Technology Centre (BTC) is an organization established at the local level, with the main task to initiate networking between economic players and technology providers. The BTC's aim is to increase the growth of regional economy. A BTC functions as an intermediary for technology transfer process. Furthermore, BTC has the task to identify local economical potentials, technology diffusion, and technology transfer as well as commercialization of R&D outputs. The focus of BTC is to push local SMEs in acquiring and mastering appropriate technologies to produce better goods based on market demand, higher productivity, more innovation and stronger competitive advantage.

2.2 History

During 2000 to 2002, the Ministry of Research and Technology, Indonesia (RISTEK) collaborated with the German Federal Ministry of Education and Research (BMBF) in the PERISKOP Project (Evaluation of the Science, Research and Technology Landscape to Strengthen the National Innovation System). The project surveyed 10 regions in Indonesia scrutinizing problems of regional innovation capacity, *i.e.* Medan, Padang, Bandung, Yogyakarta Semarang, Surabaya, Mataram, Makasar, Samarinda and Manado. One of the results of the project is the Map of Indonesia research and technology landscape as shown in Figure 13.1.

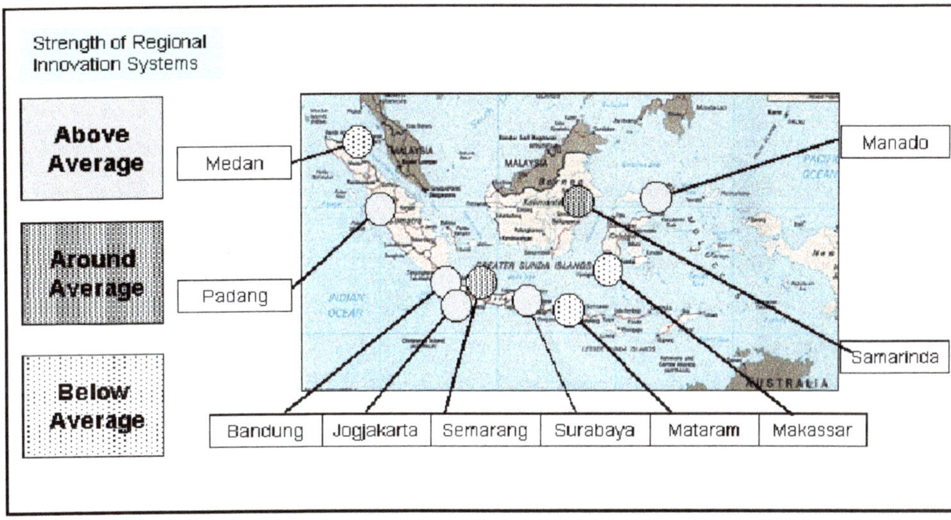

Figure 13.1: Regional Capacity Innovation[1]

Public R&D Agencies in Indonesia have many research results but many cannot be applied in industries. Therefore, PERISKOP study recommended building an intermediary institution as a hub for business and technology support/initiatives. Special attention shall be given to improve technological capabilities of the local SMEs. In the Indonesian context, the intermediary institution is known as Business Technology Centre (BTC).

2.3 The Vision and Mission of BTC

The vision of BTC is to increase regional growth by empowering SMEs, enabling them to have better access and bargaining power in entering dynamic markets, supporting them with networking power, connecting them to potential business partner and R&D institutions.

The Mission of BTC is[2]:

1. To consult and to assist SMEs on their technological needs
2. To connect SMEs to inward partners (local SMEs and related governmental institutions) and to outward partners (buyers, investor, research institution, BTCs in other regions).
3. To support SMEs in developing their business, in term of: access to market information, access to funding agencies, business management and utilization of proper and modern technology, enabling SMEs to master the required know-how to develop new and better technology-based products using local resources.

---

1 PERISKOP PROJECT, 2002.

2 BTC Guideline, The State Ministry of Research and Technology of Republic Indonesia, 2007.

4. To improve regional investment by promoting local brands and regional potentials.

2.4 Services

A BTC delivers services for SMEs, in term of the following activities

- ☆ Innovation consultancy
- ☆ Business matching and networking
- ☆ Intellectual property rights consultancy
- ☆ Conveying technology offers from universities and research institutions
- ☆ Supporting SMEs to promote their products and brands in exhibitions
- ☆ Settlement of enterprises from Indonesia and foreign countries
- ☆ Regional marketing

Innovation consulting is the main service of a BTC. The BTC shall conduct its basic services within the concept of one stop in innovation consultancy for local SMEs, in term of the following activities:

1. Visit the companies and industries
2. Analyze and optimize the value chain of SMEs
3. Find the right business partner from business, science, and research institutions.
4. Optimize the internal process, observe the actual situation for solutions, open the door to other institutions in others areas to get information/ technical assistance. If the required assistance is not available at regional level, assistance at national level could be pertained.
5. Identify risk and opportunity. One way is to identify other institutions in other areas that have potentials to offer certain technology assistance/ information especially for SMEs.
6. Access funding agencies for R&D activities.
7. Stimulate projects and find the right partner by arranging events, expert's exchange, and meeting of experts.

If a BTC has obtained the necessary capability in innovation consulting, an expansion of services is possible. Promising expansion of BTC services can include the followings:

1. Foreign trade advice
2. Intellectual Property Rights (IPR) advice
3. Initiate the establishment of start-up companies, based on the output of R&D institutions
4. Commercialize R&D results to industries
5. Arrange/initiate placement of students for internship/job training in companies or R&D institutions

6. Financial support for innovation process from designing to marketing
7. Support and organize joint-exhibition for company/SME at national or international expo event
8. Innovation audit
9. Attract domestic and foreign investors to the region
10. Regional marketing support and promote regional potential
11. Initiate, develop and/operate industrial clusters

2.5 Roles of BTC

A. Assisting Research Institution

Research institutions have many research results about new products, new processes and new technologies. They have many well trained people to do research. The problem is they cannot sell them. Most of the researches are not market oriented. Moreover, many researchers are not aware of IPR. Therefore, the role of BTC is to provide the followings assistance:

- ✰ Promote research results.
- ✰ Assist researchers about IPR
- ✰ Obtain partner from industry and finance institutions
- ✰ Train researchers about enterprises

B. Assisting SMEs

Innovation consulting is the most important service of a BTC. The BTC has to visit the SMEs which need advice in technology to obtain verifiable information about technology needs, to build trust, and to transform technology needs to technology demand. Visiting the SMEs may help to analyze the value chain and give first advice to optimize internal processes. Included to innovation consulting, BTC can help the SMEs in:

1. Risk identification.
2. Obtaining advice from other institutions.
3. Connection to financial resources: investors, connection to investors and financial support (bank).
4. Project stimulation and search for cooperation partner: arrange expert meeting for SME's and expert exchange, generate a new project and event.

BTC helps SMEs to access information about IPR, for instance, by organizing IPR workshop to show the advantages of having own patent or to describe how to use patent database.

Innovation consulting requires forum like business gathering. However, business gathering should be followed by matchmaking activities. This activity is regular meeting between researcher/university with industry/SMEs. If the goal is to generate a research project, BTC match companies which are aware about their technology needs or which are looking for new breakthroughs, with experts/researchers. If the goal is to commercialize research results, BTC gathers researchers willing to establish start-up companies and match them with potential investors and business incubators.

C. Enhancing Industrial Cluster Development

In the region, a BTC collaborates with regional science and industry clusters. The role of the BTC is to:

1. Initiate idea to build industrial cluster based on prime competency of the region
2. Perform SWOT analysis (Strength, Weakness, Opportunity and Threat) to identify growth potentials and pitfalls to avoid
3. Assist industrial clusters by conveying information about market, technology advancement and availability of experts.

D. Supporting Local Government

BTCs support local governments in regional marketing, emphasizing on the strengths of natural resources, culture, investment regulation, human resource, regional economy, existing industries, unique products, etc. In this case, BTCs have to build collaboration with other institutions in the region to commercialize and to promote research results of local institutions.

Local economy depends on its industries in the region. In the context of Indonesia, SMEs dominates the industrial structure. Therefore, SME is the most important element in Indonesia's local economy. Strong SMEs shall contribute much to the local economy. In this case, BTCs shall help local governments to empower SMEs to strengthen the local economy.

3. LOCAL BTCs

The Ministry of Research and Technology of Republic of Indonesia has developed following nine BTCs since 2003:

1. BTC – Batam

BTC – Batam was founded in Batam, Riau Island Province in 2003. The shareholders are Batam Government, Batam Authority and Batam Polytechnic. BTC Batam focuses on electronics technology.

2. BTC – BPTI Yogyakarta

BTC – Yogyakarta was established in Yogyakarta in 2003. The shareholders are Gadjah Mada University and Yogyakarta Information Technology Development Board (BPTIY). The BTC focuses on information technology and handicraft.

3. BTC–Jawa Timur (East Java)

BTC – Jawa Timur was founded in Surabaya, East Java in 2005. The shareholders are Surabaya University and Business Development Service (BDS). The BTC focuses on fish preservation technology such as frozen, canned and smoked fish/shrimp, and fish/shrimp nugget.

4. BTC – Makasar (in South Sulawesi)

BTC Makasar was founded in Makasar, South Sulawesi in 2005. The shareholders are Makasar Management Consulting (MMC) Foundation and Business Development Service (BDS). BTC Makasar focuses on agro industry and fisheries.

5. BTC – BBRC Bandung

BTC – Bandung was founded in Bandung, West Java in 2005. The shareholders are Indonesian Basic Science (LIPI) and Bandung Business Resources Center (BBRC). The BTC focuses on information communication technology. BTC Bandung is initiating ICT cluster in Bandung together with Tarina Consulting and Small Business Digital Community (SBDC).

6. BTC – UKSW Salatiga

BTC – Salatiga was founded in Salatiga, Central Java in 2006. The shareholders are Christian Satya Wacana University (UKSW). The BTC focuses on training on information technology. In 2008, BTC-Salatiga together with Ristek, Fraunhofer – Germany, and Central Java Forum for Economic and Resources Development (FPESD) are initiating to develop wooden cluster in Central Java. The project focuses on environmental friendly technology for furniture and handicraft production in Central Java and to be proposed for funding to the Commission of Europe Union.

7. BTC – JRC Jababeka Research Center

BTC- JRC (Jababeka Research Centre) was founded in Karawang, West Java in 2007. The shareholders are President University and Jababeka ltd. BTC Jababeka focuses on empowerment of SMEs in Jabebaka industrial area and sell the research results from President University.

8. BTC – Kendari

BTC – Kendari was founded in Kendari, Southeast Sulawesi in 2007. The shareholders are Kendari Government. The BTC focuses on agriculture products.

9. BTC – Jawa Tengah (Central Java)

BTC Jawa Tengah was established in February 2010. The shareholders are Research and Development Board of Jawa Tengah (Balitbang Jawa Tengah). Balitbang Jawa Tengah is a part of Jawa Tengah Local Government. BTC Jawa Tengah focuses on tourism, handicraft and agriculture.

In 2010, BTC-Network is initiating to establish a BTC in South Sumatera. The shareholders are Research and Development Board of South Sumatera. The BTC is called Serambi Difusi Iptek (SDI) and its activity will be focused on agriculture, food processing, and handicraft products.

4. ESTABLISHMENT OF THE LOCAL BTCs

The establishment of a BTC shall be based on the region's need from which the following questions shall be answered: (*i*) Who should establish BTC? (*ii*) Who can access the BTC services? (*iii*) Where should the BTC be established, and (*iv*) How to establish?

4.1 Who should Establish BTC?

BTC shall be established by the regional government, universities or community members by involving other key stakeholders. As long as the objectives of such establishment and operation are not contradictory to the existing legal framework, and in line with the concept of local SME development, anyone may initiate the establishment of BTC at regional level. In the context and application of BTC establishment itself, it is important to remember that such establishment is purely to support local SME, with the assistance from larger network of BTCs and other concern institution, in the implementation of technology in developing SMEs. The ultimate goal is to strengthen the SME competitiveness.

4.2 Who can Access the BTC Services?

BTC services, particularly delivering information on technology advancements, shall be provided for all local stakeholders (regional government, association, universities, SMEs, businessmen, research institution, etc). One shall not forget, however, that the main role of a BTC is to provide innovation consultancy. Within the establishment of institutional BTC, divisions can also be made to efficiently handle and to control different access for the benefit of the stakeholders.

4.3 Where a BTC should be Established?

A BTC shall be established either at provincial level, regency or municipality level. However, considering that SME clusters tend to be more at regency/municipality level, it is recommended that in the long run all BTCs should be established at that level. All BTCs will interact strategically within BTC-Network.

4.4 How to Establish a BTC?

There are four stages to establish a new BTC, *i.e.* need assessment, pre-socialization, socialization, and establishment. The details of the stages will be described as under:

1. Needs assessment stage

Needs assessment is necessary to determine where a BTC should be located. Important tasks to be considered are as follows: (1) clustering SMEs based on their most prominent/valuable products and compare them with the demand of national/international market; (2) reviewing the regional development plan, (3) identifying potential local partners who may assist the establishment of the BTC. Local partners must be selected carefully because they will act as the long term partner. The partners must understand and have extensive experience at the local level.

2. Pre-socialization Process

Before entering the socialization stage, it is important to hold meetings with formal leader (head of local government officials) as well as informal leaders (chairman of local business association, NGOs, craftsman association, chamber of commerce). On that occasion, presentation of the concept, the role and the function of BTC for regional development, must be made and followed by in-depth discussions. It is necessary to collect feedbacks, to identify potential obstacles, and to ensure the commitment of local stakeholders. The result of this stage is invaluable for the next step of establishment.

3. Socialization of BTC

In this stage, one has to organize formal presentations, discussions, and public exposures on the topic of how BTC may improve regional development. During those events, the following topics shall be brought forward:

- ✰ Understanding the core concept of BTC for the interest of local SMEs and all stakeholders
- ✰ Explanation on the concept, strong points, and benefits of BTC for local SMEs development
- ✰ Case study and success story of BTCs from all over the world.

Advance socialization stage can only be carried out if there has been a certain level of interest and commitment from the local stakeholders to establish BTC in their area. This socialization is more targeted at providing the stakeholders with detailed and technical information regarding BTC establishment. In this process, briefing or presentation shall cover the following details:

1. Organizational and managerial concepts of BTC
2. Scope of activity of the BTC
3. Benefit and practical issue/obstacles in establishing the BTC
4. Establishment of the BTC

4. Establishment of BTC

The founders (local businessman or local government) acting as facilitator for BTC establishment shall prepare and provide the following review papers and legal documents:

- ✰ Legal form and framework of BTC
- ✰ Articles of Association of BTC
- ✰ Feasibility study of BTC
- ✰ Working plan of BTC (short, medium and long term)
- ✰ Strategic plan of BTC
- ✰ System and SOP for daily operational BTC

Minimum Requirements of a BTC

To function well and to be acknowledged by the BTC Network, a BTC shall fulfil the following requirements:

1. Possess a definite legal entity, Limited Company (PT), Common Service Board (BLU) or Cooperative.
2. The founder must have certain capability, strong commitment, entrepreneurship and leadership characteristics.
3. Have competent and fully dedicated staff.

4. Maintain healthy transparently managed financial performance
5. Have clear annual target which includes performance indicator

5. BUSINESS TECHNOLOGY CENTER - NETWORK (BTC - NET)

In 2007, The Ministry of Research and Technology initiated the BTC-Network to coordinate the existing local BTCs. The BTC-Network is placed under the coordination of the Deputy Minister for S&T Utilization and Dissemination. BTC-Network is the coordinator for all BTCs in Indonesia and a part of the Indonesia's National Innovation System. The stakeholders consist of state ministries, industrial chambers, universities, and other institutions at national level. BTC-Network is the initiator and the driver for local BTCs. BTC-Network functions as the pioneer, facilitator, motivator and moderator for components within the supporting network. The network may initiate and carry out numerous services and maintain cooperation between the partners (also related ministers, R&D institutions, chamber of commerce, universities, etc.) while disseminating the information about the advantages of establishing a BTC. BTC-Network shall assist local government on the establishment of a BTC from the initiation stage up to the implementation and operation stages. Such assistance shall also encourage diffusion and dissemination of practical science and technology output, so that the system itself should be sustainable in the long run.

The BTC-Network ensures the exchange of information at national level and oversees cooperation agreement between individual BTCs with other partners in Indonesia. To ensure promotion and to create a stronger and more positive image for the society, a unique Corporate Identity (CI) of BTC shall be established.

The BTC network provides the following services:

1. Disseminate information to BTCs, corporate identity and brand image
2. Manage the website
3. Develop a national database of technology
4. Manage the network's funding
5. Create and promote cooperation, advocacy and support for individual BTCs
6. Develop tools necessary for BTC operations
7. Organize training for BTC members
8. Promote cooperation among the existing industrial clusters
9. Initiate projects with local BTCs to attract international funding

Additional functions and roles are open to be decided by the Steering Committee of BTC Network.

6. THE POSITION OF BTC IN THE INDONESIAN NATIONAL INNOVATION SYSTEM

There are three main elements of the Indonesian National Innovation System, *i.e.* (1) institution, (2) networking, and (3) policy. Institution includes research

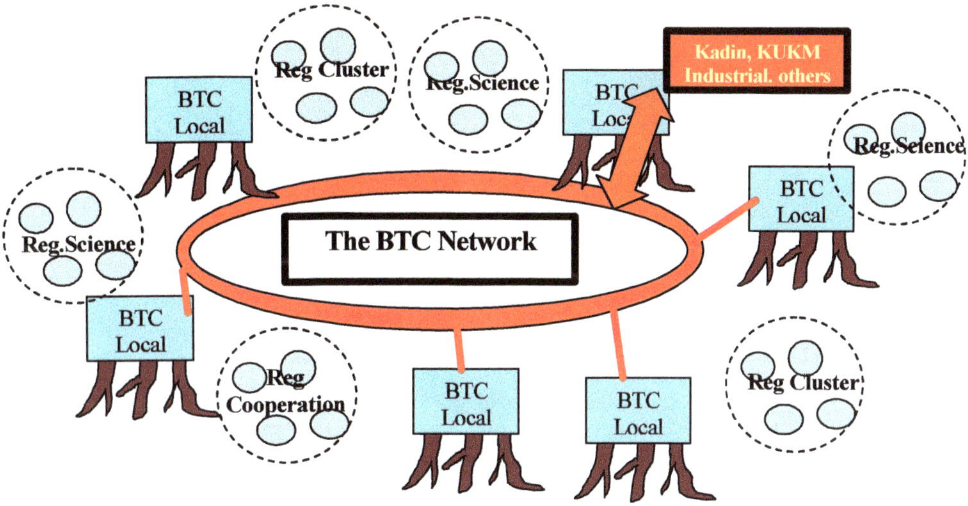

Figure 13.2: The BTC Network System[3]

institutions, universities, banks, industrial centres, science parks, Business Development Services (BDS), Business Technology Centres, incubators, technology or industrial parks and centres of excellence. Each element is under coordination by different ministries.

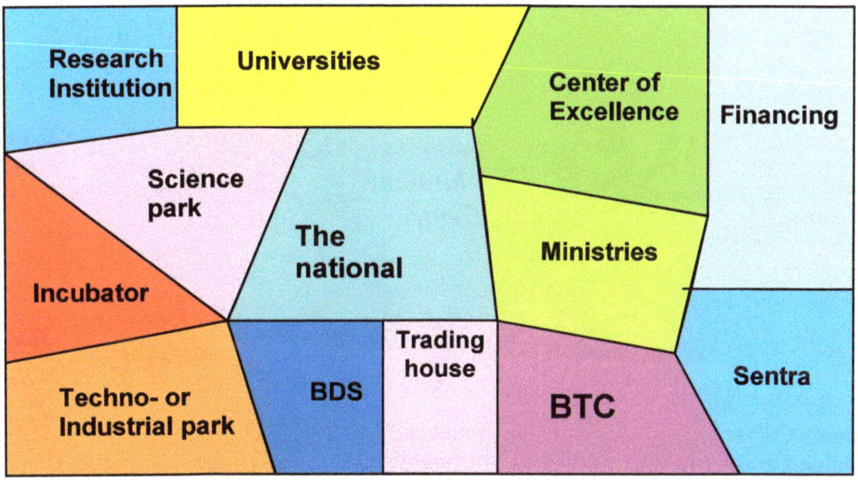

Figure 13.3: The Elements of National Innovation System[4]

---

3 BTC Structure and Task. Presentation from Franz Gelbke. Advisor Minister for BTC Development in Indonesia. November 1, 2007. Jakarta.

4 *ibid.*

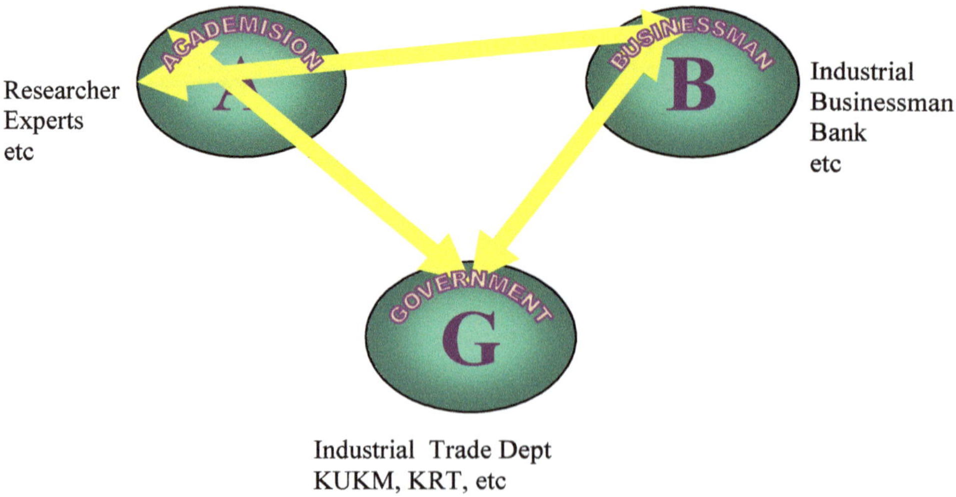

Figure 13.4: Interaction Between Academic, Business and Government in the National Innovation System[5]

Indonesian Innovation System describes the interaction between academic (*a*), business (*b*) and government (*g*). The government of Indonesia has the responsibility to help SMEs to become more competitive by generating conducive business climate through regulations and incentives such as tax regulation, incentive research programs, market information and training. The relevant governmental agencies are

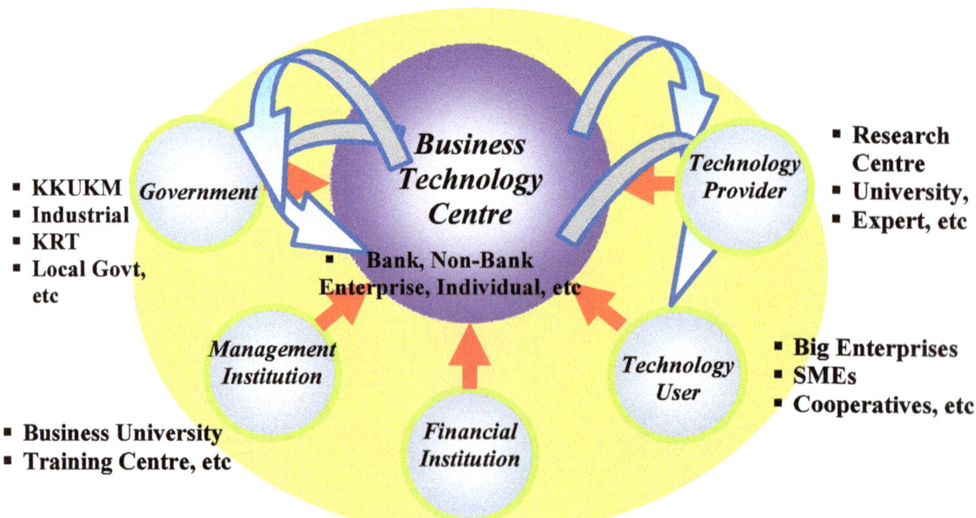

Figure 5: Interaction ABG Elements through BTC[6]

5 *ibid.*

6 *ibid.*

the Ministry of Cooperative and SME, the Ministry of Industry, The Ministry of Finance, The Ministry of Trade, and The Ministry of Research and Technology. Universities and research institutes are the technology providers. Meanwhile, SMEs and big companies are the business, which generate profit and values.

BTC is positioned between the A-B-G as follow. A strong interaction among the three elements through BTC shall build a strong innovation culture.

BTC Network to Support the Indonesian National Innovation System

In 2009, BTC Network planned to host a national workshop on Innovation System in collaboration with the Federal Ministry of Research and Education (BMBF) Germany, which aims to strengthen the Indonesian Innovation System. This workshop is called Modular Training Concept Implementation (MTCI) and consists of three modules: National Innovation System (SINAS), Regional Innovation System (SIDA) and Innovation Manager (Innoman)[7]. The workshop however could not be held in 2009 and is postponed to seek the convenience of the President of the Republic of Indonesia who will officially launch the workshop himself.

The purposes of this workshop are as follow:

1. To strengthen the Indonesian National Innovation System through the collaboration of key ministries and governmental bodies;
2. To implement the innovation system in the region based on their strength the SINAS Training

6. CONCLUSION

BTC as an intermediary institution has significant roles in developing Small and Medium Enterprises in Indonesia and may help local governments to developed regional economies. The initiative to establish a BTC has to come from the region because all BTC activities prerequisite local stakeholders' support (local government, local commerce, university, industries) since most of BTC's activities involve interaction with others institutions within the region. At the national level, BTC Network (RISTEK) coordinates and advices local BTCs. Moreover, BTC as an intermediary institution has also the role to run the Regional and National Innovation Centres. Implementation of Regional and National Innovation Centre needs intermediary to bring innovative technology from research institute to the market place (industries/businesses).

REFERENCES

Gelbke, F. 2007. Structure BTC and Task (Presentation). RISTEK. Jakarta, Indonesia.

Ristek 2002. Periskop Project. Evaluation of the Science, Research and Technology Landscape to Strengthen the National Innovation System, Jakarta, Indonesia

Ristek. 2007. BTC Guideline. BTC Network Team, Jakarta, Indonesia

Ristek. 2009. Project Description of Integrated Training for SINAS, SIDA and Innoman. BTC Network Team. Jakarta, Indonesia.

7 Project Description of Integrated Training for SINAS, SIDA and Innoman. BTC Network Team. RISTEK 2009.

Chapter 14

The Role of Science and Technology for the Development and Growth of Small and Medium Enterprises in Malawi

***Jonas F. Mwatseteza*[1], *Felix Maoni*[2] *and Lyson John Kampira*[3]**
[1]*Chancellor College, University of Malawi,*
E-mail: jfmwatseteza@chanco.unima.mw
[2]*Ministry of Industry and Trade, Malawi*
E-mail: fmaoni@yahoo.com
[3]*Department of Science and Technology, Malawi*
E-mail: ljkampira@yahoo.com

ABSTRACT

The Government of Malawi endeavors to promote, support, and facilitate local private enterprise efforts in order to make Malawi a manufacturing based economy, capable of creating and sustaining a competitive advantage in domestic and international markets. The initiatives aim to embrace the interests and needs of enterprises at all levels, which include small, medium and large scale enterprises so that they can become more competitive in their operations. Such initiatives include the formulation of relevant policies such as the Integrated Trade and Industry Policy, and establishment of appropriate technology institutions and departments. This paper examines Malawi's progress in this sector as it relates to engagement of technology for its growth.

Keywords: *Malawi, Small and Medium enterprises, Policies, Technology, Initiatives.*

1. INTRODUCTION

Malawi's economy relies heavily on agriculture, with a per capita GDP of about US$ 205 of which 60 per cent is from commercial and smallholder agriculture,

13 per cent from the manufacturing sector and 27 per cent from others. The country's population as of 2008 stood at 13 million of which 80 per cent are employed in the agricultural sector while the industrial sector accounts for less than 20 per cent of employment. As of the year 2000, agriculture provided profit based income to 25 per cent of households in Malawi, and mostly those in the rural areas.

The small and medium enterprises (SMEs) in Malawi are defined as those entities that employ 5 to 20 and 21 -100 individuals with a turn over of US$830 to 2800 and US$27580 to 68960 respectively. The SMEs are therefore a very important sector in the livelihoods and economy of Malawi.

Growth and development of Malawi's industrial sector is hampered by a number of structural issues with only agro-processing, especially tobacco, food products, tea and beverages accounting for the manufacturing sector. The high cost of transportation contributes to high costs of both Malawi's domestic and export products with the surplus labour being the only comparative advantage, which must be used in high value-added activities. This calls for targeted investment in technology by both new and existing investors.

With low GDP, the need for increasing cash income compels people to engage in businesses and for them to do this they need capital. The micro, small and medium enterprises (MSMEs) sector in Malawi constitutes an important and meaningful component of economic development. The Government of Malawi maintains the development of micro and small-scale enterprises as a core element in government strategy to support economic growth, employment and poverty eradication. The Government of Malawi Integrated Trade and Industry Policy recognizes the role of SMEs in the economic development of the country.

2. SCIENCE AND TECHNOLOGY INITITAIVES OF THE GOVERNMENT TO BOOST MICRO, SMALL AND MEDIUM ENTERPRISES

2.1

In recent years the Government of Malawi has been strengthening support to sectors other than agriculture, which have a potential for enterprise development and employment creation. The government has undertaken initiatives to embrace the interests and needs of small and medium and also large enterprises so that they can become more competitive in their operations. The support includes the establishment of:

(*a*) Micro, small and medium enterprises support institutions to provide credit and specialised training services; and

(*b*) Research; Industry and Trade (SMEs department); commerce and industry; and science and technology institutions to:

  (*i*) Enhance the development of industries in Malawi, especially the MSMEs through conducting industrial research and technology development, transfer and commercialization;

(*ii*) Direct donor programme support towards SMEs and provision of good will support towards the establishment of conducive environment and friendly legislation for the MSME sectors.

The SME sector is being recognized for its potential to promote economic growth and development through a subcontracting (out-sourcing) and partnership linkage with the large-scale industries. The financial management, commerce and industry, research, science and technology institutions have the capacity to improve the productivity of SMEs, through provision of technical support for private sector development, facilitation of the necessary industrial linkage between the SMEs and the large scale enterprises (LSEs) and provision of technology support for import substitution manufacturing.

2.2 The Framework of Operation

The following programmes for technological and other specialized services have been highly useful for achieving the objectives to enhance the development of the industrial sector in Malawi.

2.2.1 Industrial Research Services

Includes appraisal through contract research of raw materials as inputs into industrial manufacturing processes and evaluation of intermediate products to improve quality and process efficiency, and develop new products.

2.2.2 Technology Development and Assessment Services

Includes development, adaptation and assessment of technology suitable to local condition and provision of guidance for selection of technologies for the manufacture of new products.

2.2.3 Technology Information Services

Includes provision through the Technology Information Services (TIS) library, information and documents related to technology; information searches on technology in computer databases or databanks in international networks; provision of process know-how information and industrial investment opportunity profiles and related information. The following databases are being developed to enable provision of information to the industry:

(*i*) *Industrial Raw Materials Database*: contains information on local and imported raw materials used in manufacturing industry in Malawi;

(*ii*) *Project Database*: contains information on projects completed, undertaken and proposed, impact assessment reports on projects, etc;

(*iii*) *Company Database*: contains company profiles and provides information on company activities, which can be selected, for subcontracting in technology transfer projects;

(*iv*) *Database of Specialists*: contains profiles on specialists in various fields at local and foreign levels who can be consulted in specific projects of interest; and (*v*) *Database of Appropriate Technologies in Malawi*: contains information of appropriate technologies being used in Malawi.

2.2.4 Engineering Services

Includes design and fabrication of machinery, equipment and spare parts, production process and plant layout.

2.2.5 Consultancy Services

Includes technical analysis and appraisal of industrial projects, economic feasibility studies and preparation of business plans, identification and selection of appropriate technologies for micro and small enterprises, applied research to improve industrial processes, assistance in quality improvement and skills capacity building.

2.2.6 Government Schemes

These include One Village One Product (OVOP) programme in selected areas with resource that give the community comparative advantage over others, irrigation and farming systems which increase agricultural productivity, establishing cooperatives in various sectors.

2.2.7 Establishment of Management and Technical Institutions

These include Technical Entrepreneurial Vocational Education Training Authority (TEVETA), Malawi Entrepreneurship Development Institute (MEDI), Development of Malawian Enterprises Trust (DEMAT), Small Enterprises Development Organisation of Malawi, (SEDOM) where entrepreneurs can learn business management and technical knowledge for managing and establishing entrepreneurial activities.

2.3

Some of the accomplished activities to enhance development of SMEs in Malawi

These include:

(*a*) Implementation of renewable energy programmes such as sizing and installation of solar systems to improve access to electricity and reduce environmental degradation;

(*b*) Design and installation of biogas plants

(*c*) Development of national policies for the MSMEs and the national policy on Science and Technology;

(*d*) Development and dissemination of technologies such as the heat exchanger for agro-processing, treadle pump for small-scale agricultural irrigation, bakery ovens, electric egg incubators, oil expellers, ethanol production, etc.

(*e*) Development and promotion of technologies for traditional salt extraction from saline soils;

(*f*) Implementation of the Appropriate Technologies, which largely benefited the SME sector;

(g) Implementation of the Technology Programme Objective in the Enterprise Development and Employment Creation Programme;

(h) The Press Trust, which is a public trust, is currently funding some of the research and development projects in the Country.

2.4 Opportunities for SME Growth in Malawi

The SME sector in Malawi has a lot of opportunity to grow due to the favourable environment. There is appreciable demand for most products and services from SMEs on both local and international markets. In addition, the opportunity for further growth is due to:

(a) Low to medium income households being major buyers of both goods and services.

(b) Creation of linkages with industries and public sector in general to provide goods and services.

(c) Government has several bilateral, regional and international trade agreements for creating more market opportunities for locally produced items, including those produced by SMEs.

(d) Ever expanding road network and infrastructure, which is increasing accessibility even to the remote areas in the country.

(e) Increasing inflation indicates that demand is higher than supply, especially for basic goods and services (necessities), which are produced and provided by SMEs.

(f) Improved telecommunication in terms of both network coverage for cell phones and wireless landlines. The telephone network is also associated with increased internet coverage. This gives SMEs an opportunity to connect with customers or services beyond their immediate environments.

2.5 Current Threats Facing SMEs Development in Malawi

The country has supported, through funding of operations, equipment and machinery, and provision of business premises to institutions whose sectoral policies support the SME sector. Such institutions include Malawi Industrial Research and Technology Development Centre (MIRTDC), MEDI, DEMAT, SEDOM, etc. The challenges facing SMEs in Malawi include credit facility, raw materials and inputs, markets, entrepreneurship and management skills, infrastructure, fiscal policies, technology and production costs (ITIP Malawi GOVT). The policies that support SMEs in Malawi may be good but actualisation of those initiatives is not consistent with the assertion. The problems range from inherent flaws in the support units to lack of clear communication channels between the various user to the intended recipients.

Technology policy instruments that exist have been formulated and made available across the industrial sector. Due to lack of explicit affirmative policies targeting the growth of the SME sector, the assumption is that the SME sector competes with the LSEs in sourcing funding, investment opportunities, production, markets, technology acquisitions, land acquisition, licensing, etc., thereby leading to industrial competition rather than collaboration and integration. This results in most SMEs becoming uncompetitive and the trend is that the majority of SMEs invest in trading rather than in manufacturing. This situation is unfavourable for growth of the SME sector and national development. In general due to inadequate funding, there is a tendency to the supporting institutions to charge commercial fees to the SMEs, which they seldom afford.

In general, households use the proceeds from the SMEs income for households sustenance, leaving very little or nothing for reinvesting. The funding arrangements available to most SMEs sectors also provide further challenges to the growth of the SMEs. The financial arrangements may require weekly repayments to service the loans leaving very little room for further investment.

Reinvestment in SMEs growth is under pressure from loan repayment obligations and can be viewed as a risk venture by most SMEs. The arrangements for loan repayments are a de-motivating factor for SMEs to venture into additional loan once they are through paying for the initial loan. It is evident that loan repayment exerts a lot of financial pressure on the household. A survey in 2000 indicated that only 15 per cent of SMEs reinvested to boost their business.

Financiers are also reluctant to fund product research and development since once production has commenced such venture does not guarantee immediate increased sales that would in turn ensure success of loan repayment. The operators are likely to invest in raw materials for increased production and not so much in terms of improving quality. Improved quality of product or service increases added value and is a recipe for penetration of new and lucrative markets.

The financiers most often impose unfavourable conditions such as demanding collateral in form of non-movable assets, which is beyond the capability of many prospective entrepreneurs. Cash deposits have also been demanded by some lending institutions before they can lend out certain amounts of money. It is evident that the lending institution does not want to shoulder any slightest element of risk when lending. In such difficult circumstances usually entrepreneurs are not willing to go for repeat loans. The loans are sometimes disbursed in small amounts even when being loaned to large group of individuals making up the enterprise. Small loans lead to inadequate investment for any meaningful profits.

The majority of the SMEs in Malawi revolve around natural resources including crops, forestry, fishing and small-scale mining among others. Where there is no source of finance to re-invest in the growth of the activities, it is very unlikely that loans can be channelled towards environment sustainability of the resource raw material. This in the long run, threatens the very existence of the SMEs sector.

The country does not also have an internationally recognised institution that certifies its brands so that the SMEs can penetrate lucrative international markets such as the European Union (EU). This pushes more ambitious enterprises to seek such services outside the country at a high cost and may sometimes take too long to get results. Engaging outside services also risks a situation where the products can be mishandled and contaminated on the way to the laboratory leading to false results on the representative sample and financial losses are incurred.

3. CONCLUSION

There is need for deliberate approach that not only funds SMEs to improve their financial base but also direct investment for adopting cleaner and sustainable technologies for the Malawi SMEs. A model for technology investment in SMEs should be in such a way that the operators have no financial risks in order to increase rate of adaptation or adoption of technology. It is very unlikely that SME players will be willing to take a perceived risk in trying out technologies if they have not been evaluated in their business venture, as their main priority is to repay the loans. It is therefore necessary to initiate a suitable venture capital mechanism dedicated to the needs of the SMEs to take care of the risks involved in the adoption of new technologies in this sector of industry. The country also needs to set up proper communication channels and support mechanisms to encourage owners of SMEs to seek newer innovations for the development of their businesses.

REFERENCES

DFID 2000. Malawi National Gemini MSE Baseline Survey 2000, Lilongwe, Malawi.

DFID, 2002. Malawi District Level Gemini MSE Baseline Survey Mangochi and Lilongwe Districts 2002, Lilongwe, Malawi.

Government of Malawi, 1998. Integrated Trade Policy, Lilongwe, Malawi.

Government of Malawi; Ministry of Commerce and Industry, 1998. Micro and Small Enterprises Policy Statement, Lilongwe, Malawi.

Michael, A. McPherson, 1991. Growth and Change in Malawi's Small and Medium Enterprise Sector, 1986-1990.

NSO, 2008. 2008 Population and Housing Census Preliminary Report: Zomba, Malawi.

Peter, Quartey, 2001. Regulation, Competition and Small and Medium Enterprises in Developing Countries.

Chapter 15

SMEs and Technology Transfer in Nepal

Dilip Subba
Secretary
Nepal Academy of Science and Technology,
Khumaltar, Lalitpur, G.P.O. Box 3323 Kathmandu, Nepal
E-mail: dilipsubba@mail.com.np, dilipsubba2009@yahoo.com

ABSTRACT

Despite non-existence of a strong technology transfer mechanism in Nepal there are many examples of technology transfer and foreign investment in the public as well as private sector companies in the country. The technology transfer agreement in the private sector is found to have materialized through negotiations between technology recipient and provider companies. This has taken place through a variety of mechanisms such as direct foreign investment via joint ventures, technical collaboration, import of machinery and equipment, technical assistance through human resources, etc. While these are more formal modes of technology transfer, a lot of technology gets transferred informally through books, journals, promotional literature and personal contacts. There is not much data on the actual technology improvement and up-gradation in the Nepalese SMEs, but the observations show that there has been a slow transformation towards a higher technology level. The only data available relating to technology transfer agreements are of the companies for which permission is sought under Foreign Investment and Technology Transfer Act. Adaptation of technology and innovation of new technology is not happening in Nepalese industrial sector primarily due to lack of R&D facilities and capital. SMEs themselves are not in a position to set up R&D infrastructure.

The national apex bodies of science and technology like Nepal Academy of Science and Technology (NAST), which has mandate to facilitate appropriate technology transfer can be helpful in technology transfer mechanism.

Keywords: *Policies, Organizations, Status, Modality, Constraints, Technology transfer, SMEs, Nepal.*

1. INTRODUCTION

Technology transfer has become a global phenomenon for industrialization especially in the developing countries. Transfer of technology plays a prominent role in the economic development of a country and it is believed that economic growth of every nation is inextricably linked to the successful transfer of technology. Theoretically, the central idea of technology transfer is to provide a gamut of benefits to a country. Some of these benefits include: (*a*) it provides material means such as capital goods and machinery, necessary for economic development, (*b*) it is the only instrument for the transmission of technological knowledge, and (*c*) proper technology fosters healthy competition and keeps in check inefficient industries (Hug, 1997).

It is generally assumed that due to the large technological gap that exists between the developed and developing countries the latter is generally dependent on the technologies developed by the technologically advanced nations and transferred there after. For some, technology transfer takes place whenever technical knowledge, a technique or a device, which emerges from, or is developed by one group becomes taken up and used or applied by another. This covers both hardware (material devices and designs of such devices) and software (incentive ideas and the know-how needed for implementation of such ideas) of technology transfer. For some others it is a progressive and continuous effort to move technical devices, material, methods and/or information from the point of discovery or development to the users; *i.e.* it occurs whenever the users and producers of the new applied work are different. Technology is not merely the ability to use but to understand the process and improve on it whenever necessary.

The importance of technology in the modern day economy can hardly be exaggerated. It is the technological ability that makes the firms and companies competitive. Promotion of small and medium enterprises (SMEs) can be considered as a prominent approach to economic development, particularly, in developing countries. This sector not only mobilises scarce resources but also caters to the needs of fragmented domestic market. SMEs, thus, can be viewed as one of the important sectors that significantly contribute to the gross domestic product (GDP), create employment and earn foreign currency through export.

Various studies conducted have indicated that usually in the developing countries, SMEs are found to contribute 40-60 per cent of the total output or value added to the national economy. Emphasis to be given in the promotion and development of SMEs in developing countries is much more urgent where about 50 percent people are unemployed. The SMEs in developing countries produce goods and services of moderate quality largely with the application of outdated technologies and therefore have to face tough competition with imported products. This situation is expected to further worsen in the days to come, as they have to compete with international competitors due to increasing globalization, the widening free trade and phasing out of tariff barriers. The changes that have been taking place in the international economic scenario have definitely brought challenges for the SMEs. However, these changes at the same time have opened up tremendous business and market opportunities for this sector (IIDS 1994, Dawadi 2000, Nepal, 2002).

In most developing countries, problems faced by SMEs are more or less similar in nature. The major problems and constraints faced are in the fields of policy and legal framework, finance, entrepreneurship, management, socio-cultural values and technology. One of the main factors that influences the success or failure of enterprise is technology. The best use of technology no doubt enables enterprises in reducing cost of production, maintain consistency in quality, improve productivity and finally develop the competitiveness of the enterprise.

Most developing countries are aware of the importance of technology but do not give due attention while the process of technology transfer is taking place. SMEs are heavily dependent on the imported technology and always remain dependent on imported skills for maintenance and operation of any plant. This is only happening as because these countries have not given proper consideration to the technology transfer process. Adaptation of any imported technology to the local needs has always been of less priority.

If promoted on efficient lines, industrialization can hold out prospects of abundance of manufactured goods, increased employment, fuller utilization of natural, human and capital resources of the country, improved balance of payments and above all, greater efficiency and modernization throughout the economy. It is an important pre-requisite for economic development and transition from traditionalism and agrarian economy to modernization, especially in developing countries.

2. INDUSTRIAL DEVELOPMENT IN NEPAL

2.1 The Industrial Sector in Nepal

Nepal's industrial sector is characterised by a large number of small establishments, highly dispersed geographically, and dwelling between the formal and informal sectors of the economy, contributing less than 10 per cent to GDP and absorbing only one percent of the labour force. The industrial sector comprises of both modern and cottage and small-scale industries. The cottage industries plays a crucial role in generating employment for the large number of people in agriculture and in relieving population pressure on agriculture as 80 percent of the total population is engaged in agriculture. However, the contribution from cottage industries sector is merely 20 per cent to the total manufacturing output.

Industrialization is viewed as a major instrument of progress, modernization and social changes in Nepal. Industrialization in modern Nepal began only after 1951; however, this dates back to 1936 with the establishment of Biratnagar Jute Mill.

SMEs have been an important foundation of industrial development in Nepal. Even before the initiation of the process of industrialization in 1936, SMEs, especially the cottage industries were the basis of supply of processed products of the country. Considering the fragmented market, geographical situation and limited resources, Nepal has less choice other than to promote SMEs with comparative advantage especially small and cottage enterprises.

2.2 Classification of Industry

The government has classified the industries in the following four categories:

- *Large industries*: industries with fixed assets of more than 100 million rupees shall be termed as large industries (1 US $ = 73 Rs)
- *Medium industries*: industries with fixed assets between 30 million rupees to 100 million rupees shall be termed as medium industries.
- *Small industries*: industries with fixed assets up to an amount of 30 million rupees shall be termed as small industries.
- *Cottage industries*: traditional industries that utilize specific (indigenous) skills or local raw materials and resources, labour-intensive and are based on national tradition, art and culture and industries with fixed assets up to 200,000 rupees shall be termed as cottage industries.

Many organizations involved in the promotion and development of micro enterprises have defined it in different ways and forms. Some of them have defined it based on the number of employment whereas others have focused on volume of investment.

It has been mentioned that the micro-enterprise sector by contrast is less well defined; there are only rough estimates of its size. In general, the micro-enterprise and cottage and small industries sectors can be similarly characterized as owner-managed, with limited division of labour and using primarily family labour, and by informal relations with clients, including suppliers. Although the number of employees (less than 10) does not clearly distinguish the micro-enterprise sector from the small and cottage industries sector, the Government has defined the cottage industries as those including businesses with fixed assets up to 200,000 rupees with other characteristics mentioned above. Micro-enterprises are defined as those with fixed assets of less than 100,000 rupees.

Similarly, the Industrial Enterprise Act 1992 has classified industries as:

- *Manufacturing industries*: industries that produce goods by utilizing or processing raw materials, semi-processed materials, by-products or waste products or any other goods.
- *Energy-based industries*: industries which generate energy from water resources, wind, solar, coal, natural oil, gas, biogas or any other forms of such resources.
- *Agro and forest-based industries*: enterprises mainly based on agriculture or forest products such as integrated sericulture and silk production, horticulture and fruit processing, animal husbandry, dairy industry, poultry farming, fishery, tea gardening and processing, coffee farming and processing, herbs processing, vegetable seed farming, mushroom farming, vegetable farming or processing, tissue culture, green house, bee keeping, honey production, rubber farming, floriculture, and forestry-related enterprise such as lease-hold forests, agro-forestry etc.
- *Mineral industries*: mineral excavation or processing thereof.
- *Tourism industries*: tourism, lodging, motel, hotel, restaurant, resort, travel agencies, skiing, gliding, water rafting, cable car complex, pony-trekking,

trekking, hot air ballooning, para sailing, golf-course, polo, horse-riding, etc.

- ☆ *Services industries*: services industries include workshop, printing press, consultancy service, ginning and bailing business, cinematography, construction business, public transportation business, photography, hospitals, nursing homes, educational and training institutions, laboratory, air services, cold storage, etc.
- ☆ *Construction industries*: industries such as road, bridge, ropeway, railway, trolley bus, tunnel, flying bridge, industrial/commercial and residential complex construction and operation.

2.3 The Industrial Policies

In Nepal, the attempt to integrate technology policy and development planning was first initiated as late as 1980, the start of the Sixth Five Year Plan. Since then various policy reforms and new policies/acts have been added for the speedy development of Nepalese industrial sector.

The objectives of industrial policies in Nepal have been both to promote export and to develop import-substituting industries using indigenous raw materials.

Past industrial policies offered an array of concessional import duties, tariffs, subsidies, and tax exemptions, which ensured that the production cost of domestic industries were at least 30 percent cheaper than their equivalent imports.

Foreign Investment and Technology Transfer Act, 1992 classified industries according to the product scale. Various facilities have been provided to various categories of the industries.

Although the government has introduced a series of economic measures along with various industrial policies and acts to attract the private sector to modern industry, such a trend has not been noticed. The private sector, apart from dithering at the incompatible and unstable policies of the government, confronts an anomaly of raw materials and a dearth of trained manpower. A great deal of effort has been made to attract Foreign Direct Investment, however, it does not seem so promising at present (Maskey 1999).

3. TECHNOLOGY TRANSFER IN THE NEPALESE INDUSTRIAL SECTOR

3.1 Legal Provisions

As part of the modernization drive the country has opened to direct foreign investment and other types of technology transfer from abroad. Legally speaking the only law that governs technology transfer (from foreign countries) in Nepal is the Foreign Investment and Technology Transfer Act 1992, which has laid down the regulations and rules governing foreign investment and technology transfer.

This Act defines technology transfer as any transfer of technology to be made under an agreement between an industry and a foreign investor on the following matters:

(a) Use of any technological right, specialization, formula, process, patent or technical know-how of foreign origin;

(b) Use of any trademark of foreign ownership;

(c) Acquiring any foreign technical, consultancy, management and marketing service.

Looking at the definition adopted by the Act, technology transfer encompasses a broad range of know-how transfer including management and consultancy services. However, the Act appears to be more focused on foreign investment than on technology transfer as such.

Except for the legal provisions contained in the above Act, there is no other legal framework that governs and promotes technology transfer to Nepal. Notwithstanding, technology transfer has been emphasized in the sector-wise action policy of the S&T policy 2006.

A number of facilities and concessions have been provided to foreign investors as per the Foreign Investment and Technology Transfer Act. Such facilities include:

- Except for the projects listed in the negative list, 100 per cent equity participation by foreigners is allowed in almost all sectors. The industries listed in negative list are cottage industries;
- Technology transfer is allowed even in projects where foreign investment is not allowed;
- Firms established with foreign participation are treated equally as 100 per cent Nepalese-owned firms;
- Interest paid on loans obtained from foreign source is tax free;
- Income from royalty and technical management services is levied a standard tax rate of 15 per cent;
- Residential and business visa is provided for foreign investors and their dependents;
- Non-nationalization of industry is assured;
- Provision of dispute settlement between the contracting parties;
- Expatriate employees in firms with foreign equity can take back up to 75 per cent of their salary income abroad.

3.2 General Situation of Technology Transfer in Nepal

The access to technology is a major determinant of efficiency. The investment efficiency to a great extent depends on the capability of the firms to acquire new technologies and adopt them to local conditions (often called 'learning' mechanisms).The Federation of Nepalese Chambers of Commerce and Industry (FNCCI)/World Bank survey has made the conclusion that the learning mechanisms are weak in Nepal. To quote the report "In house firm training is very limited, buyers and suppliers are not coming to Nepal, there are few foreign investors or experienced local firms to serve as role models for 'benchmarking' the competitiveness of firms operations. Good public or private training sources external to firms are also limited, and government technical support services are either nonexistent or under financed or poorly managed. NGO business association, and donor programmes to assist enterprise learning are also limited, although, where they exist they are making a positive contribution".

This does not mean there has not been any technology improvement or there is no demand for technology improvement in the country. The FNCCI/World Bank survey of manufacturing establishments carried out in 2000 revealed that 77 per cent of Nepalese firms wanted assistance in finding new technologies. The demand for technology services was second only to the demand of information on foreign markets among the business services sought out by the Nepalese manufacturing establishments. There seems to be an acute dearth of supply of such services to the enterprises. The same survey revealed that less than a quarter of the business firms could get business support services, including technology services, from business support service providers like government, business organizations or NGOs.

There is no available statistical information on the actual technology improvement and up-gradation in the Nepalese enterprises, but the observation over the last few decades reveals that there has been a slow transformation to a higher technology level.

3.3 Organizations Involved in Technology Transfer and Development

There is no strong institutional mechanism for technology transfer in the SME sector in Nepal. Nevertheless, many organizations are found involved, in some way or the other, in the field of technology transfer and development in general. Such organizations can be found in the government, non-government and private sectors. But the majority of them are involved basically in skill development training for cottage and micro-enterprises level. Based on the human resources and physical infrastructures available, a few of them (such as the Royal Nepal Academy of Science and Technology, Research Centre for applied Science and Technology, Nepal Agricultural Research Council) seem to be capable to have significant contribution in this field. But these organizations are reported to be modestly functioning due to financial constraints. It is observed that national priority areas for science and technological research and development (R&D) have not yet been adequately determined. On the other hand, the science and technology related organizations are not in a position to contribute significantly even in areas indicated by the state policies due to lack of adequate budgetary support from the Government. Their yearly budget reveals that the fixed cost is far more than that of programme budget. As a result, even the capable public sector organizations appear to have involved in donor-supported small projects (international non-governmental organizations aided projects) as and when asked for, rather than working on a long-term basis on the national priority areas.

Despite the non-existence of strong technology transfer mechanism within the country, there are many examples of technology transfer and foreign investment in public as well as private sector companies. The technology transfer in public companies was arranged through government-to-government agreements, while such agreements in the private sector are found to have materialized through negotiations between technology recipient and provider companies.

Though not directly related to the SME sector, technology development and transfer/diffusion in the alternative energy sector in Nepal appears to be remarkable. These technologies particularly include the bio-gas, solar power and micro-hydro power plants, spread to several districts in Nepal.

Spread of these technologies has been made possible from various foreign assistance projects.

Short description of the activities of some national level technology related institutions/organizations is given in the following.

Ministry of Environment, Science and Technology

The Ministry of Environment, Science and Technology is the apex body within the Government in the field of science and technology (S&T) development in Nepal. The ministry has formulated S&T policy in 2006, however it is yet to develop technological infrastructure and other related support for the use of S&T policy in the SMEs.

Nepal Academy of Science and Technology (NAST)

This academy was established in 1982 for contributing to the development of the nation from the S&T sector. Human resources of NAST consist of a number of scientists and technical experts in different scientific disciplines, but the academy has not been able to contribute significantly in the field of technology development and transfer due to its own limitations.

Some of NAST's technological projects include:

- ☆ Solar photo-voltaic pump and lighting system
- ☆ Biomass briquetting technology from solid fuel
- ☆ River boat (used in Ghatbesi, Trisuli River), etc.

Research Centre for Applied Science and Technology (RECAST)

RECAST was established in 1976 under Tribhuvan University with an objective of undertaking research and development in the field of S&T. It has human resources as well as fairly developed physical facilities (laboratories) for some specific scientific work.

Despite the availability of physical as well as human resources, it is learnt that the Centre could have made significant contribution in technology transfer and development, particularly in the SME sector. It is reported that the Centre is not in a position to conduct research programmes due to budgetary constraints. Some of the Centre's work on S&T includes:

- ☆ Energy: solar, bio-mass, bio-fuel, improved cook stove, water turbines
- ☆ Food technology

- ☆ Building materials and low cost housing
- ☆ Natural dye, etc.

Nepal Agricultural Research Council (NARC)

NARC was established in 1990 for the research and development in the agriculture sector. The Council has physical and human resources (scientists) in different disciplines of agriculture science.

Obviously, the Council's research activities are focused on agriculture field, hence there is not much contribution made to the SME sector. However, the Council can possibly make substantial contribution to agro-based industry development (*e.g.* mushroom, bee-keeping, horticulture, etc.) in the country. The Council is basically a research and development organization, and the research findings and knowledge are disseminated through different departments and offices of the Ministry of Agriculture and other related organizations.

Other Organizations

Apart from the above, there are a number of organizations in the public sector, non-governmental organizations and the private sector involved in technology improvement in enterprises as well as in other areas. Giving details of all of them is out of the scope of this paper. To name few of them, the Department of Cottage and Small Industries and Cottage and Small Industries Development Board under MoICS have network all over the country, and are involved primarily in technology transfer through skill development training programmes. Similarly, a number of international agencies and non-governmental organizations are also involved in the alternative energy, food industry, rural transportation sector, bio-resources conservation, non-timber forest products-based enterprise promotion through dissemination of technical and market information to the small entrepreneurs. But most of these organizations are primarily involved in skill development training programmes, and lack adequate physical infrastructure and human resources (technologists/scientists) for technology transfer (from abroad), adaptation and innovation.

A Technology Transfer and Development Project was established under the Ministry of Industry, Commerce and Supplies (MoICS) in 1996 with an overall objective of providing services to the industrial sector in technology-related field. Due to lack of necessary budget, human resources and infrastructure, the project could not achieve its objectives effectively. The project was merged within the Industrial Enterprise Development Institute (IEDI) in 2002. Since its merger with IEDI it has conducted some awareness workshops, and a technical need assessment study of small and micro-enterprises in the food industry sector in the Kathmandu valley. It is also modestly functioning due to the similar problems mentioned above.

3.4 Modality and Mechanism of Technology Transfer

Most of the technology transferred to Nepalese industries in the past was in the form of turnkey plants to the state sector, financed through international aid and/or loans. However, in the latter years, many large and medium-scale industries have been established with foreign collaboration in the private sector as well. This has taken place through a variety of mechanisms such as direct foreign investment via joint ventures, technical collaboration, import of machinery and equipment, technical assistance through human resources, etc. While these are more formal modes of technology transfer, a lot of technology gets transferred informally through books, journals, promotional literature and personal contacts. As many Nepalese industries, particularly in the small and cottage industries sector, use Indian machinery and equipment, the informal mode of technology transfer is more prevalent in such industrial units. It is difficult to find out, thus, not only the level and extent of informal technology transfer, but even the numbers of such units, as the details regarding technology transfer are not found in the government records (Joshi, 1996).

3.5 Status of Technology Transfer

The only data available relating to technology transfer agreements are of the companies for which permission is sought under Foreign Investment and Technology Transfer Act.

Number and Category of Projects

According to the statistics of the Department of Industries, a total of 1350 small and medium scale projects with foreign collaboration (financial and technical) have been approved so far. The data in Tables 15.1 and 15.2 indicate number, category, scale (type) of collaboration and scale of the projects.

Table 15.1: Number of Projects with Foreign Collaboration

Year	*No. of Projects*	*Year*	*No. of Projects*
Upto 1990	95	1999	52
1991	24	2000	95
1992	58	2001	94
1993	55	2002	60
1994	22	2003	81
1995	39	2004	68
1996	46	2005	106
1997	88	2006	130
1998	66	2007	193

Source: DOI Statistics, 2008.

Table 15.2: Number and Scale of Projects with Foreign Collaboration

Scale	*Number of Industries*	*Per cent of Foreign Investment*	*Employment*
Large	146	32.53	36967
Medium	199	37.96	26313
Small	1151	58.02	62277

Source: DOI Statistics, 2008.

Table 15.3 indicates that around 12 percent of the projects involve technology transfer, and the rest are only financial collaboration among the concerned firms/ parties.

Table 15.3: Number and Category of Projects with Foreign Collaboration

Category	No. of Projects	Equity Only	Equity + Technology	Technology Only
Manufacturing	419	338	39	42
Agro-based	14	13	0	1
Energy-based	14	14	0	0
Construction	17	15	2	0
Mineral-based	3	1	1	1

Source: DOI Statistics, 2008.

Table 15.4: Status of Foreign Investment Projects by Country or Area (with 10 and more projects)

Sl.No.	Country	Numbers of Projects
1.	Australia	21
2.	Austria	11
3.	Bangladesh	15
4.	Canada	17
5.	China	219
6.	France	36
7.	Germany	64
8.	India	401
9.	Italy	17
10.	Japan	132
11.	Malaysia	10
12.	Netherlands	19
13.	Norway	10
14.	Pakistan	13
15.	Russia	13
16.	S. Korea	100
17.	Singapore	14
18.	UK	77
19.	USA	138
20.	Others	169
	Total	1496

Source: DOI Statistics, 2008.

The country-wise data of foreign investment projects in Nepal shows 401 projects are from India, while China stands at the second position with 219 projects followed by 138 projects from USA and 132 projects from Japan. A total of 60 countries so far have been involved in foreign collaboration in Nepal. Table 15.4 depicts the country-wise status of projects.

Operational Status of the Projects

Looking at the operational status, approximately 27 per cent are in operation.

Table 15.5 shows the latest status of foreign investment projects in Nepal.

Table 15.5: Status of Foreign Investment Projects

Status of Industries	Numbers
Operational	401
Under construction	261
Licensed	122
Approved	619
Closed	21
Cancelled	72

Source: DOI Statistics, 2008.

Sector-wise Status of the Projects

The sector-wise list of projects under foreign investment shows that the majority of the projects under manufacturing industries are textile and readymade garments. Table 15.6 depicts the number of foreign investment projects (sector-wise).

Table 15.6: Number of Foreign Investment Projects by Sector-wise

Type of Industry	*Number of Industries*	*Employment*
Agriculture and forestry	32	1780
Manufacturing	581	70633
Food, beverage and tobacco	92	10668
Textile and readymade garments	216	38166
Wood and wood products	13	1730
Paper and paper products	27	1015
Chemical and plastic products	99	7744
Non-metallic mineral products	23	3413
Basic metal products	26	2134
Fabric metal	64	4307
Other manufacturing	21	1456
Electricity, water and gas	33	6246
Construction	52	3263
Transport and communication	41	5464

Source: DOI Statistics, 2008.

3.6 Constraints

Since Nepal is striving for fast economic development, remarkable progress has not been made in the field of technology transfer and foreign direct investment (FDI). FDI can be termed as one of the means of technology transfer, especially in a developing country. But due to several reasons and basically on account of inadequate infrastructural development, and small domestic markets; multinational corporations have not yet come up in the country in a large number. Some of the reasons for not investing in Nepal may be inaccessibility of raw materials, lack of facilities with regard to power, transportation, maintenance, auxiliary materials etc., inadequate investment and credit facilities, a limited domestic market, competition of foreign products, lack of trained personnel at all levels and so on. However, the change in the government policy in the recent past has certainly encouraged increasing the numbers in a considerable amount as compared with the past years.

Nepalese industries are facing numerous problems. One study found that the lack of a workshop to repair machinery has been regarded as a constraint on the smooth functioning of industries.

Industries in the private sector are not developing because of lack of co-operation from different government agencies. Government is blamed for not providing industrial facilities in a simplified manner. Besides, many industries are still dependent on imported raw materials either from India or other overseas countries. Apart from these, due to the short supply of electric power, industries have to incur higher costs. Because of these problems many industries are not able to flourish. The main causes of failure of industries are stated as lack of raw materials, working capital and markets (Nepal 2002).

The process of importation of technology in Nepal has been going on largely without any planning and reference to the conditions and needs of the country. There is no such institution which collects relevant information on advanced technologies, monitor the transferred technology, and advice both public and private sector in the transfer of technology. Furthermore, Science and Technology Policy has not yet been successful in incorporating the technology transfer network.

In general, the choice of manufacturing technology in Nepal is usually determined by taking the cost of capital into consideration as well as the availability of spares and the ease with which repair and maintenance work can be done. Nepalese manufacturing is characterized by low-level technology. The static nature of technology is borne out by the fact that in many factories plant layout has remained fixed for decades. Many plants are noted for their vast array of obsolete technology.

4. CONCLUSION

Success or failure of technology transfer is evaluated by the capability of SMEs at both management and workforce levels. The SMEs need to learn to continuously upgrade their capacity while the government needs to encourage such capacity improvement. Besides, other private and public organizations can help to devise appropriate policy as well as encourage the firms to develop such capacity by providing information, advocacy and consultancy services.

A major way forward for promoting technological development in Nepal would call for the creation of a network of links between universities or research centres and industry. In particular, the coalition between science and technology institutions and industry is very weak. A framework for promoting university-academy-industry dialogue needs to be established, so that the desired synergy in moulding of university graduates and the development of an indigenously strong, stable and dynamic industry may be expected to come forth. This task would be facilitated if the research carried at the academy, seeks to show how science and technology based development should be promoted for the benefit of rapid socio-economic development in Nepal. Besides, industrial development is possible only when industrial services are provided efficiently to the entrepreneurs.

An industrial plan should include a comprehensive and co-coordinated programme to establish and strengthen industrial services.

Clusters of industrial activities need to be developed for international competitiveness. These include product diversification; process technology innovation; design and fabrication of plants and equipment; internal competition among national producers; closer backward and forward linkages, including those linkages with universities, associations, and polytechnics; and development of specialized industrial consultancy firms in finance marketing, engineering, management and advertising etc.

In developing countries like Nepal technology transfer seems to be obsessed with the hardware part only, *i.e.* in obtaining the machines and tools. The maximum further obtained is the documentation on the processes, however, rarely do we pay much attention to the intangible material or software part of the technology, which is the ability to make best of the available technology and improve on it.

The way forward is only with proper technology transfer that SMEs in developing countries can understand, introduce, adopt and master the knowledge of the key processes.

REFERENCES

Dawadi, H. 2000. "Technology Transfer: Capacity Building and the Private Sector of Nepal" paper presented at UNESCAP seminar, Kathmandu, Nepal

Government of Nepal, Department of Industry (DOI) 2008, (unpublished data)

Government of Nepal, Foreign Investment and Technology Transfer Act 1992.

Government of Nepal, Industrial Enterprise Act 1992.

Hug, M.M., 1997. "Transfer of advanced technologies to developing countries" paper presented at the national workshop on "Promotion of New Technologies in Nepal", Kathmandu.

Institute for Integrated Development Studies (IIDS) (1994). *Inhibitors/Barriers to the Development of Small and Medium Enterprises in Nepal* (Kathmandu).

Joshi, D. R., 1996. "Technology Transfer to Nepal", Thesis submitted for M.Sc. in Industrialisation, Trade and Economic Policy at the University of Strathclyde, UK.

Maskey, B.K. (ed.), 1999. *Small and Medium Enterprise Promotion through Local Capacity Building* (Kathmandu, Community Development Group).

Nepal, C., 2002. "Technology Transfer in SMEs: Problems and Issues in the Context of Nepal" paper presented at the National Workshop on Technology.

Chapter 16

Status of SMEs in Pakistan and its Possible Uplift by PCSIR

Nighat Afza
Chief Scientific Officer/Director (Planning and Development), PCSIR Laboratories Complex, Karachi – 75280, Pakistan
E-mail: klcdirectorp_d@yahoo.com

ABSTRACT

The Small and Medium Enterprise (SME) sector is the backbone of Pakistan's economy. The significance of its role is clearly indicated by various statistics. There is no consistent definition of SME applicable across the board in Pakistan. However, in the SME policy 2007, a uniform definition for SME has been recommended to overcome all ambiguities. There is about 3.2 million business enterprises in Pakistan out of which 90 per cent are SMEs which employ nearly 35 per cent of total and 78 per cent of the non-agriculture labour force. They contribute over 40 per cent to the GDP and account for 25 per cent of exports of manufactured goods besides sharing 35 per cent in manufacturing of value added products. A new SME Policy 2007 has been approved by the Government of Pakistan (GoP). Problem statements, policy recommendations, the impact of these recommendations on the problems and the role of R&D organizations, particularly Pakistan Council of Scientific and Industrial Research (PCSIR) in technology development and transfer for the uplift of SMEs have been discussed in this article.

Keywords: *SME in Pakistan, Overview, Economic importance, SME policy, Impact, Commercialization of R&D.*

1. INTRODUCTION

Small and Medium sized Enterprises (SMEs) are the most dynamic and vital sector and are the backbone of a national economy, particularly in developing countries. SMEs contribute to economic development of a country in the following ways:

- Create employment for rural and urban workforce
- Provide flexibility and innovation in the economy
- Help in alleviating poverty
- Contribute to exports

The benefits of SME in Pakistan have not been fully recognized till the last decade. In 1998, the Government of Pakistan, realizing the economic importance of SMEs, formed the Small and Medium Enterprises Development Authority (SMEDA) to provide support to SMEs in Pakistan (Gazette of Pakistan, Extraordinary, August, 2002)

1.1 Definition of SME in Pakistan

There is no uniform definition of SME applicable across the board in Pakistan till the adoption of the SME Policy in 2007. Different departments and organizations define SME in accordance with their functional convenience rather than market situation. Definitions used by various Government agencies have been given (Developing SME policy in Pakistan, SME issue paper; Iqbal and Farrukh, Small and Medium Enterprises in Pakistan) in Table 16.1.

Table 16.1: Definition of SME in Pakistan

Agency	*Small Enterprises*		*Medium Enterprises*	
	No. of Employees	*Assets*	*No. of Employees*	*Assets*
Small and Medium Enterprise Development Authority (SMEDA)	10-35	Productive Assets of Rs.2-20 million	36-99	Productive Assets of Rs.20-40 million
SME Bank	-	Total Assets up to Rs.20 million	-	Total Assets more than Rs. 20 million and up to 100 million
Federal Bureau of Statistics	Less than 10	-	-	-
State Bank of Pakistan	An entity, ideally not being a public limited company, which does not employ more than 250 persons (manufacturing) and 50 persons (trade/services) and also fulfils one of the following criteria: 1. A trade/services concern with total assets at cost excluding land and buildings up to Rs.50 million. 2. A manufacturing concern with total assets at cost excluding land and building up to Rs.100 million. 3. Any concern (trade, services or manufacturing) with net sales not exceeding Rs.300 million as per latest financial statements. 4. Enterprises exporting up to US$2.5 million a year.			

Contd...

Table 16.1–Contd...

Agency	*Small Enterprises*		*Medium Enterprises*	
	No. of Employees	*Assets*	*No. of Employees*	*Assets*
Punjab Industries Department	Fixed assets with Rs.10 million excluding cost of land			
Sindh Industries Department	Entity engaged in handicrafts or manufacturing of consumer or producer goods with fixed capital investment up to Rs.10 million including land and building			
Punjab small Industries Corporation	Fixed investment of Rs.20 million excluding land and building		-	

1 US$ = ~ Rs. 84.5......

Absence of a single definition makes it difficult to identify target firms, align development programme, collect data and monitor progress of SME in Pakistan. To encounter this problem, it is recommended in the SME policy 2007 that Government of Pakistan should adopt a single SME definition acceptable to all public and private agencies. However, various organizations may be allowed a two year time frame to align their current SME definition in line with the SME definition proposed in this Policy. (Saleem, 2008; SME policy 2007).

The uniform definition of SME as suggested in the SME policy is as follows (Saleem, 2008; SME policy 2007):

Table 16.2: SME Definition as Suggested in SME Policy 2007

Enterprise Category	*Employment Size*	*Paid Up Capital*	*Annual Sales*
Small and Medium Enterprise (SME)	Up to 250	Up to Rs.25 Million	Up to Rs.250 Million

2. PRESENT STATUS OF SMEs IN PAKISTAN

2.1 No. of Units

According to the latest Economic Census of Pakistan (GOP 2005, Economic Census of Pakistan) there are 2.96 million SME units in the country. The surveys usually cover those firms that are registered and a large number of very small firms go unrecorded either because they want to remain hidden or the enumerators have not been able to include them in the data, so their number, especially SE, is probably much larger than is officially reported. In general, the firms with relatively more workers are smaller in number and those with smaller number of workers are in majority. Moreover, most of the firms are owner-managed, supported by family workers. The latest Economic Census of Pakistan (2005) has distributed SMEs between Establishment unit–an economic unit under single ownership or control, and Household unit–in which the production activity and the residence of entrepreneur share the same premises.

Table 16.3: No. of SMEs in Pakistan

Total SMEs Units	*Establishment Units*	*Household Units*
2.96 million	2.8 million (93.9 per cent)	0.18 million (6.1 per cent)

The province wise distribution of SME establishments is as follows (GOP 2005, Economic Census of Pakistan):

Table 16.4: Province-wise Distribution of SMEs

Name of Area	*SMEs Unit*
Pakistan	2.96 million
Punjab	65.26 per cent
Sindh	17.82 per cent
NWFP	14.21 per cent
Balochistan	2.09 per cent

2.2 Employment

The Economic Census (2005) showed that 2.96 million SMEs employed 6.58 million persons and their distribution is as follows (Economic Census of Pakistan, 2005; SME Development in Pakistan: Issues and Remedies):

Table 16.5: Distribution of SMEs According to No. of Employees

Total SMEs Units	*SMEs Employing 1-5 Persons*	*SMEs Employing 6-10 Persons*	*SMEs Employing 11-50 Persons*	*SMEs Employing Over 50 Persons*
2.96 million (100 per cent)	2.85 million (96.6 per cent)	0.079 million (2.67 per cent)	26,000 (0.87 per cent)	1617 (0.054 per cent)

SMEDA estimates the share of SMEs in the total employment of labour force of Pakistan to be about 35 per cent.

The role of female in industrial sector of Pakistan is very negligible though they constitute nearly 50 per cent of the population of the country, however in recent years they are participating in SMEs. As regards the female work force, 0.46 million, 7.0 per cent of the total (6.58 million) SMEs employees are female and their distribution is given in the following table (SME Development in Pakistan: Issues and Remedies):

Table 16.6: Distribution of Female Workers in SMEs

Total Female Employees	*Self-Employed/ Proprietor*	*Unpaid Family Helper*	*Paid Employed*
0.46 million	15 per cent	30 per cent	55 per cent

2.3 Types of SMEs

Nearly 53 percent of all SME activities is in retail trade, wholesale, restaurants and the hotel business whereas the contribution of industrial establishments and those involved in service provision is 20 per cent and 22 per cent respectively. Among the Household units, the highest share is that of manufacturing sector (66.5 per cent), followed by community, social and personal services (20.5 per cent), agriculture, poultry farming, fishing etc. (8.7 per cent), and wholesale and retail trade, etc. (about 4.00 per cent).

Among the SMEs involved in retail, wholesale etc. 98 percent employ less than 5 persons and 99 percent less than 10 persons. Even within the manufacturing sector the trend is not different and nearly 87 percent employ less than 5 persons and 98 percent less than 10 persons. Similar pattern of employment distribution can be traced among other sectors except for mining where SMEs tend to employ more people. In the mining sector, it averages around 56 percent employing between 6 to 50 persons. On the whole, the percentage of large firms is very small. Punjab is home to more than 65 per cent or 1.9 million businesses in Pakistan. (SME Development in Pakistan: Issues and Remedies).

Approximately, half of the total SME activity is concentrated in five sub-sectors; grain milling, cotton weaving, wood and furniture, metal products and art silk. For the past three decades, the fastest growing export industries have been dominated by the SMEs. Export contribution from SMEs emanates from sub-sectors, cotton weaving and other textiles and, surgical equipment (Economic Census of Pakistan, 2005; Iqbal and Farrukh, Small and Medium Enterprises in Pakistan). Share of key SME sub-sectors in Pakistan is given in Table 16.7.

Table 16.7: SMEs' Share in Sub-Sectors

Sl.No.	*Sub-sectors*	*Percentage Share of SMEs*
1.	Cotton Weaving	13 per cent
2.	Other Textiles	6 per cent
3.	Metal Products	7 per cent
4.	Carpets	4 per cent
5.	Art Silk	5 per cent
6.	Grain Milling	16 per cent
7.	Jewelry	4 per cent
8.	Wood and Furniture	10 per cent
9.	Others	35 per cent

2.4 Economic Importance of SMEs in Pakistan

The Economic Census of Pakistan-2005 reveals that over 90 per cent of the total (3.2 million) enterprises of the country are SMEs. Their share in industrial employment according to an estimate is 78 percent and in value addition approximately 35 percent. They are contributing a 40 per cent of national GDP and 25 per cent in export earnings (Economic Census of Pakistan; 2005, World Bank Report; Economic Survey of Pakistan 2007-08). All of the SME function within the private sector and mostly operate in the undocumented informal part of the economy. Contribution of SMEs in the manufacturing sector is given in the Table 16.8.

Table 18.8: Economic Importance of SMEs

Employment	*GDP*	*Value Added*	*Export Earnings*
78 per cent (6.8 million)	40 per cent	35 per cent	25 per cent

Pakistan has a fairly developed financial system spread over most of the urban and rural areas of the country. There were 6974 branches of commercial banks, in December 2004. On top of these, a number of specialized financial institutions are working in support of SMEs: (*i*) the SME Bank, (established 2002); (*ii*) Khushali Bank, (established 2000); (*iii*) The First Micro Finance Bank; and (*iv*) Network Micro Finance Bank and Rozgar Micro Finance Bank (established 2001). The three types of micro finance banks have the mandate to work nation-wide, province-wide or district-wide with minimum capital of Rs.500, Rs.250 and Rs.100 million, respectively. In addition, all commercial banks have the mandate to provide credit to SMEs. (SME policy 2007)

2.5 SME Policy of Pakistan (SME policy 2007)

After the independence, the Government of Pakistan at various levels, formulated and to an extent implemented small-scale business promotion policies. However, these policies were limited in their scope, and are not in line with the dynamics of present time. Government of Pakistan initiated small industries promotion through formation of the West Pakistan Small Industries Corporation. Although the Corporation created an impact in its early years yet its Provincial successors could not sustain its contribution due to various limitations.

In order to devise a basic framework for the growth and development of SMEs, the Government of Pakistan constituted a Task Force for SME Policy Development in January 2004 under the chairmanship of Secretary, Ministry of Industries, Production and Special Initiatives. The policy drafted by the Task Force has been approved by the Federal Cabinet of Pakistan on January 17, 2007. The policy will be implemented in a period of 8 years *i.e.* 2007-2015.

The objective of SME Policy of Pakistan is to provide a short and a medium to long-term policy framework with an implementation mechanism for achieving higher economic growth based on SME led private sector development.

3. PROBLEMS FACED BY SMEs OF PAKISTAN AND POLICY RECOMMENDATIONS

SME policy 2007 also addressed the problems faced by SMEs. A brief account of the policy recommendations on various problems and expected impact of these recommendations to solve the problems is given below (SME policy 2007):

SMEs in Pakistan currently operate below their potential due to the constraints and discrimination on the following issues during the last two decades:

- ✰ *Business environment*–Adverse Government policies and inadequate infrastructure.
- ✰ *Access to finance and related services*–financing barriers and disincentives.

- *Entrepreneurship development.*
- *Technological constraints*–shortage of skilled workers, and need for technology up-gradation and.
- *Marketing support.*

3.1 Business Environment–Adverse Government Policies and Inadequate Infrastructure

3.1.1 Problem Statement

The financial, labour and enterprise regulations of the Federal and Provincial Governments in Pakistan do not provide any solution of the specific problems of SMEs and are not satisfying the needs of these sectors. Generally the fiscal regulations divide enterprises by income levels and labour related regulations realize only two forms of enterprises, small and large, thus, not providing laws and implementation mechanisms that are sensitive to SME needs, thus making it difficult for SMEs to access public support programs and gain attention of public authorities when competing with the large firms. This dilutes the ability of SMEs to effectively compete with large firms.

Inadequate Infrastructure

The Pakistani industry is becoming sick because of lack of infrastructure, electricity, gas, sewerage, effluent treatment etc. Other than lot of promises, no visible efforts are being made towards solving these problems. This has caused the industry to become internationally uncompetitive and non-compliant in several issues like environment and social standards. A World Bank report argues that Pakistan has to invest around Rs. 60 billion (US$1 billion) per year in reservoirs and related infrastructure over the next five years. In the energy sector, the country will face severe power shortages of around 6,000 mega watts by 2010. Similarly, inefficiencies in the transport sector cost the economy around 4-5 per cent of GDP each year. To overcome these constraints, the Government of Pakistan is tripling its annual infrastructure investment from an average of Rs150 billion (US$2.5 billion) to Rs 440 billion (US$7.3 billion). However, the report points out that mega projects in the past have experienced frequent delays and cost overruns, illustrating a lack of capacity in the industry to plan, program, and execute large projects. (World Bank Report No.41630-pk)

3.1.2 Policy Recommendations

Some of the Policy Recommendations on this issue are:

- Promulgation of an SME Act that provides identification of fiscal, registration, labour and inspection laws that may be simplified for SMEs.
- The process of SME certification is proposed to be voluntary for those firms who choose to do so, to benefit from the support mechanism and simplified regulatory regime proposed through this Policy.
- Periodic review of all fiscal laws and labour legislation with a view to facilitate and improve growth of the SMEs and relaxation in the application

of certain laws to SMEs for a certain period of time. However, provision of education/assistance to SMEs to achieve proper safety at work and compliance with international conventions shall be essential.

- Establishment of complaint cell at State Bank of Pakistan and SME desks at the Federal, Provincial, Banking and Tax Ombudsman offices for handling and addressing SME grievances.
- A minimum quota is proposed to be established for SMEs for allocation of land in the Industrial Estates and Export Processing Zones.
- All public sector procurement may encourage participation of SMEs with certain products/contracts exclusively to be competed amongst the SMEs.
- GoP may consider to support establishment of an SME Promotion Council and SME specific Trade Associations for organizing SMEs and providing them with a platform to lobby for favourable government policies.
- Government to ensure adequate provision of physical infrastructure (roads, utilities etc.) in existing SME clusters.
- Government functionaries in tax, registration, labour, etc. departments are proposed to undergo specialized training for handling SME related issues.
- Genuine SME representatives are proposed to be inducted in all federal monitoring and dispute resolution schemes with private sector participation.
- Regulatory compliance procedures will be widely communicated to SMEs. Communication will increasingly use Urdu and local languages where useful.
- Improving and enforcing Intellectual Property Rights for SMEs.

3.1.3 Expected Impact

Implementation of the recommendations of SME policy 2007 (5) for creating conducive 'Business Environment' for SMEs will result in reducing the number of and simplifying the compliance process of SME regulations. This is expected to trigger fast paced creation and growth of enterprises resulting in economic development and job creation.

3.2 Access to Finance and Related Services–Financing Barriers and Disincentives

3.2.1 Problem Statement

According to the Investment Climate Assessment 2003 (SME policy 2007), banks provide for only 7-8 per cent of the total funding requirement of SMEs.

Pakistan's Finance is a major obstacle on the development of SMEs as they are unable to access the formal sources of credit. The SMEs are unable to meet Prudential Regulations and documentation requirements. Usually the SMEs do not maintain regular account book so they are unable to get any advantage from banks and other financing agencies.

3.2.2 Policy Recommendations

Some of the Policy recommendations on this issue are:

- ✰ Incorporation of SME financing in the Annual Credit Plan of the SBP and monitoring to cater for underserved segment of the SMEs.
- ✰ Review of Prudential Regulations, periodically, in line with the SME credit demand and supply data and establishment of Credit Guarantee and Credit Insurance agencies to provide incentives and risk cover for banks, so as to provide them the relevant comfort in financing SMEs.
- ✰ Support to FIs in designing and launching industry based program-lending schemes and improvement in the regulatory procedures and fiscal incentives for Venture Capital companies.
- ✰ Promotion of Islamic mode of financing for SMEs. Introduction of Bankruptcy Laws with dedicated and effective judicial process and expansion in the role of Banking Ombudsman to include redressal process for SME complaints.
- ✰ Awareness and promotion of options for formal financing and good accounting practices amongst SMEs.

3.2.3 Expected Impact

It is expected that improvement in the regulatory framework and removal of financing barriers will result in creating an environment where SMEs will be benefited by the credit schemes of banks and FIs.

3.3 Entrepreneurship Development

3.3.1 Problem Statement

In Pakistan entrepreneurship is usually undertaken by those belonging to the existing business families. As a result a small number of new enterprises are being created and that too in traditional areas of business overcrowding the supply/product base and their markets. On the other hand, there are no limitations in the entrepreneurial capabilities in the general public. There is a need for the Government to actively promote entrepreneurship through changes in education curricula, by creating awareness amongst youth and by providing effective support to those who wish to establish new enterprises.

3.3.2 Policy Recommendations

Some of the Policy recommendations on this issue are:

- ✰ Revision in primary and higher education curricula for promoting entrepreneurship amongst the educated youth and inclusion of 'Entrepreneurship' courses in all professional degree awarding higher education, technical and vocational training institutions in Pakistan.
- ✰ Entrepreneurship competitions at university and at national level for selecting best business plans/models and providing grant for project implementation.

- ✩ Establishment of technology and business incubators in selected universities in Pakistan.
- ✩ Identification of investment opportunities offered by backward and forward linkages of successful services/products.

3.3.3 Expected Impact

Implementation of the recommendations on supporting entrepreneurship amongst the Pakistani youth will result in the establishment of a larger number of enterprises in diverse segments. This will improve income distribution and help in creation of new jobs.

3.4 Technological Constraints–Shortage of Skilled Workers and Need for Technology Upgradation

3.4.1 Problem Statement

There is always a skill shortage in Pakistan due to out-migration of labour to Middle East and other countries and because of this shortage of qualified and skilled workers increases day by day as they are in bargaining position. SMEs are not able to attract such workers because of the low wages paid by them as compared to LEs, so they always have skill deficiencies and are unable to compete with larger firms' better-qualified manpower. So a smaller firm may be out competed because of limited funds despite its need for high skilled workers. Further more skilled and qualified workers are produced either by the higher education institutions or the technical training infrastructure, both of which are not attuned to the SME needs nor are they equipped to address them. This situation limits the capacity and capability of not only SMEs but also those of LEs to innovate, add value, upgrade technology and devise new marketing strategies.

Due to the absence of knowledge and information the SMEs of Pakistan depend on depleted and obsolete technology and this is also associated with the lack of technical skills required for the production of quality goods. This obstacle acts as a major barrier in acquiring knowledge-based modern technology. They also do not have sophisticated equipment and R&D facilities. To overcome these shortcomings they adopt manual and labour intensive production methods, which results in lower productivity levels, inferior quality of products and overall economic deficiency.

3.4.2 Policy Recommendations

Some of the Policy recommendations on these issues are:

(a) Human Resource Development

- ✩ Need Assessment Survey to identify major SME needs in HRD, technology up-gradation and marketing.
- ✩ Establishment of Institutes of Small and Medium Enterprises and Entrepreneurship Development (INSMED), and specific technical training institutes in selected business schools.

- Encouraging use of the technical training infrastructure by the private sector, Business Development Service Providers (BDSPs) serving SME sector and incentives for investment in setting up SME training facilities.
- Induction of genuine SME representatives in private sector boards of the technical training institutes.

(b) Technology Up-gradation

- Introduction of SME specific research projects supporting R&D and University-Industry liaison programs by Ministry of Science and Technology (MoST), Higher Education Commission (HEC), Pakistan Software Export Board (PSEB), Ministry of Information Technology (MoIT) and others.
- Establishment of Technology Innovation Centers (TICs) offering common facility, technology up-gradation, R&D and design related services to SMEs.
- Launching of pilot technology up-gradation projects for major SME clusters on cost sharing basis.
- Incentives for Investment in new Emerging Sectors and Skills upgrade.

3.4.3 Expected Impact

It is expected that effective implementation of Policy recommendations in supporting HRD and technology up-gradation will create a human resource pool and necessary infrastructure for adding value to SME businesses by technology upgradation and innovative products and processes.

3.5 Marketing

3.5.1 Problem Statement

Limited options are available to SMEs to invest in exploration of new markets, therefore they usually operate in limited and traditional markets. Public sector organizations can not adequately provide marketing support, thus concrete and substantial government commitment is needed in this area.

3.5.2 Policy Recommendations

Some of the Policy recommendations on the issue are:

- Encourage establishment of SME sector specific export marketing companies.
- Matching grants for developing 'world-class' trade and product directories for major SME clusters.
- Establishment of SME quota in trade delegations supported by EPB.
- Compilation and dissemination of data on local markets using manufacturers, distributors and retailers data.
- Provision of support to SME associations in exploiting local market opportunities by holding domestic product exhibitions.
- Establishment of Annual SME Awards for recognizing outstanding performance in domestic and International markets.

3.5.3 Expected Impact

Effective implementation of policy recommendations in the area of marketing will create necessary infrastructure for innovative marketing. It will also increase role of private sector BDSPs in serving the SME sector in undertaking business development activities.

4. ROLE OF PCSIR ON TECHNOLOGY DEVELOPMENT AND TRANSFER

As suggested in the policy, a linkage between R&D organization/Universities and SMEs will not only help SMEs but LEs also in solving their problems related to HRD and technology up gradation. In this contest example of Pakistan Council of Scientific and Industrial Research (PCSIR) may be quoted, which is a role model for technology development and transfer and technology up gradation and Human Resource Development for SMEs.

4.1 Background and Objectives of PCSIR

Pakistan Council of Scientific and Industrial Research (PCSIR), a premier R&D organization of the country, was established in 1953 at Karachi. PCSIR is committed to execute quality research by exploitation of indigenous natural and other resources for industrial development of the country. The work is undertaken in accordance with the prescribed standards to ensure satisfaction for its customers. The following are the objectives of PCSIR:

- ☆ Optimum utilization of indigenous raw material and resources.
- ☆ Development of technologies based on local resources from bench to pilot plant scale and technology transfer, for industrial exploitation leading to import substitution and export enhancement.
- ☆ To conduct R&D on the problems faced by the industrial sector and extend test/analysis/calibration services to users.
- ☆ To undertake cooperative research with local and foreign R&D organizations and industries on the projects of national/mutual interest.
- ☆ Human resource/skill development for the organization as well as to feed the industrial sector.

At present PCSIR have five multi-functional laboratories located at Karachi, Lahore, Peshawar, Hyderabad and Quetta, and five mono-functional units of Leather, Fuel, National Physical Standards Laboratories and Human Resources Development Centres *i.e.* PSTC and IIEE at Karachi, Lahore, Quetta and Islamabad.

4.2 Human Resource Development by PCSIR

Two Technical Training Centres of PCSIR are working at Karachi and Quetta and three others are almost in the final stage of establishment. These centres are entrusted with the functional responsibility of producing technically trained manpower for industrial sector by offering 03-04 years Diploma course of Associate Engineering in Precision Mechanics, Instrumentation Technology and Dies and

Moulds. The Dimension Stone Evaluation Centre has been established in PCSIR Peshawar, giving training in Dimension Stone Evaluation to a number of workers from SMEs every year. The Institute of Industrial Electronics Engineering (IIEE) of PCSIR at Karachi is successfully feeding the local industry by producing graduate engineers in Industrial Electronics. These trained Engineer Graduates/Diploma/ Certificate holders usually serve the SMEs of the country. PCSIR is also engaged in Human Resource Development for the industrial sectors of Textile, Polymer, Food, Ceramics, Pharmaceutical, Leather, Fuel, Engineering Services, Calibration, Environment, Material Science, and Agriculture (Including Tissue Culture) etc. by providing training to their workers and qualified personnel on sophisticated instruments as well as on management system.

4.3 Development, Upgradation and Transfer of Technology through PCSIR

Knowledge and technology are professed today as the driving forces of national productivity and ultimately economic growth of the nations. In this perception, commercialization of R&D results is critical for sustaining growth. Technology commercialization is a process of converting research to successfully marketable products. PCSIR, being an R&D organization, has the main objective of technology development from laboratory scale to pilot scale and successful transfer and commercialization of technology in Pakistan. Special efforts have been made in PCSIR to enhance the potential of R&D commercialization, which in turn would help strengthen public institution, particularly SMEs, in the long run.

In Pakistan, particularly in PCSIR, development of any technology starts with the market survey of that product. The process involves any of the following routes for the selection of technology which has to be developed:

- ☆ Scientist/technologist searches the market about the need of that particular commodity which has to be developed
- ☆ Client/SME contacts the organization/scientist for the development of technology
- ☆ After the selection of technology, the following steps for the commercialization of the required commodity are taken
- ☆ Preparing a summary of the method
- ☆ Selection of a group of scientists/technology for R&D
- ☆ Discussion on the details of the method among the research group
- ☆ Start of R&D
- ☆ Completion of R&D and preparation of pre-feasibility report based on the findings of lab scale production
- ☆ Fabrication of pilot plant for the economically feasible lab scale developed technology, start of production at pilot plant and preparation of feasibility report based on the findings of pilot plant production
- ☆ Submission of report to the client/SME

Methods for the Transfer of Technology

Various modes adopted for the transfer of technology from PCSIR to the client/ SME are:

(a) Transfer of Know-How of the Technology

Following services are offered by the PCSIR to the client/SME:

- ☆ Selection of space/building, construction of building
- ☆ Selection of machinery and equipment
- ☆ Procurement of raw material
- ☆ Appointment of manpower
- ☆ Training in production and quality control

(b) Manufacturing at PCSIR Site by the PCSIR Scientists

The responsibility of PCSIR and SMEs/Clients.

- ☆ PCSIR is responsible for the production, either at pilot scale or at industrial scale
- ☆ All the cost for the setting up of the plant as well as raw material will be born by PCSIR
- ☆ Quality control will be the responsibility of PCSIR
- ☆ The client/SME will market the product

(c) Production in Technology Business Incubator Centre (TBIC)

In this method the PCSIR provides

- ☆ Space/building for the production
- ☆ One or two scientists for the supervision of work at the plant
- ☆ Help in selection of machinery and equipment
- ☆ Help in procurement of raw material
- ☆ The quality control is also the responsibility of PCSIR, however client/ SME has to pay for it

In this method the following are the responsibility of Client/SME

- ☆ Procurement of machinery/equipment and raw material
- ☆ Fabrication of plant under the supervision of PCSIR scientists
- ☆ Client is responsible for the production under the supervision of PCSIR's scientists
- ☆ Marketing of the product.

(d) Role of PCISR in Quality Control of Commodities Produced by SMEs

PCSIR provides analytical services to SMEs for their products in following areas:

- ☆ Textile
- ☆ Food

- Pharmaceutical
- Polymer
- Material Science
- Leather
- Fuel
- Chemicals
- Engineering Goods
- Ceramics
- PCSIR also provides calibration services to SMEs

4.4 Role of PCSIR in Improving and Informing about Intellectual Property Rights for SMEs

The historical purpose of the patent system was to encourage the development of new inventions, and in particular to encourage the disclosure of new inventions. Many SMEs do not get patent for their invention due to lack of awareness about rules and regulations of Pakistan Patent Office as well as of other countries. Due to this reason both economically and academically the benefits for novel patent and design is not been obtained and a lot of innovative talent of our country has been wasted. The Ministry of Science and Technology funded the project and the patent advisory cell has been set up in PCSIR Labs. Complex, Karachi. The established Patent Advisory Cell is providing a complete protection to the process/technologies developed in Pakistan. The cell helps the SMEs starting from filing the patent application form till its registration.

5. CONCLUSION

The possible strategy for productivity improvement in Pakistan through SMEs has to follow a four step approach where each step is building on one another. At the first step, SMEs need a conducive regulatory environment, balanced regulatory framework, provision of adequate infrastructure and the supporting attitude from provincial and federal government levels. On the second step, short to long-term funding and access to capital at affordable terms should be allowed to the SMEs. Pakistani society is an employee society, and to transform it to an entrepreneurship society we have to bring reforms by taking revolutionary steps in our education system and have to motivate and encourage the educated youth towards entrepreneurship. It will result in the establishment of a larger number of enterprises in diverse segments. The fourth step should target the quality competitiveness among SMEs in Pakistan. It is understood that SMEs typically suffer from weak entrepreneurial skills as well as deficiencies in accounting, production management, technical know how and business planning. As SMEs grow, they increasingly need connectivity to export markets and capacities to match up-to-date quality standards. In this regard quality training of their workers and up-gradation of technologies are the essential factors. R&D organizations and Higher Educational Institutions should come forward and help Pakistani SMEs in these respects.

REFERENCES

Economic Census of Pakistan, 2005, FBS, Statistics Division, Islamabad [GOP 2005, P.14], http://www.statpak.gov.pk/depts/fbs/publications/ec_2005/ec_2005.html

Economic Survey of Pakistan 2002-03, http://www.accountancy.com.pk/docs/Economic_Survey_2002-03.pdf

Economic Survey of Pakistan, 2003-04, http://www.accountancy.com.pk/docs/Economic-Survey-2004-05-Part-III.pdf

Economic Survey of Pakistan 2007-08, http:www.defence.pk/forums/economt-development/11995-economic-survey-2007-08-overview-economy.html

Developing SME policy in Pakistan, SME issue paper –for Deliberation by SME Task Force-, Policy Planning and strategy Department, Annex III, page IV, http://www.smepolicy.net.pk/SME_Issues_Paper.pdf

Iqbal, M. and Farrukh, M. K., Small and Medium Enterprises in Pakistan, http:www.southasianmedia.net/Magazine/Journal/9-small Medium enterprises Pakistan.htm

Pakistan Country Assistance Strategy, World Bank Report, Annex II, Page 3, http:www.sbp.org.pk/bpd/Conference/Day One/SME in Pakistan.ppt

Pakistan Infra Structure Implementation Capacity Assessment (PIICA), World Bank Report No.41630-pk, South Asia Sustainable Development Unit (SASSD), South Asia Region, November 22, 2007, http://siteresources.worldbank.org/SOUTHASIAEXT/Resources/Publications/448813-1202436185914/PIICfull.pdf

Saleem, S., 2008. SMEDA SME Policy paper 2007-A Critical Review (An Analytical commentary upon SME policy proposed by SMEDA Pakistan), http://mpra.ub.uni.muenchen.de/7329/

MPRA Paper No. 7329

SME policy 2007, SME led economic growth- creating jobs and reducing poverty, page 14, <http://www.pakistan.gov.pk/ministries/industriesandproduction-ministry/media/SME PolicyDevelopment.pdf>

SME Development in Pakistan: Issues and Remedies, http://www.gcu.edu.pk/publications/vc-sme.pdf

ABBREVIATIONS

BDSPs: Business Development Service Providers

EPB: Export Promotion Board

FI: Finance Institute

GDP: Gross Domestic Product

GoP: Government of Pakistan

HEC: Higher Education Commission

HRD: Human Resource Development
IIEE: Institute of Industrial Electronics Engineering
INSMED: Institute of Small and Medium Enterprise and Entrepreneurship Development
LE: Large Enterprise
MoIT: Ministry of Information Technology
MoST: Ministry of Science and Technology
PCSIR: Pakistan Council of Scientific and Industrial Research
PSEB: Pakistan Software Export Board
PSTC: PAK-Swiss Training Centre
R&D: Research and Development
SBP: State Bank of Pakistan
SE: Small Enterprise
SME: Small and Medium Enterprises
SMEDA: Small and Medium Enterprise Development Authority
TBIC: Technology Business Incubation Centre
TICs: Technology Innovation Centers

Chapter 17

Factors Governing the Success or Failure of Technology Transfer: State Sector Experience of Sri Lanka

Muditha P. Liyanagedara, I.Y. Kushani Jayaweera and Ranil D. Guneratne

National Science and Technology Commission, Colombo 5, Sri Lanka

E-mail: nastec@sltnet.lk

ABSTRACT

Identifying factors affecting the destiny of technologies transferred is a vital requirement for the successful commercialization of technologies. Therefore, to identify the factors that determine the destiny of the technologies developed and transferred by public sector R&D institutions, a survey was conducted. In this survey, the existence of pre-determined factors that may affect the fate of the technologies transferred was examined. The following facts were common to the majority of transfers considered in this survey. Out of the technologies surveyed, 68 per cent were transferred free of charge. These technologies were mostly improvements to existing technologies (54 per cent) while new technologies accounted for 28 per cent. In almost all transfers only technical know-how was transferred. In 48 per cent of the cases considered, transferors highlighted inadequacy of human resources as the main institutional barrier. All the transferors were involved in post-transfer activities such as further improvement of the technology and supporting transferees to obtain finances. It was found that the successfully transferred technologies bear one or more of features such as cost effectiveness, easy adaptability, low cost alternative to existing technologies, and fulfilling an urgent need of the receiver. Among the many reasons given for unsuccessful transfers the following are more prominent: poor communication between transferors and transferees; lack of sound technology transfer mechanisms; lack of producers or producers not possessing technical/financial/management/marketing capabilities; introduction of cheaper products by private sector competitors; and practical difficulties inherent in the technology itself.

Further analysis also showed that lack of success was due to lack of interest or not attempting to continuously improve the technology to make the product competitive in the market and the absence of an effective technology advertising culture in the state R&D sector. Finally, in this paper, a brief account of a case study of a successfully transferred technology will be discussed.

Keywords: *Factors, Technology transfer, Sri Lanka.*

1. INTRODUCTION

Many state sector R&D institutions of Sri Lanka are engaged in the transfer of locally developed technologies. However, the diffusion of many of these locally developed technologies has not reached the expected number of potential transferees. Therefore, it is of paramount importance to identify the reasons behind this slow rate of diffusion. Hence, a survey was conducted, to identify those factors that determine the outcomes of the technologies transferred. The issues raised in the questionnaire were designed to determine the influence of pre-determined factors that could affect the fate of the technologies transferred. The main factors included in the questionnaire were:

1. Whether technology transfer was a mandate of the R&D institutions surveyed (to find out whether existence of such a mandate would help in effectively transferring technologies).
2. Type of technologies transferred; *i.e.*, whether they were new technologies or improvements to the existing technologies (to identify which type was transferred more successfully).
3. Whether a transfer fee was charged or the technology was provided to the transferee free of charge (to find out which way gives better results).
4. Transfer mode, *i.e.*, whether only technical know-how was transferred or technical know-how was provided with financial support (again, which approach gives better results).
5. Barriers faced at the transferor and transferee ends.
6. Transfer method *i.e.*, workshops, seminars etc. (to identify the more successful transfer methods).
7. Whether transferees were assisted in obtaining loans
8. The extent of post-transfer involvement of the transferor.

This survey covered technologies belonging to fields such as agriculture/ plantation, civil engineering, electronics, energy, and food and herbal products The Sri Lankan R&D Institutions covered by this survey were the Arthur C. Clarke Institute for Modern Technologies, Atomic Energy Authority, Coconut Research Institute, Department of Agriculture, Industrial Technology Institute, National Building Research Organization, National Engineering Research and Development Centre, Rubber Research Institute, Tea Research Institute and Veterinary Research Institute.

2. METHODOLOGY

A multiple choice style survey questionnaire was sent to all R&D institutions in the country. The researchers and extension officers of these institutions were asked to fill separate questionnaires for each technology they transferred. Fifty-two completed questionnaires were received. The percentage frequency of the each answer selected under each question was calculated.

The transferors were asked to rank the transfer methodologies in the order that worked well for them and in the order in which they preferred to execute them. The percentage preference for each transfer methodology was calculated to identify the most generally successful or preferred methodologies.

As the second step of the analysis, the returned completed questionnaires were divided into two sets: successful and unsuccessful transfers. Then the percentage frequency of the each answer selected under each question for both successful and unsuccessful transfers were compared to see whether there were significant differences, with regard to each factor considered, between the two categories.

Finally the transferor and the transferees of one case of a successfully transferred technology (toddy tapping technology for the fishtail palm (*Kitul* tree)) were interviewed to see how the factors considered above behaved in this particular transfer.

3. RESULTS AND DISCUSSION

3.1 Direct Outcomes of the Survey

Technology transfer was a mandate for all the institutions surveyed. Most (53.8 per cent) of the technologies transferred were improvements to existing technologies while 27.7 per cent of them were new technologies. A large fraction of the above technologies, *i.e.* 67.9 per cent, were transferred free of charge, 20.7 per cent were transferred for a nominal fee, and 11.3 per cent were transferred for their market value. The different modes of transfer of technologies surveyed were: (*i*) transferring of know-how; (*ii*) know-how with financial support; and (*iii*) know-how along with financial support and assurance of sale of products. In most of the cases, transfers were confined only to know-how while 9.2 per cent of them were know-how with financial support. But very few of them assured the sale of products.

According to the responses given for the question on institutional barriers, many transferors (48 per cent) quoted inadequacy of human resources as the main institutional barrier. This is contrary to the common belief that it was the lack of finances. The second highest (44 per cent) quoted institutional barrier was the inadequacy of financial resources. Lack of organizational infrastructure (23 per cent) and lack of management support (18.4 per cent) were among the other important institutional barriers. The main barriers at the transferees' end as observed by the transferors were poor communication (34.6 per cent) and lack of interest (19.2 per cent). The most used technology transfer method was training programs (86.5 per cent). The other transfer methods used, in order of decreasing frequency, were field trials/prototype pilot plant demonstrations (76.9 per cent), workshops/seminars (61.5 per cent), and personal contacts with individual receivers (65.4 per cent).

The order of transfer methods that worked best for most transferors was workshops/seminars and field trials/prototype pilot plant demonstrations, followed by training programmes, brochures, and handouts, newsletters, and websites

However, there is a difference between the above order and the order of transfer methods that many transferors preferred to follow, *viz.*, training programmes, followed by workshops/seminars and field trials/prototype pilot plant demonstrations, brochures and handouts, newsletters, and websites

In 67.3 per cent of the transfers that were surveyed, the transferor did not assist the transferee to obtain financial support from a third party such as a bank. In most of the cases (48 per cent) the post-transfer involvement of the transferors was continuous monitoring. Another 42.3 per cent of transferors became involved in post-transfer monitoring only when a problem had been referred to them. A summary of the above percentages against the relevant questions and answers are given in Table 17.1.

3.2 Main inferences of the Survey

The successful and unsuccessful transfers were compared to find out factors that promote or inhibit the success of a technology transferred. Again the percentage frequency that each answer had been selected was used as the criterion of comparison.

Table 17.1: Summary of the Questionnaire and Percentage Frequency of Answers Selected

Questions	*Answers*	*Per cent Frequency*
Transfer Mandate	Yes	100.0 per cent
Technology Type	Completely new technology	27.7 per cent
	Improvement to an existing technology	53.8 per cent
	Adaptation of an existing technology of another country	16.9 per cent
	Other	1.5 per cent
Transfer Fee	Free of charge	67.9 per cent
	For a nominal fee	20.7 per cent
	For market value	11.3 per cent
Transfer Mode	Know-how only	75.9 per cent
	Know-how and arranged financial support also	9.2 per cent
	Know-how, provided/arranged financial support and assured sale of products	3.7 per cent
	Other	11.1 per cent
Main Barriers at institutions' end	Lack of Resources: Financial	44.0 per cent
	Human	48.0 per cent
	Lack of Management support	11.5 per cent
	Lack of organizational infrastructure	23.0 per cent

Contd...

Table 17.1–Contd...

Questions	Answers	Per cent Frequency
	Inflexible regulations	19.2 per cent
	Other	5.7 per cent
Main barriers at receivers' end	Lack of Interest	19.2 per cent
	Poor communication between transferors and transferees	34.6 per cent
	Negative attitude towards the technology provider	7.7 per cent
	Language	3.8 per cent
	Culture	3.8 per cent
	Other	25.0 per cent
Transfer Methods	Workshops/Seminars	61.5 per cent
	Training programmes	86.5 per cent
	Field trials/prototype pilot plant demonstrations	76.9 per cent
	Personal contacts with individual receivers	65.4 per cent
	Other	1.9 per cent
Assistance to obtain Financial	Yes	28.8 per cent
Support (to Receivers)	No	67.3 per cent
Post-transfer involvement	Continuously monitored the usage of the technology	48.0 per cent
	Did not monitor at all	1.9 per cent
	Involved in post transfer activities only when a problem is referred	42.3 per cent
	Further improved the technology and transferred to the previous users	42.3 per cent
	Other	5.7 per cent

In this comparison it was found that the factors given in Table 17.2 made a notable contribution for successful or unsuccessful transfers.

The knowledge of the success of a technology in one country may provide confidence to the potential adopters from another country and this could be one reason why the adaptation of technologies successfully commercialized in other countries is more successful. It is seen that transferring technologies for a nominal fee rather than free of charge or for market value increases the success rate. One reason behind this could be that the payment of a transferring fee results in selection of the producers or end users that have the potential to commercialize the technology they received. Another may be that the transferor has to now put in more effort to ensure success because a payment is being received. When a technology is provided free of charge, those who may not have financial capability or do not have much interest to commercialize will also get it just because it is free. Transferring free of charge may increase the number of receivers but produce few actual users (*i.e.* effective transfers) which ultimately leads to a low success rate. On the other hand transferring for the market value may significantly decrease the number of receivers, as many may not be able to pay the market value.

Table 17.2: Factors that Significantly Promote or Inhibit the Success of a Technology Transferred

Factors Promoting the Success Rate	*Per cent Frequency Successful/ Unsuccessful**	*Factors Inhibiting the Success Rate*	*Per cent Frequency Unsuccessful/ Successful***
Adaptation of an existing technology of another country	19 per cent/ 12.9 per cent	Transferring free of charge	72 per cent/ 63.1 per cent
		Transferring for market value	12 per cent/ 0 per cent
Transferring for a nominal fee	36.8 per cent/ 12 per cent	Lack of financial resources	60 per cent/ 22.2 per cent
Transferring know-how with financial support	16.7 per cent/ 0 per cent	Lack of management support	16 per cent/ 0 per cent
Transferring through Workshops/Seminars	72.2 per cent/ 56 per cent	Lack of organizational infrastructure	28 per cent/ 16.7 per cent
Assistance to obtain financial support	33.3 per cent/ 20 per cent	Inflexible regulations	24 per cent/ 5.5 per cent
		Lack of interest of transferee	36 per cent/ 5.5 per cent
Continuous post-transfer monitoring	55.5 per cent/ 40 per cent		
		Poor communication between transferor and transferee	36 per cent/ 27.8 per cent
		Post-transfer monitoring only when a problem is referred to	52 per cent/ 38.9 per cent

*: The per cent frequency that a factor was selected in successful and unsuccessful transfers respectively

**: The per cent frequency that a factor was selected in unsuccessful and successful transfers respectively

Providing technical know-how with financial support or providing assistance to obtain financial support from monetary institutions like banks and continuous monitoring were the other factors that increased the success rate. Continuous monitoring may put pressure on the technology receivers to use it and/or may prevent early failure. That also provides the transferor an opportunity to continuously improve the technology, which may need to face the market competition by smart alternatives. On the other hand, the small and medium scale entrepreneurs either cannot afford to take their problems to the transferor or they are poor communicators. Therefore, active and continuous post-transfer involvement is a key factor that makes technology transfer a success.

The other key factors that inhibited the successful transfer of technologies were inadequacy of institutional financial resources, lack of management support, inflexible institutional regulations and lack of interest of transferees. An idea of the extent of the influence of these factors on the success of the technologies considered in this survey can be obtained from Table 17.2 (see percentage frequency columns of Table 17.2). The other factors considered in this survey showed similar results in both successful and unsuccessful transfers.

Other than the contribution of pre-determined factors given in the questionnaire towards the outcome of transferred technologies, the transferors were also asked to provide the main reasons that contributed either towards a success or a failure of the technologies they transferred. The success factors they highlighted included the following: the cost effectiveness of the technology, (*e.g.* biomass rice cooker); technologies fulfilling an urgent need of transferees (vegetable/fruit drier); interest of transferor and transferee and easy adaptability of the technology (toddy tapping technology of fishtail palm trees). On the other hand, poor communication between transferor and transferee, inadequacy of financial and human resources, later invasion of the market by cheaper alternatives (*e.g.*, a fan regulator and high frequency electronic ballast), practical difficulties inherent to the technology itself (*e.g.*, tea basket designed for tea-pluckers), lack of producers (*e.g.*, baddy trailer, manual fruit harvester, multipurpose cage wheel), negligence of ergonomics and aesthetic aspects of final products, and the absence of an effective technology advertising culture in state R&D sector were cited as important failure factors.

Some transferors mentioned the facilitating role of technology transfer centres (*Vidatha* Centres), or the new technology being a more effective alternative to an existing technology as reasons for success. Others mentioned the high cost of the technology as a reason that inhibited successful transfer.

While the factors described above are important at the micro level, the purchasing power of producers/end users and a small domestic market, that made it difficult to achieve economies of scale, could be significant factors at the macro level that inhibit the success of transferred local technologies.

Some institutions tried out specific strategies to enhance the success of the transfer. These included, introducing prototype products before transferring the technology in order to create a demand in the target market, providing feasibility reports to convince financial institutions to support SME entrepreneurs and providing locally unavailable components to producers to initiate their production.

4. CASE STUDY: TODDY TAPPING TECHNOLOGY OF THE FISHTAIL PALM TREE (*KITUL* TREE)

Toddy tapping is a traditional rural industry in Sri Lanka. To obtain the sweet sap from the flower of the fishtail palm tree, villagers use a mixture of plant materials as a catalyst. This mixture has no uniformity in terms of ingredients. A researcher, after studying this traditional technology, developed a more effective new chemical mixture called kasper, which gives a higher yield than the traditional one. This new technology was successfully transferred to the rural folk who have been traditionally engaged in this industry.

This technology, while being new, can also be considered as an effective alternative to the traditional technology. In the study of the transfer of this technology, it was found that the enthusiasm of both the transferor and transferees was the main reason for its success. The technical know–how was transferred free of charge and no financial support was provided to the receiver to adopt this technology. Since the cost of adopting this technology is very small there was no need of financial support. Therefore, not providing financial support with technical know-how had no impact on this particular transfer. It was transferred through capacity building initiatives such as workshops/seminars, training, and promotion through the mass media. The post-transfer involvement was confined to monitoring only when a problem had been referred to the transferor. Again, in this particular transfer of technology, since it is an alternative to the traditional technology the need of post-transfer involvement was minimal. The involvement of the potential users of this technology during the field trials conducted at their home gardens, at the initial stages of the development of this technology, could be another reason for the success of this transfer. Since they saw the success of the technology, word of mouth publicity may have helped in the rapid diffusion of the technology. Lack of management support and organizational infrastructure were the main institutional barriers encountered by the transferor in transferring this technology.

5. CONCLUSIONS AND RECOMMENDATIONS

Based on the results of the survey the following recommendations may be made.

- ☆ Nominal fees should be charged to transfer technologies to potential users who have the capability to commercialize.
- ☆ Workshops and seminars are more effective than other transfer methods. However, a successful transfer method is technology specific and could vary from one technology to another. The transferor should be able to identify the most effective transfer method based on the characteristics of the technology.
- ☆ Active post-transfer involvement is a key to a successful technology transfer as it provides the transferor with the opportunity to continuously interact with the transferee.
- ☆ Lack of financial resources, infrastructure, management support, and inflexible regulations are the main institutional barriers that significantly impede a successful transfer.

- ☆ Poor communication between transferor and transferee is also a notable deterring factor in technology transfer, according to the present survey.

To rectify the slow diffusion of locally developed technologies, several strategies can be tried out. Like the private sector, the government should also develop an effective technology advertising mechanism. Establishing a repository of government developed technologies, such as a website, which provides all the necessary basic information of technologies that any potential transferee need and advertising the availability of such a repository will be a cost-effective solution for disseminating information on marketable technologies. It is also recommended to carry out a market study before developing a technology. The findings of such a study will help to develop a technology, which has a known market demand. Since, in Sri Lanka, we do not see a strong demand pull technology market, a supply push approach to speed up the diffusion of locally developed technologies is recommended. Finally the country must have proper technology transfer policies and mechanisms, which will simultaneously boost both the supply and demand for locally developed technologies.

ACKNOWLEDGEMENTS

We wish to acknowledge the support of the research institutions that provided us with necessary details of the technologies that they transferred.

Chapter 18

Role of Business Development Service Providers: Implementation of Technology Transfer

Lakshman Wijeyewardena, H.M.S.[1]* and Wasantha Senanayaka

[1]Director General,

Industrial Service Bureau (ISB), North Western Province, Kurunegala, Sri Lanka

**E-mail: lakshmanw@isb.lk, lakshmanwijeyewardena@gmail.com*

ABSTRACT

Small and Medium Scale Enterprises (SMEs) play a critical role in the global economy in terms of creating jobs, introducing innovative ideas, products and services. As per recent investigations, it is concluded that 80 per cent of the new jobs are created through SMEs. Reducing poverty and bringing down unequal income distribution in a country are well served by involving individuals in SMEs. SMEs are well performed when they are subcontractors of established firms which assure market and technology. SMEs have not gone beyond because of lack of awareness on the benefits of legal status and their potential for diversification and expansion of business. Growth and competition are the key targets that SMEs fail to address within their resources. Lack of SME development policies, classifications and standard for Business Development Service (BDS) providers hinders SME development. Competitiveness of the SME sector has been affected due to lack of entrepreneurship, confined to regional markets and use of low technology. SMEs being small face with the problems associated with the principle of economies of scale and are highly susceptible to market distortions. Networking among SMEs hardly exists, as negative impacts of networking are given more prominence over the much more positive outcomes from being in a network.

A Business Development Service (BDS) provider can act as a catalyst in the process of filling these gaps. BDS market is comprised of government, non-governmental organizations (NGO's), associations and societies and private sector service providers. Services that can be offered by BDS providers include planning and management; marketing, training, technical and policy support and implementation; Research and Development and financing.

As an overall strategy for improving competitiveness of the SMEs, creation of an enabling environment for SMEs, accessibility to competitive funds, technological developments and entrepreneurship development are the key factors. High cost of plant and machineries, high cost of borrowed capital, unavailability of expertise, unawareness of best technologies, personal attitudes, different policies and rules and regulations in a region are the barriers to promote technology among SMEs. In order to promote technology among SMEs there should be several policy support such as affordable interest rates, business friendly banks, affordable power cost, incentives for alternative power applications, incentives for product improvement through new technology adoption and for productivity improvement, reduced import tariffs for machineries and equipment etc.

Keywords: *SMEs, BDS, Poverty, Income, Distribution, Competitiveness, Policies, Barriers, Nucleus, Sustainability, Outreach, Impact, Quality.*

1. INTRODUCTION

1.1

The Small and Medium Enterprise (SME) sector in Sri Lanka is gaining attention of the policy makers, government, non government organizations and private sector companies. The SME sector is one of the important means in addressing economic issues such as increasing Gross Domestic Product (GDP), poverty alleviation, income distribution, employment creation etc. Large private sector establishments look on SMEs for backward integration, outsourcing, etc. for getting away from dealing with large number of employees and improving financial efficiency of large establishments.

1.2

Sri Lankan government has published a White Paper on National Strategy for Small and Medium Enterprises Sector Development in December year 2002. As per the White Paper, in Sri Lanka 96 per cent of the manufacturing units is represented by the SME sector which provides 36 per cent industrial employment and contributes only 20 per cent for the value addition. Lack of access to technology due to cost of technology, and lack of knowledge and awareness are the main reasons that have hindered the potential of SMEs in contributing towards economic development of the country. The white Paper suggests establishing Technical Service Centres at district level, establishment of technology development fund and a voucher system for motivating SMEs on research and development (R&D). It is evident that the improving competitiveness of the SME sector is only possible by facilitating the SMEs to advance with technology.

1.3

Most of the Business Development Service (BDS) providers in Sri Lanka mainly work on business planning, entrepreneurship development, training, accounts and business management. Very few BDS provides are working on technology transfer and employ technically qualified professionals for delivering technological services.

1.4 The Problem

Lack of access to technology has hindered competitiveness of the SMEs in Sri Lanka. BDS providers can act as a facilitator for accessing to technology.

1.5 Objective

The objective of the paper is to highlight the role of BDS providers in technology transfer while discussing the barriers for the technology transfer and ways of overcoming the same.

1.6 Methodology

The following methodology has been used:

1. Literature review with regard to policies for SME development
2. Referring publications with regard to SMEs and its contribution to economic development
3. Experience of Industrial Services Bureau as a BDS provider since the year 1990

2. CHARACTERISTICS OF SMES IN SRI LANKA

2.1 SME Classification

There are different classifications for SMEs in Sri Lanka. They are based on assets value and number of employment. Widely accepted classification is based on Asset value as mentioned below.

Based on Asset value (excluding land and building)

Small enterprises: Rs. 01 – 20 million

Medium enterprises: Rs. 21 – 50 million

(*Source*: Task force for SME enterprise sector development programme 2002)

2.2 Ownership Structure

Majority of the SMEs are either sole proprietorship or partnership. As per information 80 per cent of the SMEs are either sole proprietorship or partnerships (Enterplan, 2003, executive summary).

2.3 Principal Business Market

SMEs in Sri Lanka mainly focus on domestic markets. Nearly 86 per cent of the SMEs are selling for domestic market and only 14 per cent of SMEs are involved in exports.

2.4 Problems of SMEs in Sri Lanka

Since SMEs by nature are small, their production capacities are relatively small and they have limitations to go for cost reduction through the theory of economies of scale. Majority of the entrepreneurs operating SMEs do not have formal educational background. They hardly design and implement marketing plans etc. Business planning is hardly done, and even if business plan is prepared *i.e.* for the purpose of an external requirement, they hardly use it for monitoring and evaluation purposes. As indicated earlier, their production capacities are limited and hence, access to market is also limited. Therefore they are concentrating on local or regional markets (Enterplan, 2003).

Further, majority of the SMEs are from either semi-urban or rural areas. Being rural, the available infrastructure for SMEs is not at satisfactory level. Quality of available infrastructure is also questionable. Breakdown and fluctuations of power supply, bad rural roads etc. are unique characteristics of rural infrastructure. This hinders the market reach of SMEs with external environment.

SMEs lack capacities to absorb technology. They lack access to technology and also capital required for investing on technology. Therefore majority of SMEs apply simple or low cost technologies that make the product expensive and low in quality.

Entrepreneurship skills are poorly demonstrated among SMEs as most of them are not aware on the concept of entrepreneurship and benefits of entrepreneurship. Such competencies can however be acquired through training. In most of the occasions, SME operators do not realize the importance of investing time and money for acquiring entrepreneurial skills and rather they treat it as a waste.

Small entrepreneurs also lack the formal knowledge on the concepts of product development, packaging, distribution, and sales promotion.

As a consequence of all the issues mentioned above, the SMEs face disadvantages in accessing to finance. Banks will be very careful to assess the loan repayment ability by way of collateral (Enterplan, 2003) rather than the project feasibility alone. However since SMEs are having above mentioned weaknesses they are not in a position to prove the ability of repaying the loan on the strength of the business. As long as above mentioned weaknesses are persisting in a SME, lending organizations will not come forward to provide support to such an enterprise. However, SMEs do not have enough wealth for submitting as collaterals.

SMEs hardly keep accounts. They mix it up with their family affairs. SMEs are hardly aware of the monthly profits and monthly household expenditure. Lack of proper information with regard to income and expenditure makes it difficult to manage the finance in a SME.

One of the efficient ways to link with the external market is the use of information technology (IT). In many occasions SMEs do not have required human resources for operating an IT system (Enterplan, 2003) in managing the entries such as to keep stock record, preparing payrolls, recording market information, access to market information, maintaining accounts and cash flow information. Also, SME owners find it difficult to find time as they are also a part of an activity at the SME. But they do not understand that they can better utilize their time if they can give enough time on such important activities.

SMEs have less access to the policy makers, as a result the problems of the SMEs hardly get noticed by them. Therefore while taking decisions with regard to industrial development of the country, the needs of the SMEs are overlooked.

Due to lack of an effective link between each other, the SME units find it difficult to resolve problems which otherwise could have been done through a collective approach.

3. BUSINESS DEVELOPMENT SERVICE (BDS) PROVIDERS

3.1

A Business Development Service (BDS) provider can act as a catalyst in the process of filling these gaps by providing various services to the SMEs in solving their problems. The influence that a BDS provider can have on an SME unit is depicted in Figure 18.1

3.2 Types of BDS Providers

The BDS providers in Sri Lanka are represented by government ministries and departments, international non governmental organizations, associations and societies and private sector operators.

3.2.1 Government Ministries and Departments

There are many government ministries and departments providing BDS services to SMEs.

Government BDS providers are active in the fields of agriculture, livestock development, export development, coconut development, tea development, rubber development, investment promotion, business development, entrepreneurship development and technology transfer. BDSs provided by government institutions are mainly characterized by operation of incentive or subsidy schemes. They play a critical role in policy making. However, efficiency and effectiveness of the services delivery is highly varied depending on the ministry and commitment of the leadership. However, perspective of SMEs on the government BDS providers is not satisfactory. The service is characterized by delay, low quality, favoritism, etc. SMEs seek BDSs from government providers as they are mostly free of charge. However, there are few BDS providers in the government sector that provide effective BDSs to SMEs. Their service delivery is at par with private operators. Government BDS providers are permanent and sector specific. Most of the time, the quality of service delivery varies depending on the social status and remoteness of the location etc.

3.2.2 International Non-Governmental Organizations (INGOs) and Non-Governmental Organizations (NGOs)

Sri Lanka receives funds from INGOs such as UN, Mercy Corps, GTZ, ILO, Swedish Cooperative Centre (SCC), CARE International etc. for SME development. INGOs and NGOs are involved in operating these funds and they either directly or indirectly provide BDSs to SMEs. Most of the occasions they identify the need to provide BDSs and finally they hire local BDS providers for the delivery of services. As

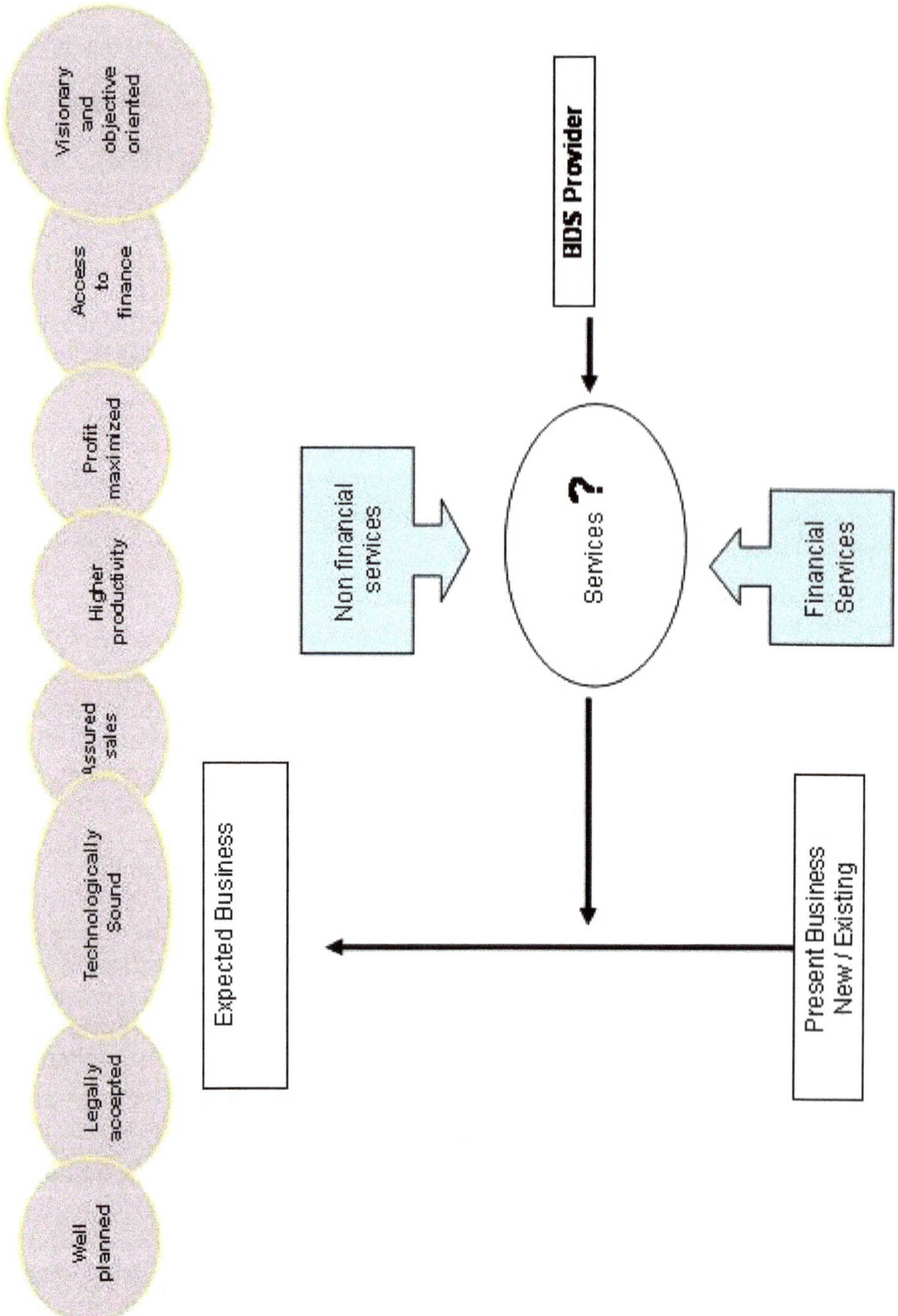

Figure 18.1: Influence of BDS on SME

most of the INGOs and NGOs operate in identified areas and work on specific problems, they have a work programme with specific objectives to be achieved. It identifies target group, geographical location and definite time period and targets to achieve during the project period. Their services are absolutely free of charge. During the project period, it was found that the BDSs provided under the direction of INGOs and NGOs are maintaining satisfactory quality standard in service delivery. As they are limited to geographic boundaries, the package is limited to a segment of the population. Also, as they are provided absolutely free in most of the cases, it leads to distort the BDS market. Also, most of the time SMEs face difficulties as the free services provided under the sponsorship of the INGOs cease after the project period is over.

3.2.3 Associations and Societies

Grouping of SMEs to form formal and informal associations, societies and chambers are evident in the SME sector of Sri Lanka. Except formal chambers of SMEs, most of the other associations and societies are formed because of an external influence. In many occasions, the external influence is to form a society for channelling donations, incentives or subsidized loan schemes etc. Their services are free to its membership. Office bearers render their services voluntary. They do not have professional capacities to deliver technical BDSs.

3.2.4 Private Institutions

Private institutions provide BDS especially in the fields of micro-finance, entrepreneurship development, business management, management development and technology transfer. They are either multidisciplinary or specialized in one or two fields. They run it as business and for profit making.

Therefore, they always implement sustainable strategies for their existence. These institutions are highly affected by free BDSs offered by government institutions, INGOs and NGOs.

3.3 Types of Business Development Services (BDSs)

Mainly BDSs can be categorized in the following sections.

1. Planning and Management Services
2. Marketing Services
3. Training Services
4. Technical Services
5. Policy Support and Implementation
6. Research and Development
7. Financing

1. Planning and Management Services

Most of the SMEs in Sri Lanka do not give emphasis on business planning and applying management techniques in their business operations. Therefore, majority of the SMEs collapse due to lack of emphasis on planning and management aspects. Followings are the main attributes of planning and management aspects.

- ☆ Resource identification
- ☆ Market trends and network
- ☆ Potential sectors vs geographic areas
- ☆ Evaluation of entrepreneurship competencies
- ☆ Matching entrepreneurs with viable businesses
- ☆ Business planning
- ☆ Feasibility studies
- ☆ Pre-investment studies
- ☆ Performance evaluations
- ☆ Financial analysis
- ☆ Development of business plans
- ☆ Business appraisal
- ☆ Business evaluation
- ☆ Accounting and auditing
- ☆ Business registration and licensing
- ☆ Management information systems
- ☆ Linking with financial institutions
- ☆ Productivity development
- ☆ Business counselling
- ☆ Formation of associations for mutual problem solving and lobbying

2. Marketing Services

SMEs do not address their marketing issues professionally and it has become the key factor for failures in SME business operations. SME entrepreneurs are practicing experience based business running. Rarely do they apply marketing and business management knowledge in their operations.

Marketing is one of the very important activities for business success. Marketing covers under mentioned areas in a business.

- ☆ Market research
- ☆ Market plan
- ☆ Market linkages
- ☆ Trade fair and exhibitions
- ☆ Advertising and sales promotion
- ☆ Export/import marketing
- ☆ e-marketing
- ☆ Marketing evaluation
- ☆ Incentive packages for market development
- ☆ International business exploration visits

3. Training Services

This is a powerful tool for imparting skills among the entrepreneurs involved in SMEs. Training institutions in Sri Lanka are providing free as well as paid services. Main activities of a training provider are as follows.

- ✰ Training need assessment
- ✰ Training program design
- ✰ Training module preparation
- ✰ Training delivery
  - Management training
  - Mobilization training
  - Enterprise development training
  - Vocational training
  - Leadership training
  - Technical training
  - Human resource development training
  - Business counselling
  - Institutional development training
  - Capacity development training
- ✰ Training program evaluation

4. Technical Services

Technology services are offered mainly to introduce and transfer technologies. Further, providing continuous services for maintenance of machineries and equipments after the technology transfer is evident in these services. Government organizations, INGOs and NGOs are active in introduction and transfer of technology. The actual operation as a result of technology transfer is mostly carried out by private sector BDS providers. Different categories of technical services are as follows.

- ✰ Technical trainings (agriculture, engineering, dress making, beautician, electrical, etc.)
- ✰ Technology transfer
- ✰ Product development
- ✰ IT services
- ✰ Packaging and labelling
- ✰ Quality control

5. Policy Support and Implementation

Policy making and implementation is mainly an activity of the government. BDS providers support identification of problems associated with SMEs and channel to the government for policy making and formulating solution. BDS providers support the government or policy makers to develop new policies and involve in policy changes.

6. Research and Development (R&D)

Mostly government owned BDS providers are involved in research and development. They carry out R&D activities related to common issues where as private sector BDS providers undertake R&D activities specific to their own businesses. Activities under R&D services include:

- ☆ Survey programs
- ☆ Need and impact assessment
- ☆ New technology identification
- ☆ New product development
- ☆ Methodology development
- ☆ Coordination with research and development organizations

7. Financial Services

Access to finance is a common problem for all SMEs. Presently the cost of funds has been sky rocketed and the whole world is leading towards a financial crisis. The Government from time to time introduces special loan schemes with a grace period for repayment and at subsidized rate of interest. Technology transfer often requires capital. The cost of finance toady has become a bottleneck for technology transfer. Different financial services are as follows.

- ☆ Finance Product,
- ☆ Credit
- ☆ Savings
- ☆ Insurance
- ☆ Money transfer
- ☆ Bank draft

4. PROBLEMS ASSOCIATED WITH BDS MARKET

4.1

The Business development services such as after sale services, registrations, and licensing are known and common, but other BDSs such as skill development, business planning, marketing, technology transfer etc. are not popular among the SMEs. However, the concept of BDS is by name itself is new to the SMEs specially based in rural areas. BDSs in Sri Lanka emerged mainly due to donor funded projects. Donor funded projects and development projects implement their project components through a service provider. Among the project components there are components to deal with livelihood development, micro and small enterprise promotion, institutional strengthening, social mobilization, technology transfer etc. In implementing such activities donors and development projects seek the assistance of a BDS provider. There are Government and semi government BDS providers. There are chambers, associations and societies providing BDS and private sector operators. Cooperatives also providing BDSs. Government and Semi-government organizations, cooperatives, associations and societies offer BDS mostly on free of charge and they involved mainly on routine services. BDSs available at Chambers and semi-government organizations involved in technology transfer and offer their services at subsidized rate and private sector providers charge their services with a profit. Service quality is varied and most of the private sector BDS providers link their BDS charges to the output. Following problems are identified as bottlenecks in the BDS market.

- There is no standardization of BDS providers
- Lack of clear cut government policy and rules and regulations governing BDS markets
- Repetitions of BDS service especially from government sector
- Free of charge or subsidized BDS services (Government. and NGOs) vs fee based BDS (private institutions)
- Reluctance of SMEs for paying BDS (Enterplan, 2003)
- Absence of coordinated effort for BDS delivery
- Distorted and discontinuation of BDS markets by projects
- Lack of technically qualified officers with BDS providers

4.2 Barriers in Finding, Adopting and Using Technology

Technology is expensive. However one should feel the need of a technology. Thereafter the information with regard to availability of technology should be canalized to industry where it is required (Enterplan, 2003). The industrialist should know the benefit that the technology can bring in for his industry and should be able to pay for the technology. As SME operators are heavily involved with day to day activities they have poor interest in accessing information.

They do not have an idea about the additional benefits that can bring in to industry after technology required being introduced. Even though the benefits of technology are known, high cost of capital (high interest rates) has hindered the acquisition of technology. In the rural context there are limited market opportunities for upgraded (quality) products and services. Technology requires expertise for operation and it will be either not available or not affordable to SMEs. Because of the communication gap between SMEs and technology generators or providers, SMEs do not know the best technology available in the market. Moreover personal attitudes (resistance to change and lack of capacity to absorb) have hindered switching to modern and advanced technologies and SMEs continue to be tied up with existing familiar technologies. Reluctance to pay for BDS services has made them trapped in to the vicious cycle of communication gap and unawareness. In addition to the points mentioned above following factors also have adversely contributed to the BDS market:

- Lack of skilled labour
- Environmental barriers
- Rules and regulations
- Different policies in the regions

4.3 Factors Affecting Competitiveness and Business Viability

Material, man, management, money and method determine the competitiveness of an SME. In case disparities are found in a region with respect to above mentioned factors, business competitiveness is affected in contrary to the competitiveness of individual business. Therefore in order to make the business competitive, first of all, the regional disparities of factors of production and cost of technology should be eliminated or else the country should bring in required protective mechanism by way of tax to regulate the effect of price disparities of factors of production. A business friendly environment, in other words, banks closely working with SME directly or through BDS provider would help all the players *i.e.* SMEs, banks and BDS providers. As a result of the advice on business planning, management and application of technology the SMEs will move forward smoothly and it would improve the business of both banks and BDS providers in the long run. Banks and BDSs can therefore lobby on behalf of the SMEs to introduce affordable bank loans. However, among the factors of production, power plays an important role. It is observed that there is a huge regional (country wise) disparity on power cost. Even within a country there are disparities among power cost depending on the user and uses. Therefore this disparity is heavily affected on competitiveness of an SME. At this power crisis, alternative power sources should be encouraged by way of giving incentive for SMEs which will then enable to improve competitiveness of SMEs. As most SMEs are established either in suburbs or rural village context, the raw material for alternative power is either readily available or accessible easily.

In the present context, quality is one of the very important factors for competing in the market. All go in hand in hand to determine the customer acceptance *i.e.* customer go for high quality and low price. So in order to meet this challenge, SMEs have to improve on quality by applying technology and getting the commitments of the personnel employed. Applying proper technology and cultivating positive attitudes among employees will enhance the productivity of the organization.

If an organization keeps operating at higher productivity levels, such organization can deal with the context of quality and competitive price (Enterplan, 2003; Task force for SME enterprise sector development programme 2002). Marketing linkages and opportunities through establishing a net work with the key players of the market concern is one of the important factors for improving competitiveness of a SME. However, SME growth is highly influenced by the level of innovativeness of the SME as compared with other SMEs locally and regionally. Innovativeness of an organization depends on the system that the organization is put in place to flow innovative ideas of the employees from top to bottom (Enterplan, 2003). Innovative ideas must be rewarded and an organization culture should be maintained for making employees falling in "*love*" with the SME. In order to arrive at that, SME should have a benefit sharing system, that the part of the profit earned should be shared with the employees by way of an incentive. This is the only way to make employees much concerned on the affairs of the organization. Therefore, SME operators need a paradigm shift in their thinking process to get more contribution from employees towards the betterment of the SME. SME who does this will eventually win the race. This will be an answer in the long run for unequal income distribution which is a burning issue in the capitalist economic system. If the industry is reluctant to do it, the government mechanism should encourage it by formulating required policies in favour of the same.

4.4 Success Factor for BDS Provider in Technology Transfer

The BDS provider should have a mechanism to sustain in long run by way of charging a fee for their services. The services provided by BDS provider should make a visible impact in a SME. At the same time BDS provider should maintain a sizable outreach for ensuring enough income as depicted in the figure below.

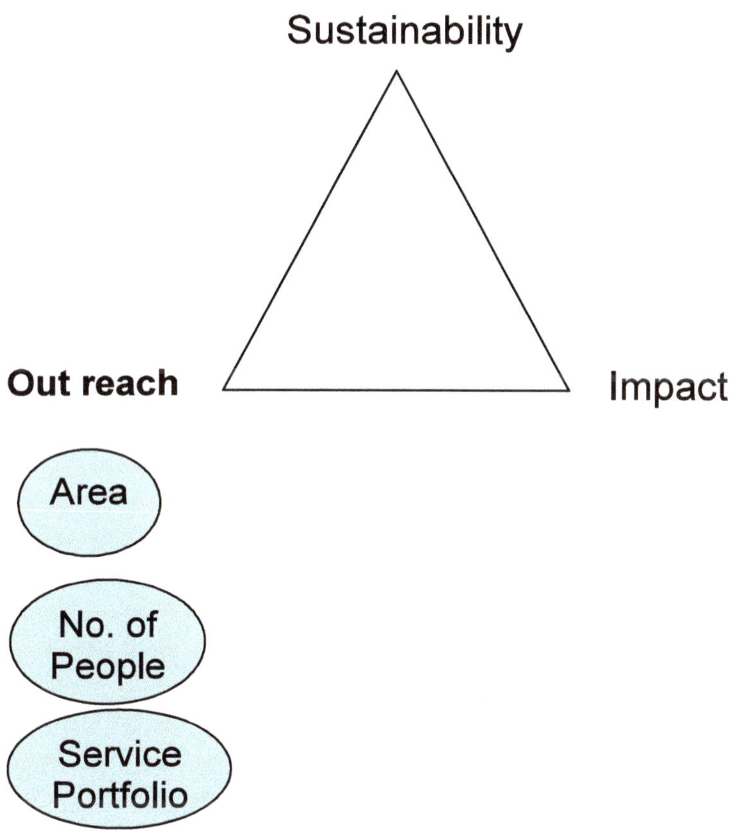

Figure 18.2: Success Factors for BDS Providers

4.5 Effective BDS for Technology Promotion

Effective business development service for technology promotion can be achieved if BDS provider gives deep thought into under mentioned aspects.

- Reputation and image
- Information with regard to technology
- Incentive schemes (loan etc.)

- ☆ Awareness campaign (international and local exposure to SMEs)
- ☆ Technical and financial feasibilities (cost benefits) with evidence
- ☆ Cluster or nucleus based approach for technology procurement
- ☆ Promotion of incubators
- ☆ Commercial approach
- ☆ Partnership or accountability (result based charging for BDS)

ACKNOWLEDGEMENTS

I acknowledge the guidance given and information provided in support of this paper by Mr. J. M. G. B. Jayasundra, Chief Secretary of the North Western Provincial Council of Sri Lanka. I extend my thanks to Mrs. Shaymali Randeni, Project Manager Entergrowth Project of Kurunegala District for giving valuable guidance and information to prepare this paper. Further my thanks go to all the employees of Industrial Services Bureau for supporting me by way of providing reading material and facts. Further, I thank NAM S&T Centre for giving me an opportunity to publish this paper.

REFERENCES

Enterplan, 2003, SME Baseline survey: Results and Key finding report, pp. 14, 19, 23 – 24, 31, 36 – 39.

Task force for small and medium enterprise sector development programme 2003, National strategy for small and medium enterprise sector development in Sri Lanka "White Paper" pp. 2, 9.

ABBREVIATIONS

BDS: Business Development Services

SME: Small and Medium Enterprises

Chapter 19

Challenges of Integrating the "Jua Kalis" into the National Science and Technology System in Uganda

Bashir Rajab Kagere
Policy Analyst
Uganda National Council for Science and Technology, Kampala, Uganda
E-mail: kagsbassh@yahoo.com, kagsbash05@yahoo.com, bkagere@gmail.com

ABSTRACT

Jua Kali is a Swahili word meaning hot sun. In the East African region, the term is used to refer to small scale artisans eking out a living through manufacturing of products and offering services in the open scorching heat of the sun. *Jua Kali* businesses are therefore small scale enterprises (SSEs) and are comprised of semi-skilled artisans operating various kinds of trade in the informal sector and have contributed significantly to Uganda's economic growth. *Jua Kali* business forms the backbone of the informal sector which is by far the most important employer in Uganda. It is estimated that there are more than 800,000 small scale enterprises in the country, employing 2.6 million people, *i.e.* about 90 per cent of the total non-farm private sector workers. Its contribution to GDP is more than 20 per cent. These enterprises also form the base for local technological innovation and transfer to the medium and large industrial establishments. Despite this contribution, SSEs are less recognized for their contribution to the economy, less facilitated, not consulted on key issues that affect them, and left out of the national planning process. The process of integrating SSEs into the national S&T system and the national development framework is quite complex. *Jua Kali* faces a number of challenges which are both institutional and national. These include, but are not limited to, unfavourable policy environment, governance and institutional challenges, difficulty in accessing finances, high attrition rate, poor enterprise management, need to deepen the concept of innovation, low levels of technology transfer, capacity of the *Jua Kali* to absorb technology, legal protection, definition of SMEs and product standardization. This paper reviews the challenges that hinder the integration of SSEs into the national S&T system and the overall development process.

Keywords: *Jua Kali, Science, Technology and innovation, Small scale enterprises, Integration, Challenges.*

1. INTRODUCTION

The political turmoil in Uganda, in the 1970s and early 1980s, shattered the country's industrial sector leading to an economic and structural breakdown. In 1986, the government's initial goal was to reduce the country's dependence on imported manufactured goods by rehabilitating existing enterprises. These efforts met with some success and, in 1988 and 1989, industrial output grew by more than 25 per cent, with much of this increase in the manufacturing sector. During the financial year 2006–07, the industrial sector contributed 21 per cent to GDP indicating an increase of 0.2 per cent the previous financial year (2005–06). The sector still focuses on light manufacturing and offering services for local consumption. Over 90 per cent of Uganda's exports are agricultural products which, at the time of export, have low technological input or value addition. Import growth has been sustained at just over 26 per cent in 2006–07 (MFPED, 2007).

The poor performance of large and medium scale industries and the resulting high levels of unemployment have led to the emergence of micro and small scale enterprises within both the formal and informal sector especially in urban areas. The informal sector acts as an employment buffer in the urban economy providing some minimum form of employment to migrants who come from the country side as well as urban dwellers. Though it is clear that vast numbers of people work in small-scale and micro- enterprises, the distinction between small scale and informal activities is not always easy to make.

The main characteristics of the informal sector include labour intensive technology, low levels of productivity, use of family labour, irregular working hours, restricted use of capital, lack of legal protection and regulation, little formal schooling, absence of book keeping procedures, few barriers to entry, personal relations with clients, a small and usually poor clientele, lack of access to credit, lack of government support, etc.

Small-scale industries in Uganda have existed in the informal sector since pre-colonial times making various tools, ornaments and cooking utensils. They have also continued to produce goods to fill in the gaps between imported products and what is produced locally either because they are too expensive to afford or not available in the market for one reason or another. Small-scale industries' operators are often good imitators and very ambitious. As expected, the type and quality of the products vary with geographical location, with the urban-based small-scale industries stretching their skills to assembling electrical and electronic appliances. They use precision instruments while the ones in villages concentrate on agricultural implements.

Some major towns have SSEs located in designated or gazetted areas. However, those processing agricultural products are scattered randomly across the country. Those in urban areas employ a broad category of people ranging from non-skilled, uneducated to skilled, well-trained certificate holders whereas those in the countryside are dominated by uneducated and low-level school dropouts or kin favoured by the proprietor/manager. They are young and energetic, some of whom acquire knowledge, experience and eventually start their own trade along the same line nearby or migrate to urban centres.

The paper is organized as follows: section two critically examines the literature on the growth path of SSEs to industrialization, and also analyses the *Jua Kali* concept and the contribution to the economy. Section three examines Uganda's science, technology and innovation (STI) system. Section four looks at the day-to-day challenges faced by the *Jua Kali* and section five gives an example of the metal fabrication industry. Section six examines the challenges faced by the S&T system in integrating the Jua Kali. The last part of the report presents conclusions.

2. SSE GROWTH: THE PATH TO INDUSTRIALIZATION

Recent literature has emphasized the role of SSEs as a source of entrepreneurship and continual innovation. New products and processes are claimed to be introduced by new and small firms which may grow rapidly if they prove to be successful. The effect of generation of employment by these firms depends on two factors–survival and growth. If the survival rate is low, then creation of new establishments will merely increase the rate of labour turnover without making any lasting contribution to employment. The survival rate is the ratio of the number of surviving new firms to the initial number of new plants.

One of the dynamic arguments frequently espoused for encouraging SSEs, is that they serve as a "breeding ground" or "seedbed" from which larger firms emerge[1]. It is often argued that virtually all large private enterprises started originally as very small entities[2]. Kilby (1988) and Biggs, *et al.* (1995) contend that this pattern of "graduation" of firms from one size category to another is not universal and that entrepreneurial and policy bottlenecks often restrict this "graduation" process.

An empirical study on small scale manufacturing in Africa[3] shows that modern small and medium scale enterprises did not primarily originate from micro manufacturing enterprises but rather started with more than ten employees. Relative growth in numbers of firms appears to be highest in 2-9 and the 10-49 worker firms and lowest in 1 person firms. The percentage of graduation of firms from micro to small scale and or medium scale varied widely, ranging from 10.7 per cent in Rwanda to 65.6 per cent in India. The graduation rates in African countries are substantially

1 See Marshall, 1920 or Bolton Committee 1971.

2 Anderson 1982.

3 Carl Liedholm and Joan Parker, 1989.

smaller than those found in Asia and Latin American. In Asia and Latin America, roughly one-half to two-thirds of the modern small and medium firms had expanded through the size structure while in no African country did even half graduate (Carl and Joan, 1989).

Evans (1987a) explored the relationship between firm growth, firm size and the firm's age for a panel of US manufacturing firms for the time period of 1976-1982. He found that firm growth decreases with firm age as well as with firm size[4], which is consistent with Jovanovic's (1982) theory of firm growth. According to Jovanovic (1982), firms uncover their true efficiencies, over time, through a learning process. But Evans' (1987b) other finding was that firm growth decreases with size, which is not consistent with Gibrat's Law[5]. Moreover, Evans also concluded that the relationship between firm's growth and size is a highly non-linear one.

2.1 *Jua Kali* and their Contribution to the Economy

With the collapse of many medium and large scale industries in Uganda in the 1970s and early 1980s, the micro and small private sector enterprises responded by absorbing large numbers of the unemployed labour force and contributed significantly to the country's GDP. Some of the micro and small scale enterprises in the informal sector have been commonly referred to as *Jua Kali*. *Jua Kali* is a Swahili word meaning hot sun. In the East African region, the term is used to refer to small-scale artisans eking out a living, by manufacturing products and offering services, in the open scorching heat of the sun.

Jua Kali businesses are therefore small scale enterprises (SSEs) and are comprised of semi-skilled artisans operating various kinds of trade in the informal sector and have contributed significantly to Uganda's economic growth. *Jua Kali* business forms the backbone of the informal sector which is by far the most important employer in Uganda. It is estimated that there are more than 800,000 small scale enterprises operating under this scheme in the country, employing 2.6 million people, *i.e.* about 90 per cent of the total non-farm private sector workers. Its contribution to GDP is more than 20 per cent. These enterprises also form the base for local technological innovation and transfer to the medium and large industrial establishments.

Most *Jua Kali* industries start off and remain as informal enterprises as their owners often cannot afford the expensive and rigorous process of registering with the relevant Government bodies. This informality means that the *Jua Kali* businesses have found it hard to access business development services or export markets as compared to the formal and larger industries. The *Jua Kali* entrepreneurs as a result remain poor and their businesses small and uncompetitive. Though the Government has recognized the potential of the *Jua Kali* sector as an employment opportunity for many Ugandans, it has failed to create an enabling and sustainable environment for these enterprises to grow.

---

4 This study controls for the exit of slow growth firms and tests for the age and size effects on growth separately.

5 The size of a firm and its growth rate are independent.

2.2 Characteristics of *Jua Kali*

Some important characteristics of *Jua Kali* include the following:

(*a*) *Jua kali* are generally sole proprietorships with the owner as the only person in a managerial position. Thus no organized structural management system or board of directors is instituted. They mainly focus on a small range of products or services which are primarily sold to the local market. Often, they employ 1 to 5 people and operate their business on trust, rather than on contracts and systems.

(*b*) Though they face fierce competition due to their small range of products, *Jua Kali* do not have well laid out future plans and the day-to-day pressures of management and resource constraints cause their vision and outlook to be restricted by the scope, skills and experience of the owner.

(*c*) *Jua Kali* are mainly dominated by micro-enterprises with low technological intensity and labour intensive activities.

(*d*) Most *Jua Kali* tend to take up business concepts based on a conviction arising from a superficial assessment of an existing business operation (copying culture). Such shallow convictions often do not stand the test of time given the risks, persistence, sacrifice, disappointments and frustrations associated with starting and running a business.

(*e*) There is easy entry and easy exit. The vision at the entry point is usually economic survival other than business growth.

(*f*) They extensively use local raw materials and scrap in production

(*g*) They lack social protection schemes such as the National Social Security Fund

3. NATIONAL INNOVATION SYSTEM

The literature on innovation systems has made important contributions to the understanding of innovation as an "interactive process, which involves the creation of qualitatively different new things and new knowledge" (Lundvall, 1992). A generic innovation system involves a network of government and non-governmental agencies, science and technology institutes, firms and education organizations, etc. However, the network dynamics and trajectory are influenced by a country's macroeconomic and industrial policies, international regulations, market governance and social innovation processes (Freeman, 1987).

Uganda, with its still developing education, research, and technology system, possesses a poorly functioning national innovation system and innovation processes for her prevailing objectives and circumstances. Viotti (2002) argues that developing countries are not organized in 'systems of innovation' but 'learning systems'. He considers developing countries as adopters of technological knowledge from developed countries, though their firms may develop incremental innovations[6] according to their capabilities to do so.

---

6 These firms do not develop innovations in the sense defined by Lundvall, 1992.

The main challenge for Uganda at present is how to build on the existing decentralized, market-oriented system and national development framework for the present-day and future generation of wealth. Total transformation is required because the old mental models, circumstances and purposes, do not work anymore today. Moreover, the established social structures and institutions, which are based on such models, theories, and rules, face major adjustment challenges. New combinations and forms of collaboration between actors in the national system of innovation also will be required.

The challenges as mentioned above can only be met if decision makers and major actors in the process are willing to fundamentally rethink existing structures and institutions in the light of the present political and techno-economic environment. The connections with the business sector are underdeveloped and there is hardly any cooperation between the universities and S&T institutions. The implication is that the Ugandan government has to upgrade its innovation system.

3.1 Components of Uganda's STI System

The STI system in Uganda ideally consists of four components namely: STI Human Resource, STI Infrastructure, STI manufacturing and Marketing and STI Policy and Institutional Framework. The system is illustrated in Figure 19.1.

3.2 Management of the STI System

Uganda's STI system is managed through a constellation of macro and sectoral policies whose major focus is on non-STI specific indices of development. This has resulted in growing public concern, especially among the scientific and research community, about the near relegation of STI as a development priority; and the sporadic and often times uncoordinated responses to STI demands at both sector and national level.

Uganda's STI system is governed by a mixture of core ministry and numerous autonomous institutions [Councils, Commissions, and Authorities] whose mandates with regard to science and technology development appear to overlap rather than

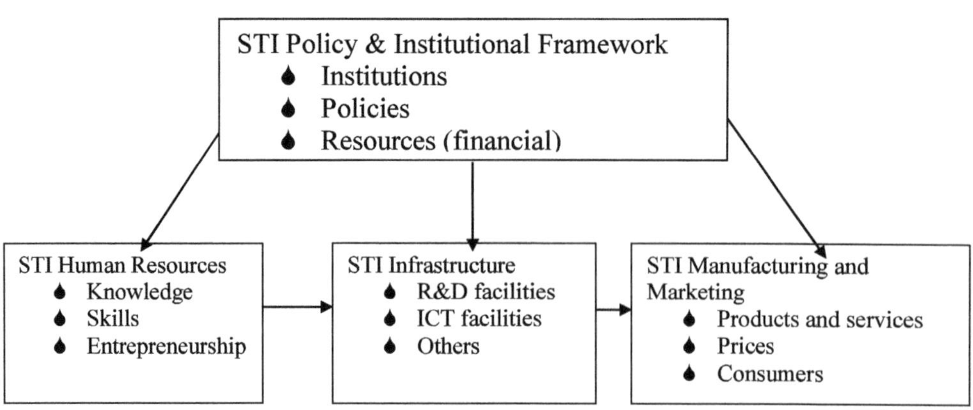

Figure 19.1: STI System

complement and enhance each other. The existence of a plethora of SETIs[7] often with parallel mandates complicates the national STI coordination function of government. Moreover, the UNCST Statute No.1 of 1990 (Cap 209 of the Laws of Uganda 2002) does not clearly spell out the regulatory and advisory functions or adequately empower the Council to undertake the co-ordination function for effectively executing its mandate. The Council therefore uses guidance and advice approaches rather than the more effective legal and regulatory approaches in management of science and technology development.

The most significant tool in the process of harmonizing STI coordination in Uganda is the National STI policy. However, Uganda does not have a comprehensive and overarching national STI policy. Though the country has since 1994 been developing and reviewing the national science and technology policy, government has never formerly adopted the resultant versions of the STI policy. The policy was reviewed in 1997, 2001 and still on going with the current version of 2007 still under review to take into account developments in science, technology and innovation (STI).

4. *JUA KALI*: A CASE OF METAL FABRICATION

The metal fabrication industry in Uganda is still in its infancy and a majority of the production takes place in small enterprises using local resources. Artisans use simple tools and locally available materials to produce consumer goods for the local market. Skills in the metal fabrication sector are basically learned on the job through an older and experienced worker teaching a novice over a relatively long period of time. This transmission of skills requires a high level of trust and a common work ethic and production attitude. The most important aspect of localized learning capacity are the common culturally based rules of behaviour, a language of engagement and collaboration, and tacit codes of conduct between firms which enable the development of trust- itself essential for innovative collaboration.

Some of the commodities produced include, among others, metallic windows and doors, staircases, mortars, vehicle spare parts, spare keys, grinding machines, tanks, source pans, chairs, and tables. The working conditions in the metal fabrication industry are tough as employees and employers have to work 6 or 7 days a week for at least 8 to 10 hours per day usually under the scorching sun. They rarely have protective gear even with the hazards at the workplace. They obtain their raw materials through collection of metal scrap or buying metallic inputs on the local market. Finished products and raw materials are kept in the open spaces since they do not own stores and therefore frequent thefts are inevitable.

The metal fabrication industry is characterized by very low technology and the rate of technological change is relatively low (Nyiira *et al.*, 2001). Cost advantages generally derive from either low labour costs or high rates of capacity utilization over long periods of time or both. Nyiira *et al.* (2001) have further indicated that 45 per cent of the metal fabricating firms did not introduce any new products over a 3 year period

---

7 Science, Technology and Engineering Institutions.

ranging from 1997 to 2000. Moreover, almost all the new products introduced in the metal fabrication industry in the year 2000 were either imitations of domestic models or applications of existing technology in the existing market. Most metal fabricating enterprises begin small (1 to 5 people) and remain in this state since the goal of the initiative is for personal survival and not growth.

However, a few modern metal fabrication enterprises have recently joined the business to cater to the middle class market. These include the following: Hwan Sung, Tumpeco (U) Ltd, Sembule Steel Mills Limited, Aluminium Holloware Manufacturers Uganda Limited, Casements (U) Limited, Steel Rolling Mills Limited, etc. These however import the raw materials as well as the technology that is used in production. Some of the few companies that have managed to follow the growth path include Seka Metal Fabricators Limited, Agali Awamu Carpentry and Welding Workshop, and Volcano Metal Works.

5. CHALLENGES OF INTEGRATING *JUA KALIS* INTO THE S&T SYSTEM

5.1 Policy Environment

A number of policies that are intended to stimulate the growth of small scale enterprises, such as the Policy on Small and Medium Scale enterprises, which have provisions for enhancing *Jua Kali* growth through establishing mechanisms for technological development as well as encouraging innovation in small scale enterprises are still being debated. The STI Policy, which is aimed at reinvigorating the STI system in Uganda, specifically through encouraging scientific and technological innovations as well as industrial development, has, since 1994, never been formerly adopted. Therefore, Uganda does not have a comprehensive and overarching national STI policy. The Industrialization Policy which has provisions on augmenting the technological capabilities of small scale enterprise – *Jua Kali* inclusive, was approved but with very little money allocated towards innovation[8] and technology transfer.

The legislations in place are therefore not conducive to SSE growth and development especially as far as effective flow of information and knowledge within the system is concerned. A sound regulatory framework is not only needed to define the role of political leadership and the civil services; it is also required to guide the operations of the SSE sector. To enhance enterprise development in Uganda, it is vital to have a suitable enabling environment for business creation, survival and growth. The enabling environment refers, among other things, to policy and regulatory issues, to infrastructure, and to financial and business development services.

5.2 Institutional Framework

The responsibility for managing the STI activities and several technology-intensive programmes is currently entrusted in various government institutions such as departments, authorities, commissions, councils and some private sector

---

8 Shs 400 million (about US$ 250,000 through the innovation fund).

institutions with very little coordination or sharing of lessons and experiences. This uncoordinated system of governance has led to duplication of effort as well as neglect of the *Jua Kali* since the absolute responsibility of managing the system is not entrusted to one particular institution.

The Uganda Small Scale Industries Association (USSIA), the institution in charge of overseeing the development of the SSEs, has very little support from the government and no policy framework. The institution does not have convenient infrastructure to help in catering to the needs of its clients. Enterprise Uganda has set up training facilities in which the *Jua Kali* and other small scale entities can be trained through the business development services. Here entrepreneurs acquire skill in marketing, pricing, management, planning, etc. However, Enterprise Uganda also lacks the required manpower to effectively train and re-orient the *Jua Kali* most of whom have little or no education.

5.3 Classification of SMEs

There is no clear and acceptable definition of SMEs. The Ministry of Finance, Planning and Economic Development (MFPED) has defined micro and small-scale enterprises, but the definition of both small and medium-scale enterprises is still lacking. Their definition considers SMEs as entities with at least 50 employees. The Uganda Revenue Authority (URA) defines SMEs as those enterprises with an annual sales turnover of at least US$ 20,000. The Uganda Investment Authority (UIA) classifies enterprises with a capital of between US$ 5,000 and US$ 50,000 as SMEs. This definition makes almost all industrial establishments in Uganda SMEs. There is therefore a need to reach industry consensus on the definition of micro and small-scale enterprises and to define medium-scale enterprises.

5.4 Deepening of the Concept of National Innovation System

Most innovation systems in developing countries are too simplistic as they only pay attention to research and development as the main engine for innovation. However, as discussed earlier, Viotti's (2002) argument of developing countries having learning systems as opposed to innovation systems appears true in Uganda's case. There is no clear system of innovation in Uganda and more so, innovation is perceived as a science based index. The level of funding allocated to facilitating innovation also further indicates the understanding and importance attached to this concept. The key challenge for the STI system is to develop a well coordinated innovation system that will enhance the capabilities of firms and individuals in all sectors of the economy. It is imperative however, that the public sector plays a major role when it comes to supplementing the self-organizing forces of the private sector in at least two respects: enhancement of the production and distribution of technology; and the reduction of transaction costs.

5.5 Financing

Access to financial services has over time been a major problem for all SMEs. Lack of financing seriously hinders the ability of small firms to increase their capacity, survive, as well as upgrade the technology in businesses. However, lack of access to formal finance is still a major element crippling the ability of a business to operate effectively, to purchase materials and services most economically, to modernize or expand business as well as maintaining or replacing machinery. When starting a new business, most of the small enterprises begin as micro businesses and are unable to obtain loans from formal institution. Thus, they decide to begin or start their businesses by investing their own savings or borrowing funds from friends and relatives. While large companies have access to subsidized credit and are supported by government, small firms cannot borrow from formal financial institutions. Furthermore, since they are not registered, they cannot offer capital, housing or fixed structures[9] as collateral. They are therefore dependent on informal sources of credit where exorbitant rates of interest are charged owing to the high risks of default. Sometimes entrepreneurs in the informal sector are actually hindered by the government.

Apart from access to formal finance, the *Jua Kali* also lacks the resources to transfer technology or exploit existing technologies. Analysis has shown that the private sector will not invest in any technology that does not promise immediate gains. It should be noted that the private sector in Uganda has not vastly invested her resources into research and development. The *Jua Kali* is risk averse with their primary goal being profit maximization. The available financial resources are sometimes cautiously and tightly managed which limits investments in projects and or technologies whose returns cannot be predicted with precision. This has limited the process of technology transfer, use and application in the country.

5.6 Technology Incubation/Clustering

Technology incubation has become increasingly popular the world over since the late 1980s. The introduction of collective solutions where firms collaborate and create technology centres and other forms of inter-firm clearing houses for the exchange of innovations (Romer, 1993) is called clustering. If firms exchange their marginally relevant knowledge via a technology centre, the rate of productivity increase will be larger than if they keep the innovative results for themselves. The innovations may be individually of minor importance to innovators but as a whole they may be of major importance to nationally located industrial sectors. However, technology centres are confronted with obvious free-rider problems and historical experience suggests that the viability of self organized centres should not be over estimated and that normally the public sector has to be heavily involved. For the innovation system to function in this kind of arrangement, the public sector has to play a role in stabilizing the competitive conditions of firms in order to allow them focus their research efforts and reduce costs of duplication of innovation efforts.

5.7 Technology Transfer and Management

The management of the flow, use and application of science and technology in Uganda's industrial sector and specifically among the SSEs, has remained weak even where it exists. The process of searching, acquisition and diffusion of technology has continued to be slow mainly due to information management constraints.

---

9 Since they do not have any.

Information relating to the required appropriate technology is on several occasions not readily available. Furthermore, with the current globalization and knowledge explosion trends, it is difficult for an innovating country to transfer new techniques of production to a country that does not have an IP regime as it underlines the fact that ethical considerations may not be appropriately taken care of (Nyiira, 1994).

Furthermore, *the Jua Kali* lack dependable structures and support mechanisms for technology adaptation, reproduction, modification, innovation and infrastructure development. There is often a strong temptation amongst *Jua Kali* to stick with an existing range of products fearing that developing new ones will be complicated and fruitless. Some feel that transferring technology could result in loss of daily revenue while others view it as a costly and complicated venture. The technologies developed by these entities are not standardized though they have the potential to feed into the industrial sector.

Jua kali also lack common facility centres and incubation centres where they can acquire and upgrade skills, form self-help groups, and share techniques and expertise related to their trade. Putting in place such facilities could facilitate access to technologies that they are unable to procure on their own.

Although it is very difficult to estimate countries' performance in developing and commercializing technology, high-tech trade is definitely one important indicator in this direction, as high-tech products are the leading edge of science-based trade and are often the result of considerable R&D investments. During the year 2006, total imports of goods accounted for US$ 2,488 million. 43.4 per cent of these were medium high technology imports[10], 33.8 per cent were medium low technology[11], high technology[12] contributed approximately 0.01 per cent and the rest was low technology[13]. Total exports for the year 2006 were US $ 1 billion of which low technological exports contributed 80.4 per cent, low medium technology 17.4 per cent, medium high technology 0.5 per cent and high technology 0 per cent (UNCST, 2007).

5.8 Standardization of Products

Though most *Jua Kali* enterprises produce unique products, their commodities are not standardized for quality and safety. Most of their products are of low quality and do not meet the minimum requirements for both local and export markets. This is attributed to general inefficiencies in quality control and assurance practices. There are 600 national standards in place and 60 locally manufactured products meet the national quality standards. The Uganda National Bureau of Standards (UNBS) is charged with the responsibility of monitoring quality and standards. However, UNBS is faced with several limitations that hinder its effectiveness.

---

10 Motor vehicles, Scientific Instruments, Electrical machinery, Chemicals, transport equipment.

11 Rubber and plastic products, non-ferrous metals, Petroleum, Fabricated metal products, etc

12 Aerospace, computers, office machinery, Pharmaceuticals.

13 Paper, textiles, wood and furniture, foods and beverages.

Although there are 600 national standards for the different products, most products do not have an agreed national standard and this makes it difficult to describe a product as either being of poor quality or substandard. The national infrastructure at UNBS is inadequate in terms of physical infrastructure and staff, and the government releases only 0.067 per cent of the total value of imports to the UNBS for carrying out its operations.

5.9 Competition and Legal Protection

There is excessive competition among the *Jua Kali* and these are at most times incapable of differentiating their products substantially from those of others. This in turn reduces the profit margins leaving fewer resources for technological innovation. Moreover, in this kind of environment, secrecy is hard to maintain and legal protection of innovation is non existent. In such a situation, knowledge spillovers may also have drawbacks as they reduce innovation incentives for the party that generates them.

Furthermore, the legal regime governing intellectual property (IP) in Uganda is obsolete; some pieces of legislation were inherited from the British colonial masters. The Government of Uganda has, since the year 2000, instituted a task force under the auspices of the Uganda Law Reform Commission (ULRC) to amend and update the IP laws. A number of them have been reviewed and tabled before cabinet and await passing. There is therefore no explicit policy on intellectual property. The process of legal protection of innovations has in turn become too expensive and unaffordable especially by the small-scale firms. It is therefore no surprise that less than 3 patent applications are submitted per year.

5.10 Link between *Jua Kali* and Industrial Establishments

The *Jua Kali* have on many occasions invented machinery that can enhance production in the medium and large scale establishments. Their technologies are very cheap as compared to imported ones. However, due to the prevailing communication gap between the *Jua Kali* and the industrial sector, these technologies have come to waste. There is also an inherent belief among the industrial entrepreneurs that imported technology is more efficient and of the best quality. The STI system therefore has a triple challenge of first of all getting the industrial entrepreneurs to appreciate the initiatives of the *Jua Kali*, improving the quality of their technological innovations to meet the demands of large scale manufacturing, and making the *Jua Kali* a hub for technological innovation for the industrial sector.

5.11 Entry and Exit

As I mentioned earlier, most *Jua Kali* enter the informal sector small business for survival. In this regard, entrepreneurs keep trying their luck at different businesses. When the market experiences a downturn they quit the business. James Tybout (1998) has indicated that firms exit when they experience a series of adverse productivity shocks, driving their expected future operating profits sufficiently low that exit is their least costly option. He further argues that young firms have not yet survived a shakedown process, so they tend to be smaller and to exit more frequently. However, since science and technology products take long to provide any tangible results, for the *Jua Kali* to be integrated in to the S&T system, they ought to have a stable business.

5.12 *Jua Kali* Capacity to Absorb Technology

Most of the *Jua Kali* enterprises employ less than 10 people who are in most cases family members or close relatives. In addition, these firms have very low levels of capitalization, literacy levels, management skills and R&D. The *Jua Kali* capacity to absorb the required technology is, among others, impeded by the small size and growth of the markets in the private sector. The *Jua Kali* lack the much needed human resource capacity to install, operate and maintain the different novel and output conscious technologies in existence on the world market today. All in all, the lack of adequate private sector capacity to access and absorb the existing relevant and appropriate technologies has continued to affect the efficient flow, use and application of modern technologies and the whole system of technology transfer in Uganda.

5.13 Tacit Knowledge

Another challenge for the STI system is the tacitness of knowledge and the difficulty it posses to transfer such technology. The *Jua Kali* usually develops innovative technologies through hands on experience. Often times they do not have the theoretical background to back their assertions and neither do they have a scientific orientation in terms of their education. It is usually true that these ideas originate from people who have a low regard and understanding of scientific oratory. Their ideas are however, not documented but rather tacit in nature and are perfected through trial and error. The fear of revealing information to competitors also adds to challenge of codifying their technologies. This kind of knowledge flow is a hindrance to the STI system in the area of technology transfer, patenting, funding and Innovation.

6. CONCLUSION

In developing countries, SMEs are the engines of growth and are responsible for most jobs, products, tax revenue and development. It is therefore imperative that governments in developing countries as a matter of necessity and survival of their economies support and fund knowledge creation and infusion into SMEs as a whole. It is clear from the discussions above, that there a number of challenges that inhibit the integration of *Jua Kali* into the STI system, most of which are beyond their control. We have also seen that the *Jua Kali* is a source of innovative ideas for industrialization if effective mechanisms are put in place to support them. Owing to the importance and size of the informal sector, the government should play a significant role in enhancing the growth of the *Jua Kali* businesses into well functioning industrial establishments through carefully thought out interventions such as a favourable policy environment, comprehensive innovation system, financial support, and literacy and skill development programs.

REFERENCES

Alain, Albagli (undated), *'Science and Technology Insights in SMEs'*

Anderson, Dennis, 1982. *"Small Industry in Development Countries: A discussion of the Issues"*, World Development

Bengt-Ake, Lundvall, 1992. 'National systems of innovation", Pinter

Biggs, Tyler, Shah, Manju and Srivastava, Pradeep, 1995. *'Technological Capabilities and Learning in African Enterprises.* World Bank Technical Paper No. 288

Bolton Committee, 1971, Small Firms: *Report of the* Committee *of Inquiry on Small Firms,* HMSO, London

Carl, Liedholm and Joan, Parker, 1989, *"Small Scale Manufacturing Growth in Africa",* Initial Evidence

Evans, David, 1987a. *"Test of Alternative Theories of Firm Growth",* Journal of Political Economy.

Evans, D. S., 1987b. *"The Relationship between Firm Growth, Size, and Age: Estimates for 100 Manufacturing Industries."* Journal of Industrial Economics

Freeman, C., 1987. *'Technology and Economic Performance: Lessons from Japan',* Pinter, London

James, Tybout, 1998. *'Manufacturing Firms in Developing Countries: How well do they do, and Why?'* Georgetown University

Kilby, Peter, 1988. *"Breaking the entrepreneurship bottleneck in the Late-developing countries: Is there a useful role for government?"* Draft paper, Wesleyan University.

Marshall, Alfred, 1920.*Principles of Economics,* 8th Edition, London

MFPED, 2007. *"Background to the budget"*

MTTI, 2007. *"Industrialization Policy"*

Nyiira, Z.M., Stella, Nagujja, Richard, B. and Lutaalo, 2001. *'Uganda's SMEs and their support systems',* Uganda National Council for Science and Technology (UNCST)

Nyiira, Z.M., 1994. *'Uganda's Initiatives to enhancing Science and Technology Capacity: Policy and Practice',* UNCST

Romer, P., 1993. *"Implementing a national technology strategy with self-organizing industry investment boards",* Brookings paper on Economic Activity

Sunil, Mani and Henry, Romijn, 2004. *'Innovation, Learning and Technological Dynamism of developing countries',* UNU press

UNCST, 2008. *"Draft National Science and Technology Policy",* STPCD

UNCST, 2007. *"State of Science and Technology in Uganda",* STPCD

Viotti, E., 2002. *'National Learning Systems: Technological Forecasting and Cultural Change'*

Chapter 20

Small Scale and Medium Scale Enterprises in Uganda: Challenges and Opportunities

Adam M. Sebbit

Department of Mechanical Engineering, Technology Development and Transfer Centre, Makerere University, P.O. Box 7062, Kampala, Uganda

E-mail: amsebbit@tech.mak.ac.ug, amsebbit@gmail.com

ABSTRACT

Uganda has achieved significant economic growth over the past decade. The economy has grown at an average rate of 6.8 per cent per annum. There is a general decline in both formal and informal manufacturing sectors. It was aggravated by the high electricity tariff and outages. Uganda is predominantly an agricultural economy with agriculture contributing over 29 per cent per cent to the GDP and employing 80 per cent of the working population. Small and medium scale enterprises play a crucial role in the development of the country. There are an estimated 800,000 Micro and Small and Medium Enterprises (MSMEs) in urban and rural areas. It provides employment to approximately 1.5 million people. Employment growth is 20 per cent per annum in this sector and it contributes to 20 per cent to GDP. This sector nonetheless forms an important part of the small business community in spite of numerous constraints. The government of Uganda has put in place intervention measures such as the Business Development Scheme for small enterprises and the master craftsmanship programme which aims at putting in place a self-sustaining, demand driven assistance programme. The Technology Development and Transfer Centre of Makerere University acts as a backbone link between industry and the faculty. The Uganda Gatsby Trust whose purpose is to support manufacturing and value adding business and the cluster initiative that is particularly advantageous to small and medium scale industries in developing their competitiveness are both based at the Faculty of Technology of the Makerere University. Small and medium scale enterprises which are crucial for sustaining high economic growth and creating jobs need long term finance mechanism to support their growth. Unfortunately, such support is not available. There are many fragmented efforts targeting small scale industries. However, these efforts are not coordinated and the cost of technology transfer and its maintenance is prohibitive. To overcome these problems there is need for international cooperation in which developed countries have crucial roles to play.

Keywords: *Uganda, Technology transfer, Finance, Constraints, Intervention.*

1. INTRODUCTION

Uganda has achieved significant economic growth over the past decade. The economy has grown at an average rate of 6.8 per cent per annum. The projected population in mid 2008 was 29.6 million. The population growth rate is 3.2 percent per annum (Statistical Abstract, 20081). The agriculture sector value added percentage growth rates remain constant at 4 per cent per annum from 2002–03 to 2006–07; it contributed significantly to the Ugandan economy. The contribution to the national economy decreased from 35.7 per cent in 2002–03 to 29 per cent in 2006–07.There is a general decline in both formal and informal manufacturing sectors. The growth rates of the manufacturing sector reduced from 3.0 per cent in 2002–03 to 1.8 per cent in 2006–07. That could be due to the influx of imported goods and the high dependence on semi-processed import input to the industrial sector. It was further aggravated by the electricity cost and outages (Ministry of Finance, Planning and Economic Development, 2008)].

Although there was a decrease in the growth rates in posts and telecommunications from 42.5 per cent to 31.5 per cent in 2007, it is the most vibrant sector. Generally the transport and communication has been registered growth rates from 14.2 per cent in 2002–03 to 20.6 per cent in 2006–07. Other sectors of the economy also registered strong growth, notably electricity (8.1 per cent), wholesale and retail trade by 10.6 per cent, construction 13.1 per cent and mining and quarry 24.2 per cent in the calendar year 2006–07. There is a general shift from the manufacturing to the service sector (Ministry of Finance and Economic Developmen, 2008).

Uganda is predominantly an agricultural economy with agriculture contributing over 29 per cent per cent to the GDP and employing 80 per cent of the working population (Bremer and Zuehlke, 2009). The country is gradually moving away from a predominantly agricultural economy, towards one emphasizing construction, manufacturing, and services, including trade, tourism, transport and mining. In recent years, manufacturing, electricity generation, water supply and transport and communication recorded substantial rates of growth. The government has launched a ten-year road development programme to rehabilitate the main trunk roads and to build and maintain feeder roads to open up rural areas (Ministry of Finance, Planning and Economic Development, 2007).

1.1 Energy

Energy is the driving force in economic development, although it is not specifically mentioned as one of the pillars in Uganda Poverty Eradication Action Plan (PEAP).

It is Uganda national framework that guides public action to promote economic growth and eradicate absolute poverty [Poverty Eradication Action Plan (PEAP), 2004]. The Millennium Development Goals, there are eight goals to be achieved by 2015, that responds to the world's main development challenges, energy role is crucial but not mentioned (http;//www.undp.org/idp/, 2010). The demand of energy exceeds supply in Uganda. This leads to load shedding during peak hours. The hydropower potential is estimated at 2000 MW. The present demand stands at about 450 MW. Power generation, in Uganda, is largely hydro-based, with an installed capacity of 300 MW but only about 160 MW is available. The thermal power installed capacity 100 MW. The government is planning to increase installed capacity to 200 MW by 2009. There is ongoing construction to add 250 MW by 2011 and there are series of large scale hydropower plants which are expected to be commissioned by the year 2012. The expected installed capacity will be 830 MW and by 2017 it will reach 1200 MW [7]. Most of the thermal power plants will be decommissioned except a 50 MW heavy fuel oil. The present shortfall is over 200 MW (Renewable Energy Policy for Uganda, 2007). The electricity tariff is about 0.28 US$ per kWh and is US$0.20 for industrial applications. Kenyan manufacturers were paying between US$ 0.16 and U.S$.0.22 cents per kilowatt (http://allafrica.com/stories/200803181293.html, http;//English. Peopledaily.com.cn, 2010).

2. NATIONAL INDUSTRIAL POLICY

Small Scale Industrial development is an integral part of national development, which can be achieved through transforming Uganda into a modern industrial country, if the right policies are put in place and implemented. Small and medium scale enterprises play a crucial role in industrial development. The National Industrial Policy sets out the strategic direction for industrial development in Uganda. It caters to all sectors of industries. In most cases large-scale industries have leverage in policy formulation because of large-scale single investments.

2.1 Policy Objectives

The principle focus of the national industrial policy is mostly on large scale industries; limited attention is given to small scale industries. However, there are issues like, engineering knowledge for capital goods, agricultural implements and construction materials and fabrication/Jual Kali operation. Some of the specific policies that will address the issues of small scale industries are to: (*i*) promote safe work practices in all industry sub-sectors and (*ii*) create support systems for sustainable micro and small industries development (National Industrial Policy a Framework for Uganda Transformation Competitiveness and Prosperity, 2007)

2.2 The Constraints

There is low level of funding for research and the institutional framework that enables identification of research priorities, greater commercialisation of research and exploitation of technologies. The major constraint with respect to growth in most small and medium scale industries is at enterprise and managerial level. Most of these firms were started as a family business.

3. SMALL AND MEDIUM SCALE ENTERPRISE

There are an estimated 800,000 MSMEs in urban and rural areas. These enterprises provide employment to approximately 1.5 million people – 90 per cent of total non-farm private sector workers. Employment growth is 20 per cent per annum in this sector and it contributes to 20 per cent to GDP (http://www.gdrc.org/techtran/techtran-sustdev.html, 2009). The Uganda Small Scale Industries Association is currently made up of the following industrial sectors: metal fabrication; food and beverages; textiles and garments; wood craft; leather products; printing and graphics products; chemicals and pharmaceuticals; handicrafts; and building materials and ceramics.

This sector nonetheless forms an important part of the small business community. The informal sector – '*Jua Kali*' – is engaged in the production and distribution of goods and services and generates employment and income in spite of numerous constraints. Small enterprises in the informal sector, in handicrafts, manufacturing, construction, trade and service industries are increasing. While some businesses are one-person units, others work on a family basis and some hire labour or train apprentices. Most *Jua Kali* acquire their talents through on-the-job training or apprenticeships before establishing their enterprises

Initiatives have been made to promote and develop this sector, but the efforts have been largely scattered, uncoordinated – if not conflicting and isolated. The enterprises continue to face many constraints which inhibit their development and growth. One of the main factors contributing to this situation is the absence of a coherent general policy framework for MSME development in Uganda.

In small scale industries the business environment is not very favourable. The challenges are different yet they employ a large number of people. They have limited say in national policy formulation and they need a different policy which will specifically address issues of small-scale industries. It is for these reasons that the government of Uganda is developing a National Micro, Small and Medium Enterprises Policy and Strategy to support the birth and growth of viable and competitive industries.

3.1 Definition

There is great diversity in the definition and classification of small and medium scale enterprises among countries. This is a reflection of the significant differences in aggregate income and distribution, in production structures and capabilities, and in industrial and technological characteristics among economies. As such, there is no sufficient reason to expect that SME definitions be uniform or even comparable across countries and through time.

Several criteria are used to define small businesses. The most widely used include labour, capital, and annual receipts. Definitions vary from using a single criterion or more than one. A micro enterprise is defined as an enterprise employing a maximum of 4 people, having an annual sales/revenue turnover of a maximum of Ugandan Shillings 12 million (US$7,500) and having total assets of a maximum of Ugandan Shillings 12 million.

A small enterprise is defined as an enterprise employing a maximum of 50 people; having an annual sales/revenue turnover of a maximum of Ugandan Shillings 360 million (US$ 22,200) and total assets of a maximum of Ugandan Shillings 360 million. A medium enterprise is defined as an enterprise employing more than 50 people, having an annual sales/revenue turnover of more than Ugandan Shillings 360 million, and total assets of more than Ugandan Shillings 360 million.

3.2 The Policy Objective

The most critical policy objective emerging is to focus on creating an enabling environment targeting the regulatory environment, providing financial support to these industries and fostering public-private partnerships including universities to deliver support to these industries.

There is a need to give attention to introducing and enforcing policies that support start-up and growth in these industries. The openness of the economy provides many opportunities, but to avoid risks like flooding of the market by imported goods mostly from China, the government must play a role so that these industries can have a fair chance to defend local markets and pursue export opportunities. These industries can play a prominent role as engines of innovation, growth and prosperity. They need coordinated institutional support and not fragmented, inefficient support.

There is a need for extensive infrastructure investment, technology upgrading and training and acquisition for these industries. In addition to non financial services, access to finance is also addressed. With the exception of micro-finance, this remains a perennial problem. There should be an established financing mechanism for these industries.

These firms are faced with various constraints such as the following; lack of managerial and entrepreneurial skills; lack of sufficient capital; use of inappropriate and inefficient technologies; limited co-operation and networking amongst the firms; lack of adequate working space; limited access to information and business support services; and inadequate infrastructure. Although the conglomeration of the Small Scale Enterprises constitutes an informal sector, their contribution to the national economy is well established.

4. TECHNOLOGY TRANSFER

Uganda like most of the least developed countries is characterised by low level of technology development. However there are various technologies available in developed and some developing countries which can be transferred to Uganda for fostering development. The priority sectors have to be identified. These technologies need to be evaluated based on national criteria. Furthermore, the technology transferred needs to be well managed for national benefit. It is unfortunate that small-scale industries are not supported. The transfer is often done by individuals who bring in cheap technologies for which the maintenance cost is high.

There are many definitions for technology transfer. Technology transfer typically involves the following steps: discovery or innovation, evaluation, acquisition, adaptation, and implementation. Transfer mechanisms span the spectrum from simple disclosure and licensing to joint ventures and research agreements.

Technology transfer is defined as a process for conceiving of a new application for an existing technology. It is also defined as a process for converting research into economic development. The term technology transfer is also used to mean licensing intellectual property to a manufacturer for production in a product, or reducing an idea to practice in a prototype, or even the process of recording concepts of technology know-how in a professional paper or patent application (http://www.gdrc.org/techtran/techtran-sustdev.html, 2009). A number of players are involved in technology transfer including governmental and related organizations, universities and research institutes, large and small companies, inventors and entrepreneurs.

If one looks at the footprint of technology development and transfer, it is vital for most of the developing countries. But, each step costs money. Most of the Sub Sahara countries can neither afford to pay nor can the other developing countries transfer the technologies to the least developed countries like Uganda at no cost. It costs to develop these technologies. There should a mechanism put in place for identification of technologies and consequently transferring these technologies. The Government in conjunction with the international financial institutions should put in place finance facility to address financial need needs of technology transfer Developed countries should come out and support these initiatives so as to improve the economic development of the least developed countries like Uganda.

The main challenges in technology transfer could be among the following:

- Lack of awareness about technologies;
- Lack of capital;
- Inadequate skilled personnel to adopt technologies;
- Lack of enabling policies initiatives, institutional mechanism, information and opportunities; and
- Lack of know-how for sourcing and efficient use of technology.

These challenges can be turned into opportunities if all stakeholders work together as a team in all stages of program development, technology identification, acquisition, implantation and sustaining the technology.

5. INTERVENTIONS

The government has put in place various intervention to address the issues relating to the development of small and medium scale industries in Uganda. These favourable policies, interventions and the good intentions of the government do not succeed, because there is no financing mechanism in place to sustain the implementation of such policies.

5.1 Business Development Scheme for Small Enterprises

The government of Uganda has put in place intervention measures in different sectors. The Business Development Scheme for small enterprises is a cost-share grants project, co-funded by the European Union. It is implemented by the Private Sector Foundation through the Business Uganda Development Scheme (BUDS) (Business Uganda Development Scheme, 2007). The project supports the acquisition of know-how through training programs to increase the capacity and performance of small-scale enterprises in the private sector.

The aim of this project is to provide support to small enterprises in order to increase the pool of technical and business qualified personnel, improve the supply and quality of teaching personnel in technical and business subjects, encourage the entry of women in technical, business and managerial activities and improve business performance. Since its inception in 2001, 300 businesses have benefited from their services.

5.2 Technology Acquisition Fund

Most of Uganda's exports have earned very little respect and value on the international market owing to their low quality and failure to meet international standards. In order to improve on the performance of Uganda's exports, BUDS is operating a new matching grant facility. The Technology Acquisition Fund (TAF) is a matching grant product, focusing on raising skills standards, and the quality of firms in export supply chains so as to meet international standards based on demonstrated export orders. The Technology Acquisition Fund is available for only few firms that fulfil the criteria, which are mostly targeted at export markets. There are very few industries which can benefit from the funds. They give priority to firms with products which can gain access to international markets like dried fruits.

5.3 Presidential Innovation Fund

Presidential Innovation Fund is a major initiative for encouraging and funding scientists and innovators to produce prototypes of their innovations; providing assistance to patent operational innovations; commercialisation of innovations and linking innovators to the private business sector. It seems to target large scale industries. Small scale industries can be encouraged to apply but often they lack capacity to document and produce good research proposals for funding.

5.4 The Master Craftsman Programme

This aims at putting in place a self-sustaining, demand driven system of assistance whereby, business advisors who are skilled technicians provide practical advice to the small-scale entrepreneurs that can be translated to immediate impact on upgrading skills, improved productivity, designs and management practices in order to increase profit to the enterprise. The Uganda Small Scale Industries Association, Northern Uganda Manufacturers and the Uganda Gatsby Trust are partner organisations in the programme as are institutions like the Nakawa Vocational Training Institution Lugogo, Uganda Leather Allied Industries Association, and Texda. Some of these industries claim the technical colleges are not doing enough, and the only polytechnic they were relying on had been turned into a university (Brief on Phase II of UNIDO, 2004).

5.5 Technology Development and Transfer Centre

This is under the Dean's Office at the Faculty of Technology, Makerere University (TDTC) and was established in 2002. The main activities of TDTC are development and application of innovative technology, research and technology transfer with the aim of uplifting social-economic development of Uganda in sustainable manner. The TDTC is the backbone link between the community, industry and Faculty of Technology.

Among the objectives of TDTC are to:

- Create capacity at Makerere University for development/adoption of intermediate technologies for rural and urban development;.
- Spearhead applied research in technology and commercialisation of research output and development of spin off companies; and
- Assess and transfer technologies to stakeholders.

5.6 Student Incubation Projects

In 2007, the projects were initiated by students from the Department of Mechanical Engineering. One of the projects is an animated digital display unit which is used to relay information at low cost. The display unit will be used for advertisement which plays an important role in modern marketing of products. This research is designed to enable small-scale entrepreneurs to advertise their businesses. Similar technology can be used for traffic control.

The second project is improvement on the Aloe Vera Gel Extractor. The objective of this research is to improve on the aloe vera gel extractor so that it can reduce the burden on women who use knives to remove gel from aloe vera. It works well with up to 25 mm thick leaves. The next stage is to design and construct a machine which can extract gel from 45mm thick leaves.

A computer numerically controlled machine was designed and constructed by a student. It can be used for engraving items, drilling and cutting. The Faculty of Technology will support the commercialisation of these projects. It is envisaged that these two students may set up their own business for manufacturing these products.

5.6 Uganda Gatsby Trust

Uganda Gatsby Trust is a Non-Government Organization based at the Faculty of Technology, Makerere University, and whose purpose is to support manufacturing and value-adding businesses with the potential to grow. It was established in 1994 with seed funding from the Gatsby Charitable Foundation in U.K. Uganda Gatsby Trust (UGT) provides a responsive institutional framework for the exchange of skills and knowledge between the Faculty of Technology and small-scale enterprises in Uganda.

The key functions are to develop a network of micro-enterprise linked to the Faculty of Technology, Makerere University are: increase the quality and value of their output; to introduce students of higher education institutions to opportunities and potential of the small scale sector and to assist them to develop and transfer technologies appropriate to it; and to assist small-scale enterprises to access, acquire credit and technology as well as overcome their problems through extension services and business development services. The Trust also implements activities aimed at informing engineering students of the potential of small scale enterprises for job creation and self-employment.

5.7 The Cluster Initiative

A cluster is an agglomeration of firms operating in the same sector and in a close geographical proximity to each other (Cluster Navigators, 2005). The cluster is particularly advantageous to small and medium scale industries in developing their competitiveness. The cluster initiatives began in 2005 with seven pilot clusters namely, Katwe Metal Fabrication; Basketry Processing; Katwe Salt Processing; Fashion and Textiles Garments, Biofuel; Pineapple Processing; and Management Consultancy. The cluster formation is based on the Triple Helix approach that involves the public/ government sector, academia and entrepreneurs. After two years, four out of the seven clusters were upgraded and fifteen new clusters were formed. These clusters are able to attract funds, provide joint training for members and win contacts.

5.8 Long Term Financing

Small and medium scale enterprises need long-term finance mechanisms which are crucial for sustaining high economic growth and creating jobs. Banks do not want to lend to small-scale enterprises because it is expensive and risky. For the same reason most banks shy away from also extending development finance to this category. The Government has taken deliberate efforts to provide access to long-term development financing. The Government has decided to: (*i*) issue an industrialisation bond for Uganda Development Bank; and (*ii*) provide a guarantee for bank lending to the agricultural sector (Cluster Navigators, 2005).

6. CONCLUSIONS

There are many fragmented efforts targeting small scale industries. All these efforts are not coordinated. Most of the small-scale industries and other enterprises have low level of education and so it may be difficult for them to differentiate the programmes. They may interpret programmes to be repetitive. If they fail in any of them, they get discouraged.

The small-scale industries have limited investment and are highly dependent on grid power for most of their activities. Frequent load shedding coupled with limited access for development finance, high interest rates, and stiff competition with imported products may see some of these industries disappearing. There is no active financial mechanism which is favourable to most small-scale enterprises.

There is lack of awareness in many areas including technology sourcing and acquisition, and access to sources of development financing. Technology transfer is left to individuals who have to bear the risk of failure of the project. There is hardly any funding, from different sources which could encourage the formation of linkages between universities and small scale enterprise linkages and the fostering of technology transfer both locally and internationally.

REFERENCES

A Framework for Uganda's Transformation, Competitiveness and Prosperity, National Industrial Policy, Ministry of Tourism Trade and Industry January 2008.

Bremer.j, Zuehlke. E, Integrating Population, Health and Environment in Uganda, Population Reference Bureau, June, 2009.

Brief on Phase II of UNIDO Integrated Programme in Uganda for Period November, 2004 – April, 2008.

Business Uganda Development Scheme, Private Sector Foundation Uganda, 2007.

Cluster Navigators, Innova, Innovation Systems and Clusters Handbook. May, 2005.

http://allafrica.com/stories/200803181293.html, http;//English. Peopledaily.com. cn accessed on 20th February 2009.

http://www.gdrc.org/techtran/techtran-sustdev.html accessed on 20th February, 2009.

http;//www.undp.org/idp/accessed on 19th February, 2010.

Ministry of Finance and Economic Development. The National Budget Framework Paper FY 2008/09-FY 2012/2013, March, 2008.

Ministry of Finance, Planning and Economic Development, Background to the Budget 2008–09) June, 2008.

Ministry of Finance, Planning and Economic Development, Background to the Budget 2008–09 June, 2007.

National Industrial Policy a Framework for Uganda Transformation Competitiveness and Prosperity, Ministry of Industry, Trade and Tourism, 2007.

Poverty Eradication Action Plan (PEAP) 2004–05 and 2007–08 Ministry of Finance, Planning and Economic Development,, December, 2004.

Renewable Energy Policy for Uganda, Ministry of Energy and Mineral Development, 2007

Statistical Abstract 2008. Uganda Bureau of Statistics, June, 2008.

Chapter 21

Technological Renovation in Vietnam's SMEs

Ngo Thi Loan
Vice Head
Project Department, Centre for Regional Research and Development, Ministry of Science and Technology (MOST), Vietnam
E-mail: ntloan@most.gov.vn, loancrd@yahoo.com

ABSTRACT

At present when science and technology development speeds up rapidly and international economic integration becomes a necessity, the economists think that renovation is actually the decisive factor for the socio-economic development of any nation. The continuous renovation of technology has become a key element in maintaining the competitiveness of an economy. Renovation also creates solutions to overcome challenges of society, health services and environment. Especially, for the small and medium enterprises (SMEs), renovation has been a very important factor, playing a decisive role in their strategic improvement and success. In recent time, industrial enterprises of Vietnam have become conscious of the importance and the need to have technological renovation. They have carried out investment in this activity although the size of such investment is not large and the methods of implementation are different for different groups of enterprises.

Keywords: *Technological renovation, Economy, Small and Medium Enterprises, Socio-economic development.*

1. SMALL AND MEDIUM ENTERPRISES OF VIETNAM

The classification of small and medium enterprises (SMEs) in Vietnam is made according to the Decree No. 90/2001/ND-CP by the Government. SMEs include different types of enterprises such as private enterprises, limited liability companies, share-holding companies, cooperatives or union of cooperatives, stated-owned enterprises (SOE) or private enterprises registered according to the Decree No.02/ 2000/ND-CP) having capital less than VND 10 billions (equivalent to USD 625, 000) and the number of permanent workers less than 300.

According to the statistics report by the General Department of Statistics of Vietnam, if classification is made according to labour-standard criteria, as on December 31, 2006 Vietnam had 127,593 SMEs, accounting for 97.16 per cent of the total number of Vietnam enterprises. And if classification is made according to enterprises' capital size, then there were 114,341 SMEs, accounting for 87.07 per cent of the total number of Vietnam enterprises (detailed in Table 21.1). The small and medium enterprises contribute about 30 per cent of GDP every year, and attract about 77 per cent of the labour force. In average, 700,000 jobs are created every year by the SMEs of Vietnam.

Table 21.1: Structure of Small and Medium Enterprises in Vietnam

Type of Small and Medium Enterprise (SME)	*Total Number of Enterprises in the Whole Country*	*Total Number of SME Classified by Labourers*	*Total Number of SME Classified by Capital Source*
	131318	*127593*	*114341*
State-owned enterprises			
Central	1744	942	181
Local	1962	1515	559
Non state-owned enterprises			
Collective	6219	6161	5757
Private	37323	37233	36292
Collective name company	31	31	30
Limited liability company	63658	62928	57398
Share-holding company having state's capital	1360	1046	387
Share-holding company without state's capital	14801	14476	12458
FDI enterprises			
100 per cent foreign-capital enterprises	3342	2555	1056
Joint-venture company	878	706	223
Per cent of the total enterprises of the whole country	100 per cent	97.16 per cent	87.07 per cent

Source: General Department of Statistics, year 2007.

In recent years, the SME sector has made an important contribution to the economic growth and structural shift, to the elimination of hunger and poverty reduction, to stabilize society, to preserve and develop traditional professions, and to increase export turnover. The small and medium enterprises make contribution of about 7 per cent to the state budget every year, which is equivalent to the amount made by FDI enterprises. SMEs also exploit the potential, intellect and manpower from the people. Small and medium enterprises involve in almost all areas of the economy:

SMEs in Industrial Sectors

The small and medium industrial enterprises account for 17 per cent of total enterprises, concentrating mainly on areas of foodstuff, wood processing, aqua-product processing. Every year, the SMEs produce 31 per cent of the total industrial output. In general, of small-scale enterprises: about 90 per cent of the enterprises have a number of workers less than 100 people, and on average each enterprise has nearly 45 workers.

SMEs in Trade and Services

Almost 55 per cent of the enterprises in Vietnam at present are involved in trade and services as it is the business that has quick rotation of capital, high profit, small amount of investment and use limited number of labourers. The small and medium enterprises being private owned account for 78 per cent of the total retail amount, and 64 per cent of the total commodity transportation. According to investigation data, the small and medium enterprises operating in trade-service area make much more effect than those in industrial area. In average, a worker of industrial enterprise makes a turnover of VND 14.6 million whilst that of trade-service creates VND 75.8 million. Enterprises dealing with trade are mainly located in big cities with improved infra-structure, favourable conditions of transportation, consumers having high incomes and demand. As there are so many enterprises in the sector, it often causes fierce competition.

SMEs in Rural Areas

The small and medium enterprises in rural areas account for 14 per cent, with the number of enterprises being 40,500, which are located in 1631 villages, of which the state-owned enterprises account for 14.16 per cent, cooperatives 5.76 per cent, and private enterprises 80 per cent. If classification is made by operational areas, about 18.62 per cent of enterprises are involved in agricultural-forestry and aqua-product processing, 32.5 per cent in small-scale industry and construction, and 49.78 per cent in service. At present, 100 per cent output of traditional products such as rush, knitting, artful handicraft is made by small and medium enterprises in rural areas. The average amount of capital of such enterprises is very small, which is mainly the owner's capital.

SMES in rural areas create 49 per cent of non-agricultural jobs, attracting about 25-26 per cent of labour force of the whole country. Labourers of enterprises in rural areas are mainly family labourers (hired labourers only account for about 32 per cent in production households). In average an enterprise in rural areas uses about 30 labourers.

2. TECHNOLOGICAL RENOVATION IN SMALL AND MEDIUM INDUSTRIAL ENTERPRISES IN VIETNAM

2.1

Due to severe and increasing competition, enterprises are forced to renovate technology for their existence. Considering the needs of the enterprises, technological renovation may include different activities such as research activity for implementation, improvement of production methods or introduction of new ones, and product innovation–design to introduce new products or modernized products. The technological renovation aims at improving capability and knowledge of technology of enterprises, to improve product quality, reduce cost price, raise labour productivity and have effective business.

2.2 Activities of Technological Renovation of Enterprises

The types of technological renovation chosen by enterprises differ largely on the basis of the scale of investment, kind of operation and nature of such operation. Most of the enterprises interviewed expressed their priority to carrying out one or more activities of technological renovation according to their financial capacity and production-business situation of the enterprise.

It can be seen that activities having direct impact on enterprises' business-production process such as the improvement of production rules/of products and renovation of production rules/of products attract much more attention from the enterprises for investment. However, research activities for implementation, to some extent, also draws suitable attention thus to effectively support the above-said direct activities, especially in groups of state-owned enterprises and FDI enterprises. Besides, through discussion with some enterprises, it is known that the technological renovation in companies mainly relates to the investment in equipments and machinery (hardware), and not in software such as systems managing information, quality, software for design, for supervision of production, etc.

Activities of technological renovation in small and medium industrial enterprises may include the following:

Research for Implementation

It is an activity that only the smallest number of enterprises opt to carry out as compared to other activities of technological renovation (67 enterprises). However, this ratio also shows that research for implementation at enterprise's level has been improved. This reflects the limitation in capacity of research for implementation of enterprises. In theory, research activity for implementation made by enterprises include research to improve products/production rules, or research to create technology for production or upgrade the imported technology for production to make it suitable to certain conditions. However, in fact, at present, enterprises carry out implementation research mainly for application purpose and for technological operation rather than for technological renovation

To Improve Existing Production Rules

To improve production rules (include: renovation of technology, of equipments, of production method, replace (change) raw materials, etc.) is an operation attracting the biggest portion of enterprises to carry out with the reason that the investment in improvement of existing production costs less than investment in new equipments and technology in production line, while it still makes active contribution to improving product-quality, to raising productivity, to production of improved (modernized) products and/or new products.

Innovate Products, Design and Produce New Products

This work is an objective requirement for enterprises, especially for enterprises producing consumer goods (such as cloth, garments, chemical cosmetics, paint, etc.) to diversify products to meet customers' requirement according to the change of surroundings such as environment, weather, fashion, etc. It is the reason the enterprises often carry out this operation.

Apply New Rules of Production

In some companies, the application of new rules of production is often made in the form of expansion of business production rather than renovation of technology. It means that the investment in technological chain (line) by enterprises either aims at increasing output of existing products or improving new products, which may be of the same types with the existing products or may be completely new ones. The application of new rule of production in the form of total replacement of existing equipments, machinery and technological line is rarely used as this requires a very big amount of investment.

2.3 Methods of Technological Renovation

The execution of technological renovation is made with the following methods:

- ✰ Enterprises themselves carry out research and design
- ✰ Cooperate with domestic scientific offices
- ✰ Cooperate with overseas scientific offices
- ✰ Reproduce, design after models (reverse engineering)
- ✰ Buy technology from domestic source
- ✰ Buy technology from overseas source
- ✰ Joint-venture, associate with domestic enterprises
- ✰ Joint-venture, associate with foreign enterprises
- ✰ Hire domestic consultant
- ✰ Hire foreign consultant

At present, the method mostly used by the enterprises for technological renovation is to buy technology from foreign sources (56 per cent of 100 enterprises investigated) and design after models (reverse engineering) (52 per cent). Enterprises also focus on cooperation with domestic scientific offices (31 per cent) and bought technology from domestic sources (22 per cent). Activity of self-research and execution within the enterprises is commonly used by many enterprises to renovate technology. However, technological renovation mentioned here mostly relates to small improvement of products and production rules (39 per cent). The method rarely used by the enterprises includes hiring domestic consultant (5 per cent). One reason for this is that the enterprises do not get used to consultant hire in various stages of technology delivery. Secondly, as domestic intermediary organizations are still weak, both in organizational and legal aspects, the enterprises have not much belief yet on them.

2.4 Factors making Impact on Enterprises' Technological Renovation

The process of technological renovation in companies is influenced by many factors. However, factors making decisive impacts on the enterprises' process of technological renovation are those originating from enterprises' requirements, from actual situation of their production (business). In another words, this is the direct motivation on the enterprises to renovate. The reason of this reality comes from two sides: the enterprises and the state

On Enterprises' Side

The enterprises have no strategic long-term vision, therefore, they have no long-term plan for technological renovation to obtain permanent development. Many enterprises' decision on technological renovation at present is only based on short-term benefit.

On the State's Side

Policies and legal regulations set out to promote enterprises to invest in technological renovation are not really suitable and satisfactory, thus not to have pressure of technological renovation on the enterprises for existence. At the same time, full propagation of such policies and regulations has not been made, and the execution is short of seriousness, therefore, it can not encourage the enterprises to renovate technology. Even sometimes it has negative effect.

For the promotion of renovation, the most effective factors are enterprises' internal requirements such as to improve quality, to reduce cost price of goods; increase productivity; to heighten power of competition, etc. Factors relating to outside environment have less impact on promotion, and most of them have no decisive nature to the process of the enterprises' technological renovation.

However, the important factors that cause hindrance to renovation are also the enterprises' internal factors such as shortage of capital, shortage of informatics/ market information, and shortage of qualified people. But the hindrance effect caused by other factors such legal environment, risks in investment; delay in process of application for support in technological renovation is rather limited and is not of decisive nature.

3. POLICIES FOR ENCOURAGING TECHNOLOGICAL RENOVATION IN SMALL AND MEDIUM ENTERPRISES IN VIETNAM

3.1 Policy Initiatives

Having recognized the important role of the SMEs in the country's economic development toward the international economic integration, in the last five years, the Government has set out many policies to develop the SMEs to a high degree of operating effect, to improve competitiveness as well as to exploit their potential to contribute to the economy. It can be seen clearly that, the system of laws, and business environment have gradually been improved and have active changes. Small and medium enterprises have enjoyed favourable and equitable policies. The discrimination in treatment in comparison to the state-owned enterprises is considerably reduced. Especially, some important factors relating to the existence and development of the small and medium enterprises such as the access to the sources of fund, and industry, land, labourers, market information are now much more open in comparison to that of previous years.

To encourage enterprises to renovate technology, the state has issued many policies and laws/regulations relating to issues such as tax policy and enterprises' finance, credit policy, transfer of technology, protection of ownership of industry, etc. Some of these initiatives are:

- ✰ *Tax*: in the 6 kinds of favoured taxes applicable to the Science and Technology operations, there are 5 kinds relating to investment in technological renovation in enterprises, which are: enterprises' income tax, value-added tax, export tax, import tax and land tax.
- ✰ *Credit*: the government has issued some favourable policies concerning credit and interest rate supporting activities of investment in technological renovation by enterprises and organizations of research for execution.
- ✰ Besides, the government has also issued decrees relating to protection of rights of ownership of industry, of information and anti-unfair competition, etc. to make base for market development of science and technology in enterprises.

3.2 Limitations

However, in reality, impact of mechanisms, policies and laws/regulations is still limited.

According to evaluation conducted on 100 enterprises which were exposed to some related laws and regulations, the impact of such legal provisions on the process of enterprises' renovation of technology is active but it is not so strong. There may be some reasons as follows:

- ✰ The lack of knowledge of the enterprises about favourable conditions applied by the state. Many investigated enterprises do not even know policies/law regulations relation to the said favourable conditions issued by the state. This prevents enterprises in availing the state's favourable provisions. The reason of such discrepancy should be considered from two sides:

- *Enterprises side*: As the enterprises have no plans/business strategy and investment of permanent, long-term and macro, but mainly the short-term ones, which directly relate to daily production, they pay no attention to finding out and understanding of the existing policies and favourable mechanism.
- *On the state's side*: as the propagation has not been made completely and timely about the state's decisions, the enterprises, especially the private small-scale ones, could not seize sufficient information of policies, and instruments of supporting encouragement from the state.

☆ The procedures to make approval of objects to enjoy favourable conditions are still complicated, difficult and some time containing negative factors of making full advantage of policies which make enterprises disappointed and not to have enthusiasm to involve in. Many enterprises complain about the administrative procedures which cause troubles and loss of time to them, "although a single window system has been applied, instead of going to several doors, the enterprises still have to come several times ".

☆ Some of the policies issued are not clear, or directions for certain execution of such policies are late, therefore it prevents process of examination of competent offices and gives opportunity for negative phenomena to occur. For example, though Decree 119 was issued in 1999, until 2002 there were no directions for implementation.

☆ Favourable policies are sometimes not suitable or not really attractive to enterprises. For example, though the objects to enjoy favourable conditions of tax are many, but this policy is not useful to those being short of finance (capital) to execute investment in technological renovation. Or the regime of favourable condition of credit from Development Assistance Fund only applies to major projects of technological renovation investment while the enterprises' latent powers are for small-scale investments only.

☆ The effect of execution of some law regulations in fact are not ensured. Although regulations on protection of ownership of industry is issued, situations of violating such rights as imitated goods, false products have not been settled completely, and this causes limitation to enterprises in investment in technological renovation.

4. CONCLUSION

At present, the renovation of technology and equipments by small and medium enterprises in Vietnam is for just purchasing new machinery, equipments and to know necessary operations to operate them. Most of the enterprises have no specialized study to master, improve and develop technology. The renovation of technology in companies is passively made under market pressure. Manufacturers only look for modern technology when there are customer demands.

Activities of equipments and technological renovation meets with many difficulties and unfavourable conditions such as: enterprises' awareness of challenges in integration is insufficient, the knowledge of the managers and workers in the companies is low; shortage of capital for technological renovation; the contents of state-administration of technology are not clear, management is overlapping and short of system method. This is the problem that the government is trying to find solutions to overcome, and thus to promote and make the enterprises' renovation of technology better.

REFERENCES

Government of the Socialist Republic of Vietnam, 1998, Decree 45/1998/ND-CP on technology transfer, issued dated 1 July 1998.

Government of the Socialist Republic of Vietnam Decree 63/CP regulating details of ownership of industry.

Ministry of Science, Technology and Environment, Vietnam, 1999, Decision 1693/1999/QD-BKHCNMT on examination of contract of technology transfer, issued dated 1999.

Government of the Socialist Republic of Vietnam, 1999, Decree 119/1999/ND-CP on encouraging enterprises' investment in technological renovation, issued 1999

Government of the Socialist Republic of Vietnam, 2000, Decree 54/2000/ND-CP on protection of anti-unfair competition rights, issued 2000.

Law on value-added tax.

Law on enterprises' income tax.

– Section IV –

The Role of Institutions

Chapter 22

ARCI: 'Translating Research into Technology'

***Arun Seetharaman*[1], *G. Sundararajan*[2] *and Shrikant V. Joshi*[3]**

International Advanced Research Centre for Powder Metallurgy and New Materials (ARCI), Balapur (P.O), Hyderabad – 500005, India

E-mail: [1]arun@arci.res.in; [2]gsundar@arci.res.in; [3]svjoshi@arci.res.in

ABSTRACT

The International Advanced Research Centre for Powder Metallurgy and New Materials (ARCI) is an autonomous R&D institute of Department of Science and Technology (DST), Government of India set up with a mission to develop unique, novel and techno-commercially viable technologies in the area of advanced materials and subsequently transfer them to industries. Consistent with its mandate, ARCI has set up a Centre for Technology Acquisition, Transfer and International Co-operation (CTATIC) which plays an important role in transferring the technologies developed at ARCI to interested small and medium scale enterprises (SME's). In addition, ARCI provides a number of key services in its areas of core competence.

Based on its experience in dealing with diverse materials based technologies (process know-how based, equipment based etc.) that have been developed by ARCI on its own or in collaboration with a partner, ARCI has developed appropriate road maps to enable transition of the technology from the laboratory to the market place. Typical examples pertaining to the Detonation Spray Coating and Nano Silver Impregnated Ceramic Water Filter Candle technologies already commercialized by ARCI will be highlighted. ARCI has established an Advanced Materials Technology Incubator (AMTI) adjacent to its R&D campus. ARCI's technology receivers are allowed to set-up their production units at AMTI for a limited period of time until their operations outgrow the incubator as a result of their commercialization efforts. This aspect will also be discussed in this chapter.

ARCI's experience during the past 11 years suggests that demonstration of a sophisticated new technology at commercial or atleast at pilot scale to highlight all the capabilities of a technology to potential users can be crucial in aiding the assimilation of such technologies by Indian industries.

Therefore, ARCI has recently taken the initiative to set up jointly operated technology demonstration centres to showcase all features of specific technologies of importance to Indian industries in collaboration with hi-tech companies abroad. Three such joint demonstration centres have been already established at ARCI and one of them will be presented as a case study.

Keywords: *Technology development, Technology transfer, Technology incubator, Demonstration centre, Detonation spray coating, Nanomaterials.*

1. INTRODUCTION

The International Advanced Research Centre for Powder Metallurgy and New Materials (ARCI) is an autonomous R&D centre of Department of Science and Technology (DST), Government of India located at Hyderabad. Operational since 1996, ARCI has been setup with a mission to develop unique, novel and techno-commercially viable technologies in the area of advanced materials and subsequently transfer them to Indian industries. A unique feature of ARCI's mandate relates to the fact that it is industry centric. ARCI is required to develop and demonstrate technologies on a sufficiently large scale, not only to prove the reliability/consistency of the technology but also to carry out effective market sensitization, so that the subsequent transfer of technology to the industry has a greater chance of success.

1.1 What Does ARCI Do?

ARCI provides a number of key services like technology incubation, technical support for developing new applications, participation in generation of new markets and upgradation of technology to its technology receivers. "Partnership throughout" is the motto in dealing with the entrepreneurs seeking to commercialize technologies acquired from ARCI. Testing and product characterization, consultancy, training as well as literature and patent search in its areas of core competence are other forms of support extended by ARCI. In addition to the above, ARCI also carries out a fair amount of its job work and limited production, as these have turned out to be crucial in providing important technological impacts to ARCI's technology development efforts.

Over the years, ARCI has developed various state-of-the-art technologies employing varied approaches on its own: (*a*) in collaboration with academic institutions/R&D laboratories within India or abroad, (*b*) through establishment of a strategic alliance within an industry. More recently, ARCI has also explored the concept of establishing jointly operated centres for demonstrating specific technologies of relevance to India in collaboration with hi-tech companies located elsewhere in the world. These joint centres, set up at pilot scale production level, have a mandate to demonstrate the reliability and consistency of the technology and develop applications for the concerned technology as a prelude to the joint transfer of the technology to the Indian Industry.

In addition, ARCI has been undertaking numerous sponsored projects with a variety of government agencies to develop specific products and/or associated technologies, besides carrying out contract research for both foreign and Indian companies. ARCI has also been carrying out basic R&D work, especially in collaboration with renowned institutes/laboratories worldwide. ARCI has made significant progress on various above fronts during the short span of 11 years since its inception.

1.2 Thrust Areas of ARCI

Nanomaterials

Considering the fact that ARCI had outstanding expertise in the area of materials, it was decided that ARCI would concentrate on the production of nanopowders and also explore on its own their utilization for atleast some applications which either cater to a large Indian market or a market unique to India. ARCI's activities in this area also include synthesis of carbon nanotubes and exploring their application potential.

Engineered Coatings

The Centre for Engineered Coatings has been playing a pioneering role in the development of state-of-the-art technologies like Detonation Spray Coating (DSC), Electro Spark Coating (ESC), Micro Arc Oxidation (MAO), Electron Beam Physical Vapour Deposition (EB-PVD) and Pulse Electro Deposition (PED) system relevant to Indian industries.

Sol-Gel Coatings

The Centre for Sol-Gel Coatings was established to demonstrate the wide ranging capabilities of the sol-based coating technology in collaboration with a high-tech company. This centre is a good example of a joint demo centre at ARCI.

Ceramics Processing

The Centre for Ceramic Processing was conceived during the formative stages of ARCI. The centre in its pursuit of developing advanced materials for niche markets has succeeded in developing a variety of products and technologies and has also been successful in effecting transfer of the concerned know-how to Indian industries for commercial exploitation.

Laser Processing of Materials

With a view to promote application of laser materials processing technologies in the Indian industry, the Centre for Laser Processing of Materials was established to serve as a job shop-cum-demonstration and training centre. The centre has been instrumental in establishing several new laser-based applications and catalyzing the adoption of laser processing by the Indian industry.

Fuel Cells

The Centre for Fuel Cell Technology specifically contributes to national efforts in the field of fuel cell technology. It is making considerable progress and is on the verge of taking up programs to demonstrate actual applications.

Apart from Centers of Excellences devoted specifically to the above mentioned thrust areas that ARCI focuses on to develop technologies for the Indian markets, ARCI also has Centres for Mechanical and Chemical Characterization to provide necessary support to the technology development efforts. Centre for Technology Acquisition, Transfer and International Co-operation (CTATIC), helps ARCI transfer its developed technologies to appropriately selected small and medium enterprises as well as coordinates the establishment of technology demonstration centres and forging of R&D linkages.

1.3 Important Highlights of ARCI

Transferred 15 Technologies to 30 Entrepreneurs

ARCI has always aimed to transfer the technologies it has developed to Indian industries and has utilized this approach to spread its technologies to various parts of the Indian market. ARCI's effort to effect technology transfers have already met with substantial success. So far, ARCI has already transferred 15 technologies to nearly 30 entrepreneurs throughout the country and these are finding use in diverse industry segments.

Ensured Utilization of the above Technologies in Diverse Industry Segments

Consistent with its industry centric approach, ARCI has utilized its expertise and infrastructure to address applications in diverse industry segments like aerospace, defence, power, automotive, steel, textile etc.

Established Fruitful Linkages with R&D Laboratories and Industries

ARCI has forged partnerships with institutes/industries in over 15 countries and also with Indian collaborators, to carry out basic R&D work and also jointly develop technologies.

Conceived and Executed Major Programs

ARCI has executed a number of major "on demand" programs for both strategic and civilian segments of the industry.

Established the Advanced Materials Technology Incubator

ARCI established the country's first Advanced Materials Technology Incubator (AMTI), which was owned and operated by it.

2. ROLE OF TECHNOLOGY TRANSFER CELL

For an organization like ARCI, successful transfer and commercialization of technologies developed by it is a crucial indicator of its performance. This is especially true in the present business scenario, which encourages intellectual property driven companies and institutes. In an effort to continuously complement existing expertise in materials technologies within its areas of core competence, ARCI has been striving to make available nationally unique technologies to the Indian industry. In addition, ARCI has also been forging alliances with R&D organizations, hi-tech companies and institutions throughout the world to develop globally competitive technologies.

More recently, ARCI has significantly enhanced the level of its international collaborations for technology development and acquisition.

Consistent with its mandate, ARCI has set up a Centre for Technology Acquisition, Transfer and International Co-operation (CTATIC), which has been playing a pivotal role in accelerating the process of transition of intellectual property rights from ARCI to the market place and strived to expeditiously transfer commercially relevant technologies developed at ARCI to appropriately selected small and medium scale enterprises.

2.1 Functions of CTATIC

CTATIC performs the following diverse functions:

- ☆ Carrying out techno-commercial analysis of existing and future technologies
- ☆ Providing a bridge between ARCI scientists and existing/prospective technology receivers as well as collaborators
- ☆ Seeking new entrepreneurs to commercialize ARCI technologies
- ☆ Understanding market needs and communicating them to ARCI scientists
- ☆ Conducting scientific exchange and joint technology programmes with collaborators
- ☆ Conducting workshops and seminars and participating in exhibitions to enhance the outreach of ARCI technologies
- ☆ Interaction with funding agencies for securing loan for technology receivers under government funding schemes
- ☆ Helping the technology receiver set up a production unit at ARCI's Advanced Materials Technology Incubator (AMTI)
- ☆ Providing him marketing assistance till the time the technology receiver completely sets up his own production unit outside AMTI and starts commercialization

The centre also performs other activities such as preparing Non Disclosure Agreements, Technology Development and Transfer Agreements, Contracts for establishing joint demo centres, Memorandum of Understanding etc. Costing of projects for sponsored and contract research and their Coordination to ensure timely implementation are other important responsibilities undertaken by the centre.

3. TECHNOLOGY DEVELOPMENT AT ARCI

3.1 Technology Development Models Adopted by ARCI

ARCI has been continuing its already proven methods of developing and commercializing technologies through following approaches:

1. Identifying an appropriate laboratory scale technology from an institution/ R&D centre located abroad, incorporating suitable modifications to suit Indian conditions and setting up a pilot unit to prove the technology before finally effecting technology transfer to interested entrepreneurs. The above approach has led to transfer of many technologies, which have met with varying commercial success.
2. In another technology transfer model adopted by ARCI, technologies have been developed at the specific request of the company. Some of the technologies under this category have already been developed and transferred to user industries while, in some other cases, technologies are under development and will be transferred shortly.
3. In yet another approach to develop and transfer technologies, ARCI has developed on its own keeping in view such technologies' market potential and their uniqueness in the national context.

3.2 Elements of Technology Transfer Strategy Developed by ARCI

ARCI has developed a proactive technology transfer strategy, the main elements of which are as follows:

Intellectual Property Protection

This is developed to secure and maintain the intellectual property rights like patents, trademarks, copyrights etc. so that no infringement in the technologies developed by ARCI takes place.

Market Assessment through Application Development

ARCI develops technologies based on applications which have a potential for commercialization. Such a strategy acts as a confidence booster to the technology receiver when he plans to take this technology to the market.

Technology Transfers on Exclusive Basis in Case of Niche Market Technologies

Some of the technologies developed by ARCI are for niche markets and are useful for a unique segment of customers or applications. ARCI transfers technologies on exclusive basis to technology receivers for them to commercialize in these niche markets when the anticipated market size is small.

Milestone-linked Payment of Technology Transfer Fee to ARCI

ARCI links the technology transfer fee instalment that it receives from the technology receivers to major milestones that need to be achieved for completing the project. This, apart from aiding in completing the technology transfer process expeditiously, also enthuses confidence in the technology receivers.

3.3 Current Status of Technologies Developed at ARCI

During its 11 years of existence, ARCI has entered into technology development and transfer agreements with 33 companies for either the creation of new businesses or for the diversification of businesses. The following table shows the wide range of technologies that have either been already transferred, or are in the process of transfer.

Table 22.1: Wide Range of Technologies Developed by ARCI

Sl.No.	*Technology*
1.	Electro Spark Coating Technology (transferred to 8 industries)
2.	Magnesia Aluminate Spinel (transferred to 1 industry)
3.	Ceramic Crucibles for C and S Analysis (transferred to 1 industry)
4.	Ceramic Honeycombs for Energy Efficient Air Heaters (transferred to 1 industry)
5.	Detonation Spray Coating Technology (transferred to 4 industries)
6.	Reinforced Graphite Sheets and Seals (knowledge transfer to industry completed)
7.	Heat Pipes Heat Sinks (knowledge transfer to industry completed)
8.	Evaporation Boats (knowledge transfer to industry completed but not commercialized)
9.	PM Grade Iron Powder (knowledge transfer to industry completed but not commercialized)
10.	Micro Arc Oxidation (MAO) Technology (transferred to 3 industries; ongoing with one industry)
11.	ESC Equipment Manufacturing Technology (knowledge transfer to industry completed)
12.	Calcium Aluminate Cements and Furnace Sealants (transferred to 1 industry)
13.	Ceramic Honeycomb Molten Metal Filters (Transfer to industry ongoing)
14.	Anti-bacterial Nano-silver Suspensions or Nano-silver Powder for Catheter Applications (Transfer to industry ongoing)
15.	Development and transfer of technology to manufacture nano-titanium dioxide based textile finishes for self-cleaning applications (Transfer to industry ongoing)
16.	Development and transfer of technology to manufacture nano-silver based textile finishes for antibacterial applications (Transfer to industry ongoing)
17.	Nanosilver impregnation of ceramic water filter candles to impart anti-bacterial function (knowledge transfer to industry completed)

4. TECHNOLOGY COMMERCIALIZATION

4.1 Advanced Materials Technology Incubator (AMTI)

ARCI's commitment towards ensuring proliferation of novel technologies in the Indian industries is exemplified by ARCI's establishment of an Advanced Materials Technology Incubator (AMTI) adjacent to its R&D campus. The AMTI is owned and operated by ARCI. ARCI's technology receivers are allowed to set their own production units at AMTI for a maximum period of upto 5 years after commencement of production. ARCI's experience in managing AMTI has also made it abundantly apparent that the tenants can potentially derive multiple benefits from their presence in the incubator in an effort to establish commercially successful ventures. The benefits that the technology receivers would gain from AMTI is that they can have day to day interaction with experts at ARCI incase they are having any queries regarding the technology or the process at AMTI. ARCI's technology receivers can also utilize facilities such as characterization, performance testing and library within the ARCI campus.

4.2 Road Maps Adopted by ARCI

Based on its experience in dealing with diverse materials based technologies (process know-how based, equipment based etc.) that have been developed by ARCI or in collaboration with a partner, ARCI has followed appropriate road maps to enable transition of the technology from the laboratory to the market place. Depending upon the type of technology, the road maps could involve the following steps:

Agreement with Collaborator

ARCI identifies an appropriate laboratory scale technology from an institute or R&D centre abroad, gauges whether the applications of this technology are suitable to the Indian market and then enters into an agreement with the foreign collaborator providing the technology.

Establishment of Prototype

ARCI then acquires a prototype unit pertaining to the technology from the foreign collaborator.

Technology Based R&D

Various R&D programs for optimization of the technology, understanding the process fundamentals and improving the technological capabilities of the system are initiated. Important inputs are obtained from these programs for deciding on any required modifications to the equipment or the process.

Job Work/Field Trials

A job work phase is crucial in ascertaining the robustness of the technology. During this phase, several potential user industries satisfy themselves about the efficacy of the technology and this significantly contributes to the market sensitization efforts.

Indigenous Fabrication

The prototype unit is then fabricated indigenously in collaboration with the foreign institute or R&D laboratory and is open to be transferred to interested entrepreneurs after thorough testing.

Business Workshop

Business opportunity workshops are then conducted by ARCI to launch the technology in question by educating potential and existing entrepreneurs on the social/economic/environmental benefits of the new technology.

Agreement with the Technology Receiver

Once an entrepreneur is interested in the technology, a technology transfer agreement is signed with him and the comprehensive technology transfer package is provided to him. The technology receiver is provided with all the technical expertise, installation and commissioning support and marketing assistance to establish the facility for commercial exploitation.

4.3 Case Studies

The above mentioned steps can change according to the type of technology (process based, equipment based) being transferred. ARCI has successfully transferred technologies belonging to both the above categories. Typical examples pertaining to the Detonation Spray Coating and Nano-silver Impregnated Ceramic Water Filter Candle technologies already commercialized by ARCI are highlighted below as case studies:

4.3.1 Commercialization of Detonation Spray Coating (DSC) Technology

The emergence of ARCI as a leading coatings' technology provider in India started with the successful commercialization of Detonation Spray Coating (DSC) technology. Incubation of the DSC technology by ARCI, and its eventual transfer to the industry, was made possible by the foundation laid by simultaneous pursuit of process-related R&D, technology demonstration, application development and market sensitization prior to technology commercialization. In many respects, the methodology adopted to introduce the DSC technology into the industrial mainstream represents a model case study in the Indian context and merits elaboration.

It was as early as 1990 when the DSC technology was first identified as being of significant promise for commercial exploitation in India in view of its ability to deposit high-quality coatings capable of enhancing the durability and performance of industrial components. Soon thereafter, a DSC system was acquired from the Institute of Problems for Materials Sciences (IPMS) in Ukraine and various R&D programs focusing upon optimization of coating quality, understanding the process fundamentals and improving the technological capabilities of the DSC system were initiated. At the same time, the above mode of operation provided an opportunity for several potential user industries to satisfy themselves about the efficacy of the detonation sprayed coatings and this contributed significantly to the market sensitization efforts. In addition to enabling substantial revenue generation, this only served to further reinforce the original conviction that the DSC technology was well-suited for commercialization in India. As a consequence, ARCI decided to indigenize the DSC technology in collaboration with IPMS and signed an Agreement to formalize the cooperation.

Fabrication of five DSC units based on the above design was completed in August 1999 and, following extended operational tests, the units were made available for acquisition by private industries as part of a comprehensive technology transfer package. The DSC technology has already been transferred to four private enterprises. Based on ARCI's assessment of the commercial potential of the DSC technology, a conscious decision was made to transfer the technology to the above entrepreneurs on a regionally exclusive basis.

Apart from limiting the number of DSC systems established in the country, this also ensured that no unhealthy competition was created in any particular region. This approach has already borne fruit as it has ensured significant commercial growth of all the DSC-based job shops. In each case, soft loan from Government funding schemes played a crucial facilitating role. Because of high quality coatings that they offer at competitive prices, all four technology receivers have been executing orders from prestigious clients in diverse industry segments including aerospace, oil and gas, thermal power, steel, textile etc. With the considerable technical expertise developed by ARCI during the DSC technology's gestation period, ARCI has been able to provide a wide rage of services to its technology receivers. This has included extension of complete technical support for suitable coating material selection for any targeted application, process parameter development, coating characterization and certification of coating quality to potential customers, demonstration of coating on prototype components etc. This is in addition to routine assistance in installation and commissioning, rectification of operation related problems and quick supply of spare parts.

While all the four technology receivers have registered impressive growth, the commercial success of the DSC is exemplified by the manner in which one of the technology receivers has been able to leverage its full potential. The company established its DSC job shop in AMTI in November 2000.

With its active marketing and periodic technical assistance from ARCI, the company could establish its presence in a large number of industrial sectors by providing a wide spectrum of ceramic, metal and alloy coatings, which were well accepted by the Industry. To meet the increased business volume generated as a result of encouraging market response, the company bought a second DSC unit from ARCI in the year 2005-06. It was also the first company to graduate out of ARCI's incubator to establish its own production facility with significantly augmented infrastructure.

4.3.2 Commercialization of Nano-silver Impregnated Ceramic Water Filter Candle Technology

Drinking water is very scarce in India due to severe contamination of water sources and lack of purification facilities. Majority of population in India depends on water from open wells, rivers and lakes for their living. Water borne diseases are very common among the population dependent on such water sources. Although purification technologies such as chlorination and UV irradiation are commonly used for water purification in urban areas, even here the water is not safe due to contamination during its distribution through pipelines. Often, people in urban areas depend on domestic UV based purification systems which are expensive for a common man. However a larger population depends on cheaper ceramic candle based stainless steel water filters for obtaining drinkable water.

The advantages of these filters over UV filters are that they do not require power and pressurized water (tap water). These filters are also convenient for use in villages, where tap water or power supply is not available. The one disadvantage of these filters is that disinfection is not possible with this kind of candle filters since they can remove suspended matter, but not bacteria or chemical contamination. It observed that modifications of such candle filters with certain antibacterial agents will make these filters on par with UV based disinfecting filters without adding much cost.

ARCI developed a nano-silver synthesis technology which allowed coating of ceramic candles with nano-silver throughout the surface of the candle. Although

Figure 22.1: Components Coated by Detonation Spraying

silver is known for its anti-bacterial action for over 100 years, nano-silver, due to the extremely large surface area, allows a more intimate contact of bacteria with it and thereby improves the antibacterial action. ARCI as a part of this process based know-how technology; tested nanosilver deposited ceramic candle filters for their anti-bacterial action and found that the candles are efficient for several months under continuous use with tap water. Furthermore, each candle requires only 1-2 grams of silver; hence, the cost increase is not significant. These candles are also useful for all kinds of water including tap water, well water and lake water.

In order to ascertain the quality of these nano-silver coated ceramic water filter candles, ARCI teamed up with a Non- Governmental Organization (NGO) for supply of water filter candles treated with nanosilver for use in rural areas for disinfection of drinking water. The scope of the project was (*a*) To address problems of drinking

Figure 22.2: Cross Section of Nano-silver Coated and Uncoated Ceramic Candles

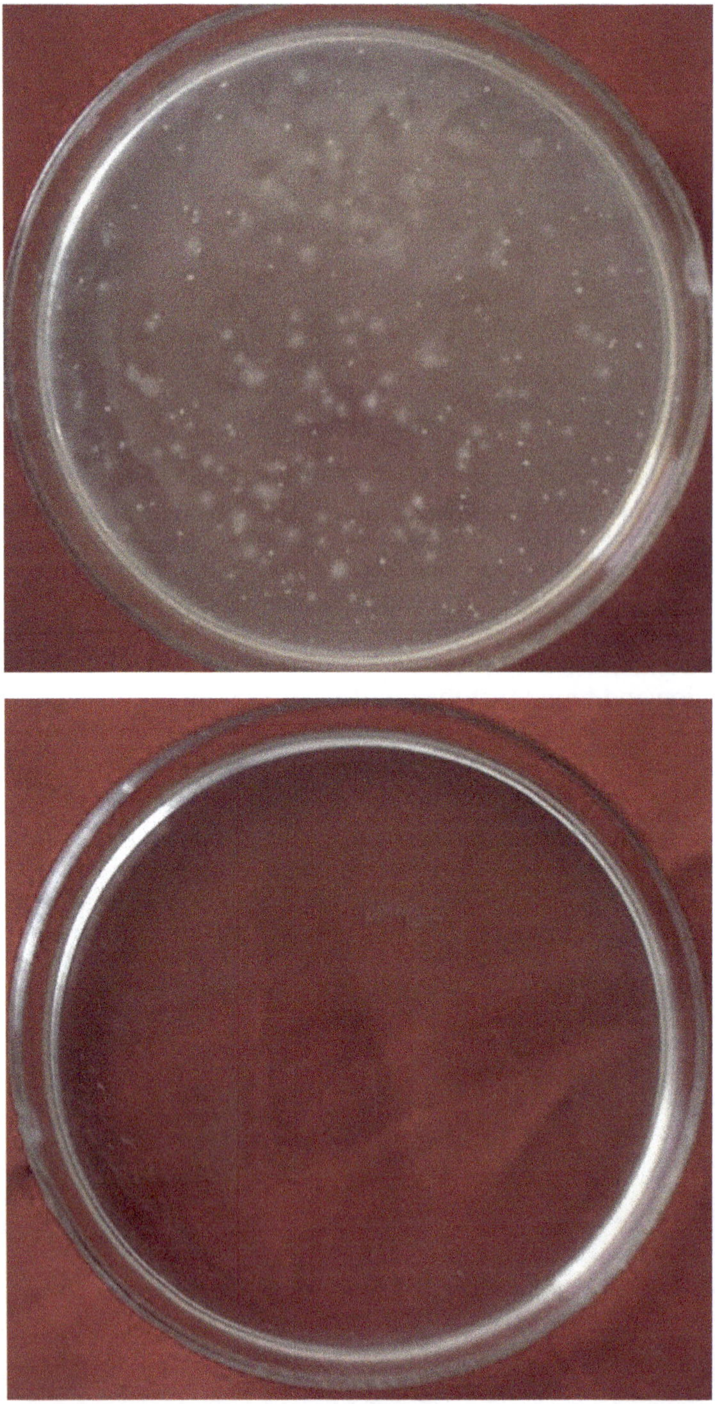

(a)

(b)

Figure 22.3: Anti-bacterial Activity Shown
(*a*) Before Filtration (10^5 cfu/ml) (*b*) After Filtration (0 cfu/ml)

water disinfection in rural areas employing nano-silver coated candle filters using the reach of the NGO and (b) generate long-term field data. Until now, 400 anti-bacterial candle filters along with stainless steel housing were supplied by ARCI to the NGO. These filters were placed in medical centres in selected rural areas "adopted" by the NGO for their ongoing programs. The performance of these filters was closely monitored at each location and anti-bacterial action of the filters and their efficiency was demonstrated through long-term user trials, for more than a year. Now, ARCI is in process of replacing these candles because of their usage for more than a year and also because the process route has been slightly modified to improve the quality. At present, evaluation is ongoing in over 40 villages and continuous data is being generated to monitor the performance.

After obtaining positive results from the field trials that ARCI conducted at various villages in cooperation with the NGO, ARCI was confident that it could transfer this technology to a company for commercialization. ARCI then found an interested entrepreneur who was engaged in the business of manufacturing ceramic water filter candles and housings. Promoters of the company identified were convinced that ARCI's technology involving nanosilver impregnation of ceramic water filter candles can add value to their erstwhile business by imparting anti-bacterial function to these filters for disinfection of drinking water. To capitalize on their knowledge of candle filters market, the company then decided to acquire ARCI's know-how of nano-silver impregnated ceramic candles.

A Know-How Transfer agreement was signed between ARCI and the company. ARCI then demonstrated the process for production of nanosilver impregnated ceramic candles at pilot scale to the promoters of the company. Subsequently, after requisite facilities as recommended by ARCI were set up by the company, a team from ARCI successfully demonstrated the process for production of 1000 nano-silver impregnated ceramic water filter candles per day and imparted comprehensive training to personnel from the company. The product was released in a few markets from March 2008 onwards and is shortly to be scaled up further.

5. CONCEPT OF JOINT TECHNOLOGY DEMONSTRATION CENTRES

During the past 11 years of its existence, ARCI has achieved significant success in its efforts to become a reliable source of frontier technologies in its core areas of competence for the Indian industry. To augment its portfolio of materials technologies and capitalize on its experience in transferring several process and equipment based technologies, ARCI has taken a pioneering initiative in the Indian context of establishing Joint Technology Demonstration Centres (JTDCs) in collaboration with high tech companies abroad to develop/co-develop, demonstrate and transfer globally competitive technologies to Indian industry. Usually, the foreign partner is the technology provider and contributes to the joint centre in the form of know-how and consultancy, while ARCI's contribution is in the form of funds for land and building, infrastructure, manpower, operation etc.

This initiative has been taken to highlight all the capabilities of the associated technology to potential Indian user companies, to catalyze the assimilation of such technologies and to shorten the time to market for Indian companies interested in commercializing cutting edge technologies.

5.1 Motivation for Establishing a JTDC

ARCI had always felt the need to demonstrate technologies at a reasonable scale of operations before it is transferred to interested technology receivers. Majority of the technologies transferred by ARCI has passed through such a demonstration phase. The demonstration phase serves a varied set of objectives, including (*a*) Demonstrating reliability and consistency of technology (*b*) Carry out limited production/job work to sensitize market and for field trials (*c*) Modifying technology based on feedback (*d*) Ascertaining market potential (*e*) Assessing techno-commercial feasibility to know whether the technology would lead to better profits and market growth (*f*) Increasing confidence levels of technology receiver (*g*) Quantum jump in technology transfer fee receivable by ARCI and collaborator. This experience of ARCI has been the main motivating factor in deciding to establish joint technology demonstration centres on its premises.

5.2 Technology Commercialization through JTDC

JTDCs can be initially used to upscale and fine tune the chosen technologies with the selected partner institutions or organizations. As a first step, a prototype unit is acquired from the partner in case of an equipment-based technology; in case of a process-based technology, basic design of equipment, process details etc. are obtained by ARCI from its partner. ARCI then procures sub systems with assistance from its partner to set up a fully operational demo facility. Operations are started to demonstrate the process and also provide samples developed for field trials for not only establishing their reliability and consistency but also to estimate the market size. Even job works can be undertaken once the developed samples have a good market. During the initial phase, JTDCs established at ARCI are planned to interact with user industry segments to evaluate the feasibility of technology for targeted applications. Once ARCI is convinced that technology being developed at a JDTC is techno-commercially viable, ARCI shall identify an appropriate entrepreneur to commercialize the technology. This is done by conducting business opportunity workshops where the processes related to the technology are demonstrated to the invited entrepreneurs and after one to one discussions with these entrepreneurs appropriate entrepreneurs are selected for commercialization.

5.3 Case Study

ARCI has signed contracts to establish a joint technology demonstration centre with three foreign collaborators. One of the three the Sol-Gel nano-composite coatings technology centre has been fully established and is highlighted below as a case study:

5.3.1 Demo Centre for Sol-Gel Coating Technology

Even though knowledge-base on sol-gel technique for synthesis of powder/ bulk is available in India, the technology for sol-based "coatings" is not yet available in the country. In the above backdrop, ARCI embarked on establishing a comprehensive facility, which will be unique in the country, to demonstrate the wide ranging capabilities of the technology. An institute in Saarbrücken, Germany was recognized as the world leader in the field of sol-based coatings technology and a spin-off company of this institute obtained license rights to a wide portfolio for sol based coating technologies developed there. ARCI then signed a contract with this company in July 2005 to jointly set up a demonstration centre at ARCI for sol-based coatings.

The Centre is jointly operated by ARCI and the company, with all required state-of-the-art facilities for sol preparation, coating deposition, pre- and post-treatment etc. available at the centre to enable demonstration of all aspects of the sol-based coating technology. Both ARCI and the company will contribute to establishment of the joint centre. ARCI will contribute in terms of meeting all expenses towards its establishment and operation, besides providing for necessary infrastructure (land and building, technical staff, characterization support, marketing efforts etc.), whereas, the company's contribution will be in terms of technical specifications, installation and commissioning, process know-how, documentation, training of staff etc. Surplus funds that accrue to the Centre from technology commercialization shall be shared by ARCI and the company.

The centre will demonstrate all aspects of the sol based coating technology and initially focus on the following areas namely: (*a*) Glass-like coatings on metal substrates (*b*) Anti-microbial, easy-to-clean and photocatalytic coatings on textiles, glasses, metals and ceramics (*c*) Interference multi-layer coatings on glass, plastics and sapphire (*d*) Fire-protection applications (*e*) Decorative coatings on glass and ceramics

6. CONCLUSION

ARCI has met with substantial success in transferring equipment/process based technologies. This can be observed from the case studies of Detonation Spray Coating (DSC) Technology and Nanosilver impregnated ceramic water filter candles technology. Transferring these equipment/process based technologies can be achieved by adopting a multi pronged approach like adopting various road maps for technology development and commercialization. ARCI has also shown that it has been able to achieve significant success by showcasing frontier technologies in the form of technology demonstration centres, which have considerable promise. "Partnership throughout" has been ARCI's motto in dealing with small and medium enterprises to ensure that they flourish. The Centre for Technology Transfer and Marketing has played a key role in this context and coordinated all essential activities for providing wide ranging support to its entrepreneurs. ARCI has always felt important to maintain a cordial and friendly relationship with its technology recipients and constantly exchange views and opinions regarding the technology with them in a systematic manner.

Chapter 23
Technology Transfer: Tanzania Experience

Matheo L. Raphael
Director
Centre for the Development and Transfer of Technology,
Tanzania Commission for Science and Technology, Dar es Salaam, Tanzania
E-mail: rmatheo@costech.or.tz

ABSTRACT

In the early 90's the Government of Tanzania pulled out of the productive activities such owning and running of manufacturing and mining firms and opted for the liberalized market economy. Since then the country has witnessed the growth of the private sector in the form of micro, small and medium scale enterprises. The sector employs 20 per cent of Tanzania's labour force and contributes about one third to the national GDP. As a result, there has been a growing demand for various technologies to support their sustainability through process and product improvements. As in any other developing Country, there are two sources of technologies, from external sources and indigenous technologies. Generation of technologies through invention and innovation is almost negligible. Irrespective of the source, Tanzania Commission for Science and Technology (COSTECH) has observed that the main obstacles to the effective transfer and sustainability of technologies are, lack of adequate information, taboos, non-engagement of the recipients by the technology developers, lack of sense of ownership by the recipients, lack of adequate preparation before acquisition, lack of two-way communication, limited resources, low level of absorption capacity, low level of capability including skills to sustain and develop further the acquired technologies, and managerial problems. To overcome some of these obstacles several initiatives have been undertaken in the country by various stakeholders. These include: the establishment of various policies by the Government that are geared towards poverty reduction strategies, establishment of a database of all SMEs in the country by the College of Engineering and Technology of the University of Dar es Salaam; establishment of incubator and cluster systems by the Small Industries Development Organization and the College of Engineering and Technology to promote innovation; and the establishment of the Association of Technology Developers, the Patents Information and Advisory Services Centre, and the Local and Indigenous Knowledge Systems Trust by COSTECH. This paper discusses these initiatives and experiences in the development, acquisition and transfer of sustainable technologies for poverty reduction in Tanzania.

Keywords: *Technology transfer, COSTECH, Micro, small and medium enterprises, Innovation systems.*

1. INTRODUCTION

The Tanzania Commission for Science and Technology (COSTECH) is the principal organ responsible for all matters pertaining to Science and Technology (S&T) in Tanzania. The Act of Parliament No. 7 of 1986 (United Republic of Tanzania, 2001), which established COSTECH stipulates under Part IV the establishment of the Centre for the Development and Transfer of Technology (CDTT). The CDTT became operational in 1994.

According to the Act that established COSTECH the Centre has the following mandate: As the principal organ of the Commission responsible for matters relating to the transfer, adaptation and development of technology including the assessment and choice of imported technologies;

- ☆ To identify, within the framework of national, social, economic and political constraints, the technological needs in different sectors of the economy;
- ☆ To acquire and analyse information on alternative sources of technology and its delivery to users;
- ☆ To make an evaluation and selection of technologies with a view towards developing the capacity for decision making in the area of science and technology;
- ☆ To play a major role in the unpackaging of imported technology including the assessment of the suitability of the technology;
- ☆ To maintain a register of imported technologies and a register of domestic technological resources and manpower;
- ☆ To act as a catalyst for the development of indigenous technologies;
- ☆ To provide training for technical personnel in various fields of analysis of transfer of technology;
- ☆ To register all technology transfer agreements; and
- ☆ To prepare short plans for identification and transferring critical technologies in each sector of the economy in collaboration with the stakeholders.

Among the functions of CDTT is to source, select, acquire and internalise (adapt and adopt) technologies from both outside the country and those that are locally developed. Since its establishment, a number of activities and initiatives have been carried out to ensure smooth and effective transfer of technologies in the country. The activities discussed in this paper are those concerned with: capacity building, documentation, promotion and facilitation of technology transfer issues in Tanzania.

2. TECHNOLOGY TRANSFER

Technologies evolve as a result of: research, innovation, or invention. In order for the technology to be used it has to be transferred. That means it has to come out of its "birth place" to the end-users. There are two forms of technology transfer–horizontal technology transfer and vertical technology transfer, as under:

Horizontal Technology Transfer

This occurs when technology is transferred from one geographical location (technology supplier) to another (technology recipient). This form of technology transfer forms the major part of the international technology market.

Vertical Technology Transfer

This involves translating local R&D results into usable products or processes that is: dissemination and diffusion of results to the society, which is at the lower level through extension services.

Through CDTT a number of activities and initiatives have been designed and carried out towards achieving these processes.

3. CAPACITY BUILDING IN TECHNOLOGY TRANSFER MANAGEMENT

The Commission for Science and Technology (COSTECH) is a secretariat that caters to all sectors in the areas of technology assessment and choice. It uses affiliated institutions, both Government and private, to carry out its planned activities. This is done through:

- Facilitation and coordination of technology transfer and training of trainers for a specific activity,
- Organizing in-house training for target groups. Examples include workshops on:
  - Technology transfer by UNIDO, 1993
  - Technology transfer management,1997
  - Technology transfer stakeholders, 2003
  - Technology developers "Sister Institutions," 2003
  - Technology needs assessment for Tanzania, 2006.
- Hosting, updating, and distributing a database on indigenous technologies available for dissemination;
- Conducting of awareness creation seminars at workplaces on patents and intellectual property rights (IPR);

- ☆ Facilitation of the establishment of technology transfer offices and institutional IPR policies for R&D and S&T institutions;
- ☆ Conducting technology needs assessment for poverty reduction.

4. CHALLENGES AND LESSONS LEARNT IN TECHNOLOGY TRANSFER

Technology transfer is not a single action but involves a number of issues which are related not only to technology but also to social and economic aspects. It therefore necessitates doing a thorough analysis before embarking upon the dissemination process. In order to promote vertical technology transfer from our R&D and S&T institutions, information regarding proven technologies is essential to encourage end users to accept and assimilate these.

COSTECH's experience in technology transfer for some technologies is as follows:

4.1 Dissemination of Mushroom Farming Technology

Scope of the Activity

Mushroom farming in Tanzania was initiated through a Zero Emission Research Initiative (ZERI) program in 1997. Since then, COSTECH has been involved in the capacity building, promotion and facilitation of transfer of mushroom farming technology in the country. Mushrooms are a good source of protein (healthy eating) and can be sold to earn cash (poverty reduction). COSTECH has facilitated the transfer of this technology to two R&D institutions (for spawn production and as training centres) and to two villages, namely, Mkuranga (a village in Coast region) and Arumeru (a village in Arusha region). In both cases it was the community-based groups, which requested for the training.

Responsibilities of the Community

In transferring this technology, communities were required to construct the demonstration of mushroom farm (small hut), mobilize the substrate (agricultural waste) and participate actively during the training.

Collaborators

In order to transfer the technology successfully, COSTECH liaised with Tanzania Industrial and Research Development Organization (TIRDO) to provide the expertise, while FAIDA-MALI (a CBO in Arusha) provided the logistics for the training.

Approach

Training of mushroom farming technology to the community was through theory and practical sessions. The trainees had hands-on practice in mushroom farming under the supervision of the trainer. After the training the trainees were required to spawn mushroom on their own. The trainer made a follow-up visit and discussed the difficulties the trainees had encountered. Solutions to the problems encountered were provided onsite.

Challenges

Being a new technology, mushroom farming requires careful sterilization (in order to avoid growth of unwanted fungi). Most farmers find it hard to maintain cleanliness due to their surrounding environment being used for other activities such as poultry and animal husbandry. Also the presence of rats that eat the mushrooms is a nuisance. Therefore some trainees give up when they experience such problems.

The issue of mushroom marketing is also posing a challenge to the transfer of mushroom farming technology. Different types of mushrooms grow in different climates. Currently we are facilitating the farming of oyster mushrooms (pleurotus species), which grow at temperatures between 25-30°C. However, customers seem to prefer button mushrooms (imported), which grow at lower temperatures.

Mushroom farming is indoor. Sometimes the construction costs of a mushroom house constrain some of the trainees and they give up mushroom farming.

Lessons Learnt

Mushroom farming is not as simple as farming of other crops and is not for everyone. Prospective farmers need to be encouraged through frequent visitation by the experts to assist them on problems they encounter. The issue of marketing is beyond the trained farmers. Institutions dealing in marketing should assist the farmers in finding markets for their mushrooms and encourage farmers in the selection of varieties to grow.

The Way Forward

In order to promote mushroom cultivation technology in the country, COSTECH has taken the following initiatives:

- ✰ Sensitise the society on the benefits of eating mushrooms – to create demand;
- ✰ Facilitated collaboration between farmers, spawn producers, buyers, and consumers;
- ✰ Facilitate the development of spawn production units at different zones in the country;
- ✰ Ensure frequent visits to the trainees until the technology takes root; and
- ✰ Involvement of a sociologist particularly in the selection and motivation of a target group. There should not be any imposition of technology to any target group.

4.2 Dissemination of Biogas Technology

Scope of the Activity

In the early 1980's COSTECH promoted biogas technology in Dodoma region as an alternative resource of energy (Kayumbo and Asman, 1981; Lushiku, Nkonoki and Sambali, 1985). Dodoma is a dry and semi-arid area threatened by desertification. However, the region is endowed with a large number of animal populations. Women and children spent several hours a day walking long distances fetching firewood and water. It was COSTECH's hope to reduce this drudgery for women and improve their health by reducing smoke inhalation. The spent manure was to be used to recondition soil.

This biogas technology originated from India and China and is known to have high initial investment costs. As a result, individuals were reluctant to invest on this new technology. Therefore, COSTECH took the initiative and constructed four biogas plants for demonstration at four different sites in Dodoma.

Demonstration of this technology took place without prior involvement of the communities. It was thought that the communities would adopt the technology after seeing its advantages.

Social and Economic Constraints (Mahgary and Biswas, 1985; Schneiders, Mkalata and Mmakasa, 1985)

Capital investment for a biogas plant, even for a family size digester is relatively high for most villagers. Furthermore, a biogas plant requires someone to spend 2-3 hours per day to attend to it. For dry areas like Dodoma most of the time is spent fetching water to be mixed with cow dung.

Before the biogas plants were installed, villagers believed that once cow dung is digested in a biogas plant it loses its strength as a fertilizer. Also, it was perceived by many that emissions from the burning gas would stink. These ideas were later proved false. Eventually, the villagers enjoyed using the gas for lighting and cooking. Villagers started local fabrication of biogas plant parts, including burners, which were previously imported from India.

During the dry season, it was noted that due to the wandering of cows there was not enough cow dung to charge the digesters. This meant that the traditional way of keeping cows was to be abandoned in favour of "zero grazing." However, it is not the culture of the Gogo tribe to fetch water and grass for cattle.

All this time personnel from COSTECH or trained personnel operated the biogas plants. After six months the four digesters were handed over to the communities and institutions. A year later they were all non-functioning.

Reasons for failure were attributed to the following.

- ✰ *Maintenance*: High maintenance cost of the selected technology was beyond the income of the villagers. The technology used required regular welding of the gasholder tank. Villages had no electricity or the means to carry those tanks to town for welding.
- ✰ *Culture*: It was taboo for the Gogo people to hold fresh cow dung and mix it with water. Therefore, the digesters were not re-charged.
- ✰ *Sabotage*: Digesters built for institutions were sabotaged because they blocked the lucrative tenders of supplying charcoal and firewood at the institutions.

Lessons Learnt (Mkwawa, 1986; Schneiders and Mkwawa, 1985)

- ✰ The project should have included social scientists to study the culture and taboos of the recipients.

- ☆ Communities should have been involved in the project, through seminars, active participation in construction and running of the plants so as to inculcate a sense of ownership.
- ☆ Due to high maintenance costs, the technology was not appropriate at the selected locations at that time.

4.3 Dissemination of Low Cost Housing Technology

Scope of the Activity

Shelter is one of the basic necessities for decent human life. The government persistently sensitizes and encourages its citizens to have good habitat both in urban as well as in rural areas. A decent shelter is one of the strategies in poverty reduction. Over 80 per cent of the population lives in rural areas where houses are constructed using mud, mud bricks, burnt bricks, wood, grass (bamboo) or sand-cement blocks.

The development of an interlocking brick press for making soil-cement bricks offers one of the best options for the construction of affordable (low cost) houses in rural and urban areas, where wood for burning bricks is either scarce or no longer available.

The interlocking soil-cement bricks have various advantages:

- ☆ They do not require binding mortar during construction;
- ☆ They do not need fire treatment and therefore save forests;
- ☆ They require less water for moulding and curing;
- ☆ They use very low ratios of cement per brick; and
- ☆ They have comparable strength and are simple to use.

Popularisation of Interlocking Soil-Cement Bricks Technology

Developers of the interlocking brick press, the Technology Development and Transfer Centre (TDTC) and the National Housing and Building Research Agency (NHBRA) are engaged in the development and popularization of this technology. Of late, two villages in the outskirts of Dar es Salaam and a District Council in the Singida Region have been beneficiaries of this technology through COSTECH's facilitation as explained below.

Dissemination of the Technology at Boko and Segerea Villages and Singida District Council

The dissemination of this technology is currently demand-driven, whereby individuals or groups that need the technology request COSTECH for assistance. In this particular case, the Chairpersons of Boko and Segerea villages requested COSTECH to provide them with the technology for constructing low-cost houses. However, the Singida District Council was approached and sensitized on the benefits of this technology.

COSTECH facilitated the training in collaboration with the NHBRA. NHBRA provided the training to trainers in soil selection, mixing ratios of soil and cement, pressing of bricks, curing of bricks and construction. The demonstration units were the offices of the village Chairpersons and two classrooms for the Singida District Council.

Responsibilities of the Community

COSTECH requested the beneficiaries of the programme to contribute partially to the training in form of labour, construction site and materials. All this was to create a sense of ownership to the communities. In all cases the trainees were responsive and cooperative. However, it was observed that the Community leaders rarely attended the training sessions. Later it was learnt that they expected to be paid allowances for attending training sessions, which was not in the budget.

Capacity to Fulfil the Needs of the Target Group

The target groups were individuals, private, and public institutions. The aim was to impart technology which, if adopted, could improve their shelters as well as reduce poverty by enabling the making and selling of bricks to others. At the end of each training course COSTECH provided one interlocking press to the trainees for their own use.

Interaction with NGO, CBO, and Government Institutions

There are several NGO's and government institutions, which are working on this technology. However, before COSTECH selects an agent it looks into several factors, such as: rates, competence, schedule of activities, previous records, and physical distance between the agent and the recipient. In this particular technology COSTECH collaborated with NHBRA because of their vast experience in this field.

Lessons Learnt

- ☆ In order to have meaningful results, the financial commitments on part of the beneficiaries must be effected before execution of the activities.
- ☆ As a facilitator, there is a need to have a legal contract with the trainer to ensure quality of the work and observe time schedule.
- ☆ Recipients need to be sensitized on the advantages of the technology that will be provided to them rather than expectations of getting paid allowances.
- ☆ Provision of one press at the end of the training has resulted in a continued diffusion of this technology to other areas.
- ☆ The demand for this technology has increased tremendously; the manually operated cinvar rams can no longer meet the demand. This has led to the development of the diesel powered presses.
- ☆ After the demonstration in Singida, other district councils are now adopting the technology.

4.4 Dissemination of Institutional Smokeless Stoves at a New Fish Market

Scope of the Activity

The Dar es Salaam Ferry-fish-market is one of the modern fish market in Tanzania, the other being in Mwanza, in the Lake Zone.

(a) (b)

Figure 23.1: Low-cost Housing Technology
(a) A wall (b) Demonstration of class rooms

The modernization of the Ferry-fish-market was realized through the support of the Government of Japan. The Ferry-fish-market activities include fishing, sales and processing.

The new-fish-market has enhanced fish business in the area. However, the business was marred by the problems caused by smoke and heat emissions surrounding the fish frying Zone No. 6 which resulted in environmental pollution.

The problems of smoke and heat emission were major concerns to all stakeholders of the fish market including the Ferry management, Ilala Municipal Council, development partners, users and visitors.

Approach

The Tanzania Commission for Science and Technology took up the initiative to look into alternative options to fuel wood to solve the problem of smoke pollution in Zone No. 6. Several measures were taken including the use of the coal briquettes instead of firewood and the introduction of an improved institutional-smokeless-wood-stove at the frying zone. The Technology Development and Transfer Centre (TDTC) of the University of Dar es Salaam designed and manufactured the stove.

The installation of a demonstration smokeless-wood-stove at Ferry by COSTECH attracted a lot of interest from the stakeholders. They considered the smokeless-wood-stove as a possible solution to the heat and smoke problems in Zone No. 6. For that matter, COSTECH donated one smokeless-wood-stove to the management of Ferry to assess its performance and acceptability by the users. Eventually, after testing the stove for one month all stakeholders accepted the stove.

The advantages of the smokeless stove were elimination of smoke and heat at the work place, fuel savings, and enhancement of the quality of deep fried fish.

Challenges

Before the new stoves were mounted there were three-stone set-ups with open fires. They could not control the burning fires. Upon introduction of the demonstration stove with a water heater, users complained of slow heating, inadequate size and depth of the frying pan and the use of only dry wood. It was pointed out that the chimney was very hot and not safe.

All technical complaints were taken into account in the second design. After satisfactory performance and acceptance by the users, the Municipality decided to order the remaining 47 institutional stoves to replace all the three-stone set-ups.

After installation of all the smokeless stoves, the biggest challenge came from the firewood vendors, who complained that their sales had dwindled from approximately US$ 30 to US$ 4 a day because the new stoves consumed less firewood.

Lessons Learnt

For the technology to be accepted there should be a working prototype and demonstration of the technology. A two-way communication between the developer and the recipient should be established.

All queries raised by the recipients should not be ignored. Frequent visits and monitoring of the system is also essential. Otherwise sabotage by other affected groups can make the technology be seen as a non-starter.

Figure 23.2: Newly Installed Smokeless-Wood-Stoves with Water Heaters

4.5 Dissemination of Energy Efficient Domestic Charcoal Stoves

Scope of the Activity

It is a fact that most Tanzanians use charcoal and firewood for their domestic cooking. This will continue to be so for many years to come. Therefore, it is imperative to design and disseminate stoves that conserve charcoal.

Fuel wood and charcoal are the most exploited energy resources in Tanzania accounting for 92 per cent of the total energy bill. A survey (2004) by COSTECH on the usage of charcoal in Dar es Salaam (the largest city in Tanzania) showed that about 80 per cent of the households in Dar es Salaam use charcoal as the major fuel, 12 per cent use firewood, LPG (7 per cent) and electricity (1 per cent). As a result over 60 per cent of the country is now threatened by desertification leading to severe soil erosion, drought, and destruction of water catchments as well as the ecosystem.

The major benefit from this project was poverty reduction through reduced bills of charcoal, job creation for artisans and women pottery groups, reduced deforestation rates, and reduced health hazards caused by heat and smoke inhalation.

For the past two decades, several stove programmes and projects have been initiated in Tanzania. Most of these programmes/projects are funded by the Government, Parastatal organizations, Development Partners or by NGO's.

The current popular charcoal stove was adopted from Kenya. It is made of metal clad with a ceramic liner inside. The cooking pot rests on top leaving a side clearance for air circulation. This clearance is the major source of heat loss from the burning charcoal. The improved version has no clearance, the pot rests on the stove and a small chimney on the side is used as a vent. Therefore, the user cannot overfill the stove with charcoal, does not see the burning charcoal, the excess air in-flow is limited and the heat loss by convection is reduced. With these improvements, the stove cuts down the charcoal usage by half.

Popularisation of the Energy Efficient Stove

Popularization of this stove has been through exhibitions, demonstrations and training of artisans. COSTECH in collaboration with the Small Industries Development Organization (SIDO), which has a network throughout the country, has been training artisans in the regions on how to convert the improved Kenyan ceramic stove into an energy efficient stove.

COSTECH has been enticing users of charcoal to switch over to the energy efficient stove. The benefits of this new design include: financial savings of about 50 per cent, less heat intensity to the user and environmental conservation. All these benefits are geared towards individual poverty reduction strategies.

Challenges

Several problems were experienced during the initial stages of dissemination of this technology, from both the artisans and the end users.

Artisans made the following complaints.

- ✰ The new stove was not moving fast because their customers did not know about it.

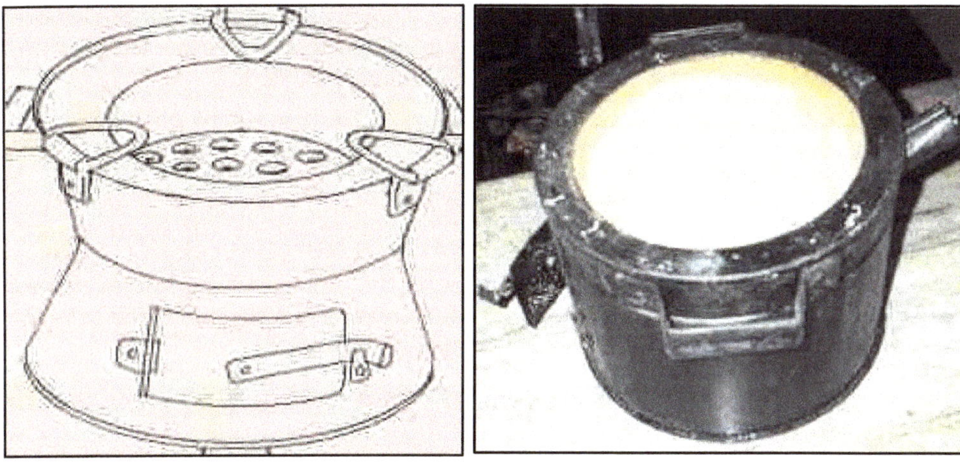

(a) (b)

Figure 23.3: (a) Improved Kenya Ceramic Stove
(b) Improved Energy Efficient Stove

- ☆ The new stove took a longer time to make.
- ☆ The new stove required more material.

The last complaint was noted and the design revised. Instead of being cylindrical, the stove is now conical in shape.

On the other hand, the end users were hesitant to buy the new stove because of the following reasons.

- ☆ They could not see the burning charcoal and therefore thought that it was not burning.
- ☆ They could not believe the amount of charcoal used was sufficient to cook their meal without adding in between.
- ☆ To them the fire would be extinguished (suffocated) once the cooking pot is placed on top of the stove.
- ☆ The new stove was cooking slowly.

Lessons Learnt

- ☆ In order to have the stove accepted by the end users, comparison demonstrations were necessary;
- ☆ Incentive for the artisans who opted to make new stoves was necessary. Otherwise they would close their shops.
- ☆ The cost and the lifetime of the new stove had to be comparable to the conventional one. Therefore, manufacturing time and material used were to be optimized.

- ✫ Education on the cooking technique and energy saving with simple lessons, such as adding more heat to already boiling water is a waste of energy and what is required is to sustain boiling, being imparted.
- ✫ Those who have used the new stove did not want to go back to the conventional stove.

The Way Forward

There are several opportunities associated with this dissemination of improved stoves for domestic and institutional uses. Currently, there is a big demand of these stoves in the country. It is an opportunity for creating employment for artisans, reducing environment degradation, and reducing family poverty.

4.6 Information and Communication Technology (ICT) Access for Rural Areas

Scope of Activities

Rural areas often suffer from neglect and isolation as they are perceived as areas of high risk to investors. However, this is where the vast majority of the Tanzanian population lives. In order to enable rural areas to participate in a knowledge-based economy, ways should be sought to improve ICT access for rural areas. Two initiatives used by COSTECH are the Community Telecentre to provide affordable computer based telecom and telecommunication services to rural communities and Village ICT projects for good governance.

Popularization of Community Telecentres

The philosophy behind the rural community Telecentre initiative is affordability, rural access and the fostering of private-public-community partnerships.

Figure 23.4: The Sengerema Telecentre (www.sengerema.or.tz)

The Sengerema Telecentre is currently operating from a recently completed building at a strategic cross road in Sengerema (Western Tanzania). The services being rendered at the moment include computer training, Internet access, fax, telephone, radio broadcasting, conference facility and meeting, secretarial and consultancy services. A glimpse of Internet usage for the month of December 2002 showed that a total of 576 email users visited the telecentre and utilized the Internet service. The gender balance was in favour of men by 51:525 attracting clients such as students, businessmen, farmers, teachers, nurses, doctors, physiotherapists, dentists, radiographers, policemen, priests, agriculture officers and artisans.

The Telecentre income has been growing steadily over the two years of operation with the main reasons being increased awareness and diversification of services provided.

After successful implementation of the first Telecentre, there have been other initiatives to setup Telecentres in other regions in collaboration with Development Partners such as:

- ☆ Ngara in West Lake region where ITU, UNESCO, UNHCR, TCC, World Space Corporation, Volunteers in Technical Assistance and The Interactive Health Network have set a Telecentre for health network and refugees
- ☆ Kasulu in Kigoma region Global Catalyst Foundation (GCF), Schools Online (SOL), UNHCR, UNDP, and RET have set a Telecentre network for the refugees camps;
- ☆ Lugoba and Dakawa in the Coast region IICD and the Commission have jointly set Village ICT Access Initiatives programme.

Challenges

Challenges faced during the diffusion of this technology were of two kinds, related to technology and those related to the target group. These include the following.

- ☆ Running and maintenance of ICT equipment required full-time personnel.
- ☆ The communities expected to use the facilities for free.
- ☆ Although the District Council provided the building, they had no sense of ownership of the project.
- ☆ Lack of software and programs in Swahili.
- ☆ Lack of management with vision to run the Centre.
- ☆ Lack of knowledge on how to use ICT as a development tool.

Lessons Learnt

- ☆ Community participation is essential at all stages.
- ☆ A detailed business plan is essential.
- ☆ Rural people have a large knowledge base that can be tapped.
- ☆ Rural communities respond fast to broadcasts directed to them from the local station than the national broadcasts.

- ☆ Rural community capacities should not be underestimated. They have the capacity to contribute for projects targeted to them.
- ☆ The cost of establishing a Telecentre is very high and public-private-partnerships should be encouraged.

5. CONCLUSION

There could be several approaches that can be adopted in technology transfer, but for the cases mentioned above the following approach has been adopted:

1. Problem identification;
2. Identification of possible technological solutions;
3. Assessment of the need to ascertain if it is demand driven or requires prior awareness creation and demonstrations;
4. Engagement of a social scientist to assess the target group;
5. Identification of the capacity in terms of funds and expertise needed to fulfil the needs of the target groups;
6. Engagement of the target group at all stages of technology transfer;
7. Addressing target group queries, no matter how small they are;
8. Willingness to revise the design to incorporate suggestions;
9. Inculcation of a sense of ownership by making the recipient contribute to the programme;
10. Organization of scheduled visits to assess the progress;
11. Identification of collaborators in the dissemination of the technology to other target groups.

Given level of the poverty in our Communities, lack of skilled manpower to support the transferred technology, taboos and myths, lack of financing, poor rural infrastructure (energy, accessibility, communication) and low participation of the private sector (especially the financial institutions) in supporting technology acquisition, the Commission has adopted the above eleven-step participatory method when transferring technologies to the rural Communities. After several successes the Government has eventually set some incentives for the private sector to participate fully in technology acquisition and transfer. The Commission sees this as a positive move for sustainability which will eventually lead to the growth of economy and poverty reduction.

REFERENCES

United Republic of Tanzania, "The Tanzania Commission for Science and Technology Act, No. 7 of 1986 (rev) –Principal Legislation" Govt. Printers, Dar es Salaam, 2001.

Kayumbo, H. Y., and Asman, S.J., 1981. "Bulletin of Research and Information" Vol. 1.No.2.

Lushiku, E., Nkonoki, S.R. and Sambali, J., October 1985. "Research, Development and Policy, Issues in the Energy Sector in Tanzania", COSTECH.

Mahgary, Y.E. and Biswas, A. K., 1985. "Integrated Rural Energy Planning", Butterworths Publishing.

Schneiders, H.H., Mkalata, K. and Mmakasa, S.E., Nov. 1985. "Sociological and Scientific assessment of demonstration/Research Biogas plants in Villages around Dodoma", COSTECH.

Mkwawa, F. J., Sept. 1986. "Dodoma Rural Energy Project (DREP) progress report", COSTECH.

Schneiders, H.H., and Mkwawa, F.J, Dec. 1985. "Design and Performance of Floating Type Biogas Plants in Dodoma", COSTECH.

Sengerema Telecentre Website: *www.sengerema.or.tz*

ABBREVIATIONS

CBO:	Community Based Organisation
CDTT:	Centre for the Development and Transfer of Technology
COSTECH:	Tanzania Commission for Science and Technology
IPR:	Intellectual Property Rights
NGO:	Non-Governmental Organisation
NHBRA:	National Housing and Building Research Agency
R&D:	Research and Development
SIDO:	Small Industries Development Organisation
SME:	Small and Medium Enterprises
S&T:	Science and Technology
TDTC:	Technology Development and Transfer Centre
TIRDO:	Tanzania Industrial and Research Development Organisation
UNIDO:	United Nations Industrial Development Organization
ZERI:	Zero Emission Research Initiatives.

Chapter 24

South-South Cooperation for Technology Transfer and Development of Small and Medium Enterprises: VIT Experiences

D.P. Kothari
Vice Chancellor, VIT University
Vellore – 632014, Tamil Nadu, India
E-mail: vc@vit.ac.in, dpk0710@yahoo.com, dkothari@ces.iitd.ac.in

ABSTRACT

This paper deals with the experience of VIT University, Vellore, India on South-South Cooperation for technology transfer. It also dwells on the experience gained at VIT in dealing with managing and building micro, small and medium enterprises.

Keywords: *South-South Cooperation, Technology transfer, Small and medium enterprises.*

1. INTRODUCTION

South-South Cooperation (SSC) is a term historically used by policymakers and academics to describe the exchange of resources, technology, and knowledge between developing countries, also known as countries of the global South. Such activities include developing mutually beneficial technologies, services and trading relationships. It is important to the developing world for two reasons. First, SSC contributes to economic advances in them especially in Africa, Southern Asia and South America. Second, SSC lacks the overtones of cultural, political and economic hegemony that is sometimes associated with traditional (Agliatello Oscar, 2006) North-South aid from US, Russia and Western World.

Over the years it was realized that within the group of developing countries also there were lot of differences. This required people to think of cooperation among the developing countries (BOND Networker, 2006) themselves thus giving rise to the concept of 'South-South Cooperation'. The use of 'South' has often been used in the realm of international interaction to highlight the fact that most of the developed nations of the world lie to the north of the developing nations, excepting Australia and New Zealand.

The UN (Brown Mark Mallach, 2001) created a special unit for SSC in the late 1970's which supported academic research and voluntary cooperative efforts between developing countries to promote S-S trade and investment.

Worldwide the micro, small and medium enterprises have been accepted as the engine of economic growth and for promoting equitable development.

In this paper we will discuss in details the contributions made by the author in particular and VIT, Vellore in general towards South-South Cooperation for technology transfer and development of SMEs.

2. SOUTH-SOUTH COOPERATION FOR TECHNOLOGY TRANSFER

All technology transfers have a definite goal that is converting the laboratory information into a commercial product. The sharing of technical information needs education and training. The arena of SSC has come to address a wide range of issues based on the areas of concerns in the developing countries. Owing to the vulnerabilities of the South to natural disasters and combining their inherent lack of adequate coping mechanisms, the need for concerted action on this front has increasingly been felt.

The UN has continually been supporting various activities to promote SSC in the form of policy support and capacity-building activities among others. UN agencies such as FAO, UNIDO, UNDP (formed in 1965) and WHO have marked out proper south-south policies and activities.

The developing nations have nurtured human resource and refined it through their mode of education into a growing asset.

The Bandung Conference in 1955 was the pioneering major south-south conference of developing countries that paved the way for the rise of the Non-Aligned Movement (NAM) and the Group of 77 (Doha, 2005).

A working group (David, Nash and Pray, 2002; Karki and Narula) on Technical Cooperation among Developing Countries (TCDC) was created in 1972 by the UN General Assembly. SAARC (South Asian Association for Regional Cooperation) with Bangladesh, Bhutan, India, Maldives, Nepal, Pakistan and Sri Lanka was formed in 1985. Now Afghanistan is added to this also. The Association of South-East Asian Nations (ASEAN) was constituted in 1967 and was a stepping stone towards bringing nations into a framework of cooperation to bridge the economic gap between developed and developing countries.

3. SMALL AND MEDIUM ENTERPRISES (SMEs)

The promotion of SMEs (Sinha, 2003) is considered as a prominent approach to sustainable development. SMEs in Least Developed Countries (LDCs) are heavily dependent on the imported technology and remain always dependent on imported skills for the maintenance and operation of the plant. All this is changing slowly but surely in the right direction. Proper thought has to be given to appropriate technology transfer process to suit the local needs.

Technology transfer (Amsden, 1989) is use of knowledge and its transfer. The concept of technology has to be made clear among concerned stakeholders through organizing workshops and seminars and this can be achieved with the help of higher education institutions/ universities

In some countries such as South Africa, data on SME suggests that these enterprises contribute about half of total employment, more than 30 per cent of total GDP. Also one out of five units exported is in SME sector.

The President under Notification dated 9th May, 2007 has amended the GOI (Allocation of Business) Rules, 1961. Pursuant to this amendment, Ministry of Agro and Rural Industries and Ministry of Small Scale Industries have been merged into a single Ministry, namely, "Ministry of Micro, Small and Medium Enterprises".

Worldwide, the Micro and Small Enterprises (MSEs) (Maurice, Wang, and Olarreaga, 2002) have been accepted as the engine of economic growth and for promoting equitable development. The MSEs constitute over 90 per cent of total enterprises in most of the economies and are credited with generating the highest rates of employment growth and for promoting equitable development. The MSEs constitute over 90 per cent of total enterprises in most of the economies and are credited with generating the highest rates of employment growth and account for a major share of industrial production and exports. In India too, the MSEs play a pivoted role in the overall industrial economy of the country. It is estimated that in terms of value, the sector accounts for about 39 per cent of the manufacturing output and around 33 per cent of the total export of the country. Further, in recent years the MSE sector has consistently registered higher growth rate compared to the overall industrial sector. The major advantage of the sector is its employment potential at low capital cost. As per available statistics, this sector employs an estimated 31 million persons spread over 12.8 million enterprises and the labour intensity in the MSE sector is estimated to be almost 4 times higher than the large enterprises.

4. ROLE OF VIT UNIVERSITY ON TECHNOLOGY TRANSFER AND DEVELOPMENT OF SMEs

4.1. Transfer of Technology

As an effort on south-south cooperation on technology transfer, several SAARC countries such as Bangladesh, Nepal and Pakistan were visited in connection with the development of local engineering colleges there *e.g.* BUET in Dhaka, Kathmandu Engineering College, Nepal and University of Engineering and Technology, Lahore, Pakistan. The VIT University is helping them in changing/modifying/updating the course curricula/structure. Also helping them in identifying good faculty and raising their standards to the best in India *i.e.*, IIT/NIT/VIT system. This is nothing less than technology transfer.

At VIT University around 400 Chinese students come from China to take education in various disciplines such as computer, management. Thus we are contributing towards the creation of knowledge hubs in the form of human resource and opening up avenues for SSC for exchange of technology. India is already ICT capital of the world and has second largest scientific and technological manpower in the world.

The Centre for Disaster Mitigation and Management (CDMM) was started in 2006 at VIT University. It conducted a roundtable meeting on landsides in March 2006 in which the recurrent problems of landslides in SAARC countries like India, Bhutan and Nepal were discussed and a Vellore declaration had been brought out. This roundtable meeting had acted as a platform to exchange views, experiences and lessons learnt from natural disasters among these countries. The centre is publishing a Newsletter on Natural Disasters and Development. It is also available on net and so is available to other countries. Books written by the author of this paper are now simultaneously published in USA, Singapore and New Delhi and hence are available throughout the world and are being followed in several developing countries. They are being translated in Spanish and Chinese. CDMM had participated in the International Roundtable meeting on Lightning Protection for Non-Aligned Nations held at Sri Lanka in 2007. South-to-South information exchange was discussed. CDMM is among the ten institutions in India to be awarded a project by Prevention Consortium of Geneva under the research and active grants for disaster risk reduction for the year 2007-08 for developing countries. The project is to create self-education knowledge products on Geomatics in Natural Disaster Management.

4.2 Establishing and Managing Entrepreneurship Clubs (E-Clubs) in Southern India

In the Union Budget 2007-08, a policy decision was announced by the Finance Minister for supporting Universities/Educational Institutions to have industry-institute linkage for promotion of Micro, Small and Medium Enterprises (MSME). Accordingly, MSME Development Institute, Chennai and VIT University have signed an MoU on 3rd July 2007. Thus VIT University has become the first University in India to undertake this prestigious project.

The primary objective of this project is to support Micro and Small Enterprises of South India through establishment of Entrepreneurship Clubs. The entrepreneurs will form a forum for discussing their common problems to find solutions. This will create a base to entrepreneurs for coming together to solve their problems; a shift from lobbying mode to facilitation mode.

Expected outcomes of this initiative are:

- To provide solutions for common problems
- Enhancing the productivity and competitiveness of entrepreneurs
- Capacity building of micro and small enterprises
- To facilitate substantial economies of scale
- To provide management training
- To provide incubator services
- Representation to industry associations/other agencies.

In achieving all these objectives, MSME Development Institute and VIT University will be establishing around 200 E-Clubs in all the four States and a Union Territory in Southern India. The first of such clubs are already working at Vellore and Coimbatore. Government of India has sanctioned Rs. 25 million (US$ 0.5 million) for this.

It is suggested that this concept should be extended to all developing countries as cooperation so that they are all benefited.

4.3 Support to SMEs through Business Incubation at VIT Technology Business Incubator (VITTBI)

A Technology Business Incubator has been established in the VIT University (VITTBI) with the support from the National S&T Entrepreneurship Development Board of the Department of Science and technology, Government of India for the promotion of knowledge based small enterprises. The activities and achievements of VITTBI include:

(*i*) Pre-incubation Services: This includes counselling the entrepreneurs to assess the interest level and the viability of the business idea. The entrepreneurs also will be guided to prepare a proposal.

(*ii*) VITTBI has established three product innovation centres in the "Auto Component", "Biotechnology" and "Information Technology Sector" with the support of DST, Government of India and infoDev (the World Bank Group) at a cost of USD 0.5 million. Entrepreneurs can also have access to the VIT University facilities through the VITTBI.

(*iii*) Continuous support is provided to prepare a business plan and vet the same by some experts. Midcourse correction support is also available.

(*iv*) VITTBI has provided a structured training named "Accelerated Business Coaching Program" (www.knowledgebiz.org).

VITTBI has a close working relationship with over 70 corporate and industries clients. This helps it to identify business opportunities available and connect them with the entrepreneurs of VITTBI. A conscious attempt also was made to publish the business opportunities available from industries like BHEL, Honeywell and other incubation systems in Europe. VIT also has a database of external experts and professional consultants who will render their services. VITTBI also helps in networking with other entrepreneurs, customers in formal and informal ways. This helps them to strike deals, partnerships and identify opportunities.

Apart from offering support to SMEs, VITTBI also has put in considerable efforts in promoting South-South Cooperation among business incubators. The main aim is to promote entrepreneurship, incubating start-up ventures and offering technology development services to firms in the developing world.

A three-day Capacity Building Program for Business Incubation Managers under SSC[11] was organized by VITTBI during May 2007. Over 19 participants from China, Mauritius, Mozambique, Philippines, Sri Lanka, Thailand and Vietnam derived immense benefits through this event. This event also paved way for networking and building alliances. Obviously this will lead to extension of Incubator services like

Market Information, Soft landing and other Custom requirement of their respective incubates.

Failure Issues

Despite such active facilitation and support, some entrepreneurs may fail. Such experiences may be shared with other developing countries so that failures are minimized (National Knowledge Commission, 2008).

(*i*) Lack of will and perseverance on the part of the entrepreneurs.

(*ii*) Peer and family pressure

(*iii*) Fear of failure

(*iv*) Lack of a failure tolerant eco system

5. THE CHALLENGES IN TECHNOLOGY TRANSFER IN MEDIUM AND SMALL SCALE INDUSTRIES

5.1

Unless there is a compulsion and required support for the researcher to commercialize his/her findings there is no desire to scale up the process and develop into a commercially viable project. The technology realization in commercial scale involves the information from different disciplines for integrating sub-systems and components, selection of suitable equipment and suppliers, erection and commissioning of all equipment and realizing quality product at affordable cost. This is also true in case of small and medium scale enterprises.

A few examples of successful transfer of technologies from Indian academic institutions and grass route innovators are given below:

(*a*) Conversion of Kerosene Stove to Increase Efficiency

Mr. Babubhai Vasoya from Ahmedabad in Gujarat has developed a stove that uses a 6V electric coil that heats the kerosene, converting kerosene into gas, which burns with a blue flame. It saves 70 per cent of fuel compared with conventional stoves running on LPG. This stove has no smell, no smoke and it burns like LPG.

(*b*) Conversion of Motorcycle to Tilling Machine to Replace Tractor

In Gujarat Mr. Mansukhbhai has converted his Enfield diesel motorcycle with a difference by removing the back wheel and replacing it with a spiked cylinder.

(*c*) Low-cost Drip Irrigation System

An old farmer in Karnataka developed a low-cost drip irrigation system to fight water crisis in his village. Now it is successfully adopted by the farmers of Tamil Nadu, Kerala and Karnataka (Kothari, Singal and Ranjan, 2008).

(*d*) A Manual Milking Device–Inventor J.S. Milker

It is a simple vacuum driven portable machine, which can be used to milk cows effortlessly. This innovation is supported by the industries and is being marketed.

Many small, medium and large scale industries are interacting with the academic institutions for developing new process, increasing efficiency of the old plant and for solving the problems faced in the industries.

Birla Institute of Technology and Science (BITS), Pilani Practical School model is extremely successful for the students to have industrial experience under faculty supervision. VIT is working on number of projects from Indira Gandhi Centre for Atomic Research, Indian Space Research Organisation, Defence Research and Development Organisation (DRDO), Ministry of New and Renewable Energy, TVS Motors, Whirlpool Washing Machines etc. Now big academic institutes have specific business development wings such as Foundation for Innovation and Technology Transfer (FITT) at IIT Delhi (set-up in 1992). National Research Laboratories such as Council of Scientific and Industrial Research and DRDO are promoting new technologies and transfer of know-how. Technology Business Incubator (TBI) is helping in building capabilities of Small and Medium scale Enterprises (SME's) in promoting technologies. TBI at VIT is one of the most successful centres supporting a number of small-scale entrepreneurs and training them in many aspects such as leadership, financial management and control and marketing.

5.2 Failures

The failures in technology development are common in large and small-scale industries. The possible reasons are (World Bank, 2001):

(*a*) Lack of funds to establish sophisticated and advanced facilities for the technology development and transfer

(*b*) Interdisciplinary work. Number of experts available is fewer, which are required to scale-up the process.

(*c*) Lack of funds and interest by the funding agency result in half backed process know-how causing failure.

5.3 Suggestions

☆ Adequate funding should be provided

☆ Joint work, better cooperation and coordination among different interest groups such as academic institutions, research departments and entrepreneurs.

From above, it is clear that any process developed in academic institutions or small and medium scale industries will be accepted for technology transfer only if government policy support with sufficient funding, training infrastructure and strict quality control.

6. CONCLUSIONS

This paper has dealt with South-South Cooperation for Technology Transfer and also discussed at length the experience obtained in dealing with SMEs at VIT University, Vellore. Light is also thrown on problems and issues in this context. Along with numerous success stories, some failures are also discussed along with the reasons and means to avoid them in future. It is envisaged that this paper will help the interest groups in dealing with the topics of technology transfer; small and medium enterprises will be immensely benefited by the VIT experience.

ACKNOWLEDGEMENTS

The author is thankful to Dr. G. Viswanathan, Founder-Chancellor, VIT University, Vellore, Mr. A. Balachandran, General Manager, VITTBI, VIT University, Vellore and Prof. Arun P. Kulshreshtha, Director, NAM S&T Centre for their cooperation and encouragement to write this paper.

REFERENCES

References from Journals

Agliatello Oscar, 2006. "Is South-South Trade the Answer to Bringing the Poor into the Export Process?" Paper presented at International Trade Centre Executive Forum 2006 Global Debate, Berlin, Germany.

Brown Mark Mallach, 2001. "Fostering South-South Cooperation", UNDP.

BOND Networker, July 2006. "North-South versus South-South", www.bond.org.uk.

Gisselquist David, John Nash, and Carl Pray, 2002. "Deregulating the Transfer of Agricultural Technology: Lessons from Bangladesh, India, Turkey and Zimbabwe", World Bank Research Observer, pp. 17:237-65.

Nepal, S.C., Karki, B.R., and Narula, K.P., "Technology Transfer in SMEs: Problems and Issues in the context of Nepal", Trade and Investment Publication, pp. 29-43. www.unescap.org/TID.

Sinha, Anil, 2003. "Experience of SMEs in South and South-East Asia", World Bank Research Paper, pp. 1-18. www.sedf.org

Schiff Maurice., Yanling Wang., and Marcelo Olarreaga., 2002. "North-South and South-South Trade-Related Technology Diffusion: An Industry Level Analysis", World Bank, Mimeo.

References from Books

Amsden, A., 1989. "Asia's Next Giant: South Korea and Late Industrialization". Oxford University Press.

Entrepreneurship in India, National Knowledge Commission, 2008.

Kothari, D.P., Singal, K.P., and Rakesh Ranjan., 2008. "Renewable Energy Sources and Emerging Technologies". Prentice Hall of India, New Delhi.

VIT Technology Business Incubator Activities Reports, 2007.

World Bank., 2001. "Global Economic Prospects and the Developing Countries 2002: Making Trade Work for the World's Poor" (Washington DC:The World Bank).

References from Websites

http://www.vittbi.com/

Chapter 25

Participation of COSQC in Technology Transfer and Development of Iraqi Enterprises

Abdulilah Mohammed Saleh
Senior Chief Chemist
Ministry of Planning and Development Co-operation,
Central Organization for Standardization and Quality Control (COSQC), Iraq
E-mail: cosqc@yahoo.com, surounded22@yahoo.com

ABSTRACT

The Central organization for standardization and quality control (COSQC) was created in July, 1979 by merging of Iraqi Organization of Standards (IQS) and the Directorate of Research and Industrial Control (DRIC). COSQC is the central agency in Iraq that is responsible to undertake tasks like development of standards, calibration and measurement, quality control, inspection, ISO certification and patent activities through specialized laboratories

All above activities are governed by special Acts issued by the Iraqi government. Activities of COSQC cover all Iraqi as well as imported goods and tests are carried out according to the Iraqi standards or international standards; so as to protect the Iraqi consumers.

About 4000 small, medium and other enterprises are serviced by the activities of the COSQC including the industries such as chemicals and petrochemicals, food, engineering, textile and construction. Thus COSQC helps the enterprises in Iraq in transfer of technology and improvement of product quality

Keywords: *COSQC, Technology transfer, Development, Iraq, Small, industry, Enterprise.*

1. INTRODUCTION

Iraq is one of the richest countries in the South West Asia due to its vast reserve of petroleum and natural gas. Further, the country also has significant quantities of other resources in its different territories such as Sulphur, Phosphates etc. In addition to that, Iraq has enough skilled manpower.

Iraq developed several industries at the beginning of the last century, most of which were based on petroleum and natural gas, and also industries on plants, and local and imported raw materials. Iraq has now progressed from a producer of raw materials into a progressive and fastest growing developing country in the Middle East. The process of development has however suffered due to the huge problems faced by the country in the last few decades, such as repeated wars, economic blockade and the worst problem facing all over the world, terrorism. These problems have affected the development of Iraq very badly.

2. STATUS OF THE IRAQI INDUSTRY

The industry in Iraq started at the beginning of the last Century, when only manual or handmade industries were found. These industries included very simple goods required for day-to-day life of the people at that time. However, the modern industries were created after the First World War. During the British rule, the industry manufactured only the simple products without much long term planning, which continued until the end of the Second World War. With the increasing World necessity of oil, Iraq became one of the most important producers of this important material. Also with the improving level of education of the people, and diplomatic relationships established with different countries, a good scientific planning was made for the country.

The State encouraged the people to establish many different types of industries by lending them to form enough capitals for working of the mills and factories, and also assisted them with different fiscal benefits such as waiver of taxes and provided support by science and technology experts.

Due to the above, several industrial complexes have been built up at different places in the country according to the availability of raw materials in each region. Most of the industries established were Petroleum and Gas complexes and industrial products based on the same, such as petrochemicals, construction industry, food processing, engineering industry, chemicals, textiles etc.

Most of the large industrial establishments are owned by the State, whereas, the small and medium enterprises are mainly owned by the private sector.

Iraqi industries during the 7th and 8th decades of the twentieth Century were in their best periods and the quantity and quality of their products had been good. But afterward, gradually due to the repeated wars followed by blockade (siege) and terrorism, the industry suffered badly. In the last few years, a new challenge to Iraqi industries has been from the imported goods from all over the world. Especially after the great changes in the country, and due to various problems of the Iraqi industry, the consumers started favouring imported goods because of their lower price and better quality.

The Iraqi Government is trying to change this situation by encouraging the Iraqi and foreign entrepreneurs and companies to invest their capitals in many different projects. A new law has been enacted for this purpose that provides for various favourable conditions, due to which it is expected that Iraq will overcome most of the problems within a short time in future and make the country industrially competitive.

3. CENTRAL ORGANIZATION FOR STANDARDIZATION AND QUALITY CONTROL (COSQC)

3.1 Origin

The Central Organization for Standardization and Quality Control (COSQC) was established in July 1979 by the law No. 54 by merging of Iraqi Organization of Standards (IQS) and Directorate of Research and Industrial Control (DRIC). The COSQC is one of the most important government institutions that belong to the Ministry of Planning and Development Cooperation.

3.2 Activities of COSQC

The activities of COSQC are governed by special Acts issued by the Iraqi government, and all bodies of the government and private sectors are covered and serviced by the organization for their industrial activities and any related task. Activities of COSQC also extend to cover all imported goods. The most important activities are as follows:

1. Finding reliable Iraqi references for measurement criteria and methods of calibration
2. Providing Iraqi standards
3. Protecting consumers, producers as well as the environment, health and public safety
4. Developing specialization in the field of standardization and quality control as well as industrial property.
5. Preparing scientific methods for rational utilization of natural resources
6. Supporting technological progress in production and services
7. Protecting national and foreign inventions by registration and granting patents
8. Analysis and hallmarking of gold, silver and other precious metals
9. Professional training in metrology, quality control and quality managements.

3.3 Functions of COSQC

COSQC is a relevant and independent government agency having many functions that acts as a consultant on quality control and related issues for the government ministries and state owned enterprises, and maintains relationships with international product standards organization. It is a central body in Iraq which is responsible for the following:

1. Development and adoption of Iraqi standards that is compulsory and applicable to both Iraqi products and imported goods. Developing Iraqi standards is based mainly on international standards.
2. COSQC has issued about 2250 Iraqi standards.
3. Quality control activities: COSQC carries out technical visits to more than 4000 Iraqi companies and factories to inspect the products and the process of production, testing and checking the quality of domestic products and imported goods according to Iraqi standards
4. In case Iraqi standards are not available, tests and analysis are carried out conforming to international standards, and conformity certificates are issued for those products and goods accordingly.
5. Auditing quality management systems for different Iraqi organizations and granting Iraqi quality certificates according to ISO 9000.
6. Metrology activity: COSQC is the only national body in Iraq which is responsible for maintaining and improving national measurement standards, calibration of measurement standards for different industrial and service sectors in Iraq, calibration of all types of measuring instruments (industrial metrology and legal metrology), calibration of imported instruments before putting them in service and authorizing other laboratories for calibration activities.
7. Accreditation of laboratories: COSQC is the only national body in Iraq which has authority to accredit testing and calibration laboratories according to ISO standard (17025); the department which is responsible for this activity is newly established.
8. Testing analysis and marking of gold and silver jewellery and inspection of goldsmith shops and granting licenses for goldsmiths under specification and certain conditions.
9. Registration and granting of patents

3.4 Organizational Structure

There are about 500 employees in COSQC. All the work is done by specialized laboratories with physicists, engineers, chemists, biologists and administrative staff.

The employees work in three main Directorates under the President of the organization, each headed by a General Director. These Directorates are as follows:

A. Directorate of Standardization

It has three departments, these are:

A-1 *Department of Standards*: Consists of six sections and their staffs is specialized in establishing Iraqi standards. These sections are for chemicals and petrochemical, foods, engineering, textiles, construction and documentations

A-2 *Department of Metrology*: Has five sections, for calibration and measurements on mechanical, physical, electrical, legal, and measurement and calibration of systems.

A-3 *Department of Hallmark.*

B. Directorate of Quality Control

The Directorate has six sections for chemical, food, construction, textile and engineering industries and quality managements.

C. Directorate of Technical Administration

D: Another five departments directly report to the President of COSQC, which are for planning and development; accreditation laboratory; legal, information and international co – operation and audit.

4. CONCLUSIONS

With its activities and functions on the development of standards, inspections, testing and quality evaluation of the domestic and imported goods, COSQC is providing a valuable service to the SMEs in Iraq to keep them competitive and motivate them to introduce new technologies in their products and processes.

Appendix

Colombo Declaration

on South-South Cooperation for Technology Transfer and Development of Small and Medium Enterprises (SMEs)

WHILE EXPRESSING gratitude to the Government of the Democratic Socialist Republic of Sri Lanka, the hosts of the International Conference on 'South-South Cooperation for Technology Transfer and Development for Small and Medium Enterprises (SME's), jointly organised by the National Science Foundation (NSF) and the National Science and Technology Commission (NASTEC) of Sri Lanka and the Centre for Science and Technology of the Non-Aligned and Other Developing Countries (NAM S&T Centre) during 18-22 August, 2008.

HAVING DELIBERATED on various issues related to knowledge, approaches and experiences related to the problems and prospects on the development and transfer of SME's,

The Specific issues encompassing:

- ☆ Implementation of Technology Transfer, Laws and Policies and Market Development
- ☆ Implementation of Technology Transfer, Partnership Development, IPR Issues and Quality Assurance
- ☆ Strengths and Barriers for Effective Technology Development and Transfer
- ☆ Technology Assimilation and Networking.

The participants EXPRESSED unanimity that the Conference has been a resounding success in coming out with focused recommendations and actions as below, that would help NAM and other developing countries in formulating their plans for the strengthening of the existing SME's and equivalent structures as well as in the creation of new entities to promote the adoption of technologies developed within their countries by the concerned users and industries aimed at their economic growth;

- ☆ It is desirable that every country should establish a National Technology Transfer Centre (or an equivalent body) to act as a bridge between technology developers (transferors) and users/investors (transferees), maintain databases of technology offers and technology requests and provide necessary services to manage the process of technology transfer, including technical and managerial consultancy, IP services, assistance with obtaining credit, etc.
- ☆ The persons engaged in technology transfer, including Technology Transfer (TT) managers, developers and users should be trained in the necessary managerial techniques, as well as the best practices in the field of TT, based on the experience of successful examples from various countries.
- ☆ Countries should streamline their laws and regulations to facilitate all aspects of SME development, thereby bringing their regulations into compliance with SME friendly policies, and eliminating contradictions between such policies and actual regulations governing the availability of credit, customs, standards, taxation, etc.
- ☆ Persons who have participated in successfully transferring technologies in R&D institutions, universities, etc. should be given appropriate incentives (*e.g.* promotions).
- ☆ Countries should set up financial institutions (*e.g.* the Grameen Bank in Bangladesh) to assist poor communities in technology transfer.
- ☆ Governments should fund R&D organizations for improvement, modernization and refurbishment of infrastructure that enable them to help SMEs.
- ☆ Governments should initiate the development of Technology/Business Incubators, where not available. For capacity building the resources available with infoDev of World Bank could be utilized. IDISC.net portal could be linked with the proposed national technology transfer facility.
- ☆ The developing countries having SME's and those desirous of establishing SME's shall network among each other with expertise of UNESCAP-APCTT.
- ☆ A database on technology offers and requests, training institutions, R&D institutions, machinery, consultancies etc., shall be facilitated through the expertise available with the Department of Science and Technology of the Republic of South Africa, Belarus Republican Centre for Technology Transfer and the National Office for Technology Acquisition and Promotion (NOTAP) of Nigeria.
- ☆ A quarterly electronic newsletter providing information on the available technologies developed within the developing countries, the available experts and any technical requirements that may be met through interactions among the developing countries themselves shall be brought out by VIT University, Vellore, India with the facilitation of the NAM S&T Centre.

- ✰ Fellowship scheme of the NAM S&T Centre shall be made use of for training of personnel in technology development and transfer from one member country to the scientific institutions and industry in another member country of the Centre.
- ✰ Workshops and training courses in the areas related to the issues concerning technology transfer and SME's, as well as other areas, shall be considered for organisation by COMSATS in consultation with the NAM S&T Centre.

THUS DECLARED AND ADOPTED IN NEGOMBO, SRI LANKA ON THIS DAY, 21st AUGUST, 2008.

www.ingramcontent.com/pod-product-compliance
Ingram Content Group UK Ltd.
Pitfield, Milton Keynes, MK11 3LW, UK
UKHW022215051026
15624UKWH00001B/170

9 789351 242024